Lecture Notes in Artificial Intelligence 9673

Subseries of Lecture Notes in Computer Science

LNAI Series Editors

Randy Goebel
University of Alberta, Edmonton, Canada
Yuzuru Tanaka
Hokkaido University, Sapporo, Japan
Wolfgang Wahlster
DFKI and Saarland University, Saarbrücken, Germany

LNAI Founding Series Editor

Joerg Siekmann
DFKI and Saarland University, Saarbrücken, Germany

More information about this series at http://www.springer.com/series/1244

Richard Khoury · Christopher Drummond (Eds.)

Advances in Artificial Intelligence

29th Canadian Conference
on Artificial Intelligence, Canadian AI 2016
Victoria, BC, Canada, May 31 – June 3, 2016
Proceedings

 Springer

Editors
Richard Khoury
Lakehead University
Thunder Bay, ON
Canada

Christopher Drummond
National Research Council Canada
Ottawa
Canada

ISSN 0302-9743   ISSN 1611-3349 (electronic)
Lecture Notes in Artificial Intelligence
ISBN 978-3-319-34110-1   ISBN 978-3-319-34111-8 (eBook)
DOI 10.1007/978-3-319-34111-8

Library of Congress Control Number: 2016938377

LNCS Sublibrary: SL7 – Artificial Intelligence

Printed on acid-free paper

This Springer imprint is published by Springer Nature
The registered company is Springer International Publishing AG Switzerland

Preface

The 29th Canadian Conference on Artificial Intelligence (AI 2016) built on a long sequence of successful conferences, bringing together Canadian and international researchers, presenting and discussing original research. The conference was held in Victoria, British Columbia, Canada, from May 31 to June 3, 2016, and was collocated with the 42nd Graphics Interface Conference (GI 2016), and the 13th Conference on Computer and Robot Vision (CRV 2016).

AI 2016 attracted 97 submissions from Canada and internationally. Each submission was reviewed in double-blind mode by at least two Program Committee members for the conference and the proceedings 12 long papers and 27 short papers were accepted, i.e., 12.4 % and 27.8 % of the total number of submissions, respectively. Regular papers were allocated 12 pages in the proceedings, while short papers were allocated 6 pages.

The conference program was enriched by three keynote speakers and three tutorials, all selected from Canadian universities. The keynote speakers were B. John Oommen (Carleton University), Michael Bowling (University of Alberta), and Froduald Kabanza (Université de Sherbrooke). The tutorials were organized by Atefeh Farzindar and Diana Inkpen (respectively, NLP Technologies Inc. and University of Ottawa), Nathalie Japkowicz (University of Ottawa), and Ted Kirkpatrick and Oliver Schulte (both Simon Fraser University).

We want to extend our warm thanks to all the individuals who contributed to the success of the conference: Brian Wyvill and Melanie Tory (both University of Victoria), the general chairs of the three collocated conferences, Artificial Intelligence, Graphics Interface, and Computer and Robot Vision (AI/GI/CRV); Ji Li (University of Victoria), the local arrangements chair for AI/GI/CRV; Gabriel Murray (University of the Fraser Valley) and Lars Kotthoff (University of British Columbia), the chairs of the Graduate Student Symposium; and the members of the Program Committee, who provided timely and helpful reviews.

AI 2016 was sponsored by the Canadian Artificial Intelligence Association (CAIAC). Cory Butz (University of Regina), the president of CAIAC, and the CAIAC Executive Committee, provided essential advice and guidance based on their experience from previous Canadian AI conferences.

June 2016

Richard Khoury
Christopher Drummond

Organization

The 29th Canadian Conference on Artificial Intelligence (AI 2016) was sponsored by the Canadian Artificial Intelligence Association (CAIAC), and held in conjunction with the 42nd Graphics Interface Conference (GI 2016) and the 13th Conference on Computer and Robot Vision (CRV 2016).

General Chairs

Brian Wyvill   University of Victoria, Canada
Melanie Tory   University of Victoria, Canada

Program Chairs

Richard Khoury   Lakehead University, Canada
Chris Drummond   National Research Council Canada, Canada

Local Arrangements Chair

Ji Li   University of Victoria, Canada

Graduate Student Symposium Chairs

Gabriel Murray   University of the Fraser Valley, Canada
Lars Kotthoff   University of British Columbia, Canada

Program Committee

Ebrahim Bagheri   Ryerson University, Canada
Eric Charton   Yellow Pages, Canada
Gabriel Murray   University of the Fraser Valley, Canada
Sheela Ramanna   University of Winnipeg, Canada
Fred Freitas   Universidade Federal de Pernambuco, Brazil
Marie-Jean Meurs   Université du Québec à Montréal, Canada
Diana Inkpen   University of Ottawa, Canada
Marina Sokolova   University of Ottawa and Institute for Big Data
                          Analytics, Canada
Jian-Yun Nie   Université de Montréal, Canada
Yang Xiang   University of Guelph, Canada
David Mitchell   Simon Fraser University, Canada
Vangelis Karkaletsis NCSR Demokritos, Greece
Frank Hutter   University of Freiburg, Germany
Lars Kotthoff   University of British Columbia, Canada
Grzegorz Kondrak   University of Alberta, Canada

Xinghui Zhao	Washington State University, USA
Fazel Keshtkar	Southeast Missouri State University, USA
Hao Wang	IBM, USA
Seyednaser Nourashrafeddin	Dalhousie University, Canada
Christel Kemke	University of Manitoba, Canada
Jimmy Huang	York University, Canada
Orland Hoeber	University of Regina, Canada
Qigang Gao	Dalhousie University, Canada
Alan Mackworth	University of British Columbia, Canada
Bruce Spencer	University of New Brunswick, Canada
Malek Mouhoub	University of Regina, Canada
Fred Popowich	Simon Fraser University, Canada
Jian Pei	Simon Fraser University, Canada
Xiaodan Zhu	National Research Council Canada, Canada
Nathalie Japkowicz	University of Ottawa, Canada
Xiangdong An	York University, Canada
Fatiha Sadat	UQAM, Canada
Yannick Marchand	Dalhousie University, Canada
Weiming Shen	National Research Council Canada, Canada
Gordon McCalla	University of Saskatchewan, Canada
Kate Larson	University of Waterloo, Canada
Virendra Bhavsar	University of New Brunswick, Canada
Dan Wu	University of Windsor, Canada
Guy Lapalme	Université de Montréal, Canada
Enrico Francesconi	ITTIG-CNR, Italy
Yong Gao	UBC Okanagan, Canada
Andrea Tagarelli	University of Calabria, Italy
Harry Zhang	University of New Brunswick, Canada
Julien Velcin	Université de Lyon 2, France
Samira Sadaoui	University of Regina, Canada
Wagner Meira Jr.	UFMG, Brazil
Christian Jacob	University of Calgary, Canada
Lyne Da Sylva	Université de Montréal, Canada
Roger Nkambou	Université du Québec À Montréal, Canada
Simone Ludwig	North Dakota State University, USA
Abidalrahman Moh'D	Dalhousie University, Canada
Michel Gagnon	Polytechnique Montreal, Canada
Alejandro Lopez-Ortiz	University of Waterloo, Canada
Angelo Di Iorio	University of Bologna, Italy
Stefano Ferilli	University of Bari, Italy
Aminul Islam	Dalhousie University, Canada
Luis Enrique Sucar	INAOE, Mexico
David Chiu	University of Guelph, Canada
Daniel L. Silver	Acadia University, Canada
Leila Kosseim	Concordia University, Canada

Keynote Talks and Invited Tutorials

Sequence-Based Estimation of Multinomial Random Variables

(Keynote Talk)

Abstract. This talk deals with the relatively new field of sequence based estimation in which the goal is to estimate the parameters of a distribution by utilizing both the information in the observations and in their sequence of appearance. Traditionally, the Maximum Likelihood (ML) and Bayesian estimation paradigms work within the model that the data, from which the parameters are to be estimated, is known, and that it is treated as a set rather than as a sequence. The position that we take is that these methods ignore, and thus discard, valuable sequence-based information, and our intention is to obtain ML estimates by "extracting" the information contained in the observations when perceived as a sequence. Our earlier results introduced the concepts of Sequence Based Estimation (SBE) for the Binomial distribution, where the authors derived the corresponding MLE results when the samples are taken two-at-a-time, and then extended these for the cases when they are processed three-at-a-time, four-at-a-time etc. This current work generalizes these results for the multinomial case. The strategy we invoke involves a novel phenomenon called "Occlusion" that has not been reported in the field of estimation. The phenomenon can be described as follows: By occluding (hiding or concealing) certain observations, we map the estimation problem onto a lower-dimensional space, i.e., onto a binomial space. Once these occluded SBEs have been computed, we demonstrate how the overall Multinomial SBE (MSBE) can be obtained by mapping several lower-dimensional estimates, that are all bound by rigid probability constraints, onto the original higher-dimensional space. In each case, we formally prove and experimentally demonstrate the convergence of the corresponding estimates. We also discuss how various MSBEs can be fused to yield a superior MSBE, and present some potential applications of MSBEs. Our new estimates have great potential for practitioners, especially when the cardinality of the observation set is small. This is a joint work with Prof. S-W. Kim, from Myongji University, in Yongin, South Korea. Being a Plenary/Keynote talk, we will concentrate and survey the new paradigm of Binomial "Sequence-based Estimation". Thereafter, we shall concentrate on the multinomial aspects of SBE.

B. John Oommen
Chancellor's Professor; Fellow: IEEE; Fellow: IAPR
Carleton University, Ottawa, Ontario

Dr. John Oommen was born in Coonoor, India on September 9, 1953. He obtained his B.Tech. degree from the Indian Institute of Technology, Madras, India in 1975. He obtained his M.E. from the Indian Institute of Science in Bangalore, India in 1977. He then went on for his M.S. and Ph.D. which he obtained from Purdue University, in West Lafayettte, Indiana in 1979 and 1982 respectively. He joined the School of Computer Science at Carleton University in Ottawa, Canada, in the 1981–82 academic year. He is still at Carleton and holds the rank of a Full Professor. Since July 2006, he has been awarded the honorary rank of Chancellor's Professor, which is a lifetime award from Carleton University. His research interests include Automata Learning, Adaptive Data Structures, Statistical and Syntactic Pattern Recognition, Stochastic Algorithms and Partitioning Algorithms. He is the author of more than 425 refereed journal and conference publications, and is a Fellow of the IEEE and a Fellow of the IAPR. Dr. Oommen has also served on the Editorial Board of the IEEE Transactions on Systems, Man and Cybernetics, and Pattern Recognition.

The author is also an Adjunct Professor with the University of Agder in Grimstad, Norway.

von Neumann's Dream

(Keynote Talk)

Abstract. Chess has long served as the measure of progress for artificial intelligence. However, at the very beginning of computing and artificial intelligence, John von Neumann dreamt of a different game: "Real life is not like [chess]. Real life consists of bluffing, of little tactics of deception, of asking yourself what is the other man going to think I mean to do. And that is what games are about in my theory." The game von Neumann hinted at is poker, and it played a foundational role in his formalization of game theory. Shortly after launching the field of game theory, he practically abandoned his new discipline to focus on the budding field of computing. He saw computers as the way to make his mathematics workable. Now, over 70 years later with both significant advances in computing and game theoretic algorithms, von Neumann's dream is now a reality. As announced in our paper in Science, Heads-up limit Texas hold'em poker, the smallest variant of poker played by humans, is essentially solved. In this talk, I will discuss how we accomplished this landmark result, along with the substantial scientific advances in our failed attempts along the way.

Michael Bowling
Professor
University of Alberta, Edmonton, Alberta

Michael Bowling is a Professor of Computing Science at the University of Alberta. His research focuses on artificial intelligence, machine learning, and game theory; and he is particularly fascinated by the problem of how computers can learn to play games through experience. Michael is the leader of the Computer Poker Research Group, which has built some of the strongest poker playing programs in the world. In 2008, one of these programs, Polaris, defeated a team of top professional poker players in two-player, limit Texas Hold'em, becoming the first program to defeat poker pros in a meaningful competition. He also started the Arcade Learning Environment, a research testbed for investigating artificial intelligence techniques capable of general competency using Atari 2600 games. His research has been featured on the television programs Scientific American Frontiers and National Geographic Today; in the New York Times and Wired; and twice in exhibits at the Smithsonian Museums in Washington, D.C.

Natural Language Processing for Social Media

(Invited Tutorial)

Abstract. This tutorial provides an overview of the state-of-the-art in Artificial Intelligence (AI) algorithms and challenges of natural language processing (NLP) tools, and of the methods used in processing non-traditional information such as social media content. AI and NLP techniques are critical to the analysis process and how these can be applied to real world challenges created by the social emergence of Big Data. We will focus on innovative applications of NLP such as opinion mining and emotion analysis, geo-location detection, event and topic detection, entity linking and disambiguation, summarization and machine translation for social media. We believe that AI methods and information diffusion in social networks need to be combined with the linguistic and content analysis. This provides the opportunity for new applications that use the information publicly available in various fields, such as social media monitoring for healthcare, finance, security and defence, business intelligence, and politics. We emphasize the importance of this dynamic discipline and the great potential of NLP in the coming decade, in the context of changes in mobile technology, wearable technology, cloud computing and social networking. The content of the tutorial will be based on the recent book, co-authored by Dr. Farzindar and Dr. Inkpen, in Natural Language Processing for Social Media, Synthesis Lectures on Human Language Technologies published by Morgan and Claypool publishers, 2015.

Atefeh Farzindar

NLP Technologies Inc. and Université de Montréal

Dr. Atefeh Farzindar is the CEO of NLP Technologies Inc., Adjunct Professor at University of Montreal and Lecture at the Viterbi School of engineering of University of Southern California (USC). She has served as Chair of the technology sector of the Language Industry Association Canada (AILIA) (2009–2013), vice president of The Language Technologies Research Centre (LTRC) of Canada (2012–2014) and a member of the Natural Sciences and Engineering Research Council of Canada (NSERC) Computer Science Liaison Committee (2014–2015). Recently, she co-authored with Dr. Inkpen a book in Natural Language Processing for Social Media, Synthesis Lectures on Human Language Technologies published by Morgan and Claypool publishers, 2015 and a book chapter in Social Network Integration in Document Summarization, Innovative Document Summarization Techniques: Revolutionizing Knowledge Understanding, IGI Global publisher January 2014.

Diana Inkpen

University of Ottawa

Dr. Diana Inkpen is a Professor at the University of Ottawa, in the School of Electrical Engineering and Computer Science. Her research is in applications of Computational Linguistics and Text Mining. She organized seven international workshops and she was a program co-chair for the AI 2012 conference. She is in the program committees of many conferences and an associate editor of the Computational Intelligence and the Natural Language Engineering journals. She co-authored with Dr. Farzindar a book on Natural Language Processing for Social Media (Morgan and Claypool Publishers, Synthesis Lectures on Human Language Technologies), 2015; 8 book chapters; more than 25 journal articles; and more than 90 conference papers. She received many research grants, including intensive industrial collaborations.

Performance Evaluation for Learning Algorithms

(Invited Tutorial)

Abstract. Data Analytics is a mature field with many sophisticated learning approaches commonly used in a variety of applications. Because of its practical relevance, it has become of critical importance that researchers and practitioners alike be aware of both the proper methodologies and the respective questions that arise when evaluating learning approaches in either an experimental or a practical setting. This tutorial aims at educating as well as encouraging the data analytics community to get involved in the discussion of these important issues. The tutorial will discuss major aspects of machine learning evaluation with a focus on classification algorithms. It will highlight the different questions, assumptions and constraints involved in the evaluation process. In particular, it will examine a number of techniques in great detail and discuss their relevance and shortcomings in different contexts. It will also present R and WEKA tools that can be used to assist with the process. Issues that arise in the Big Data age will also be surveyed. The tutorial will span four areas of machine learning evaluation:

Performance Measures (Evaluation Metrics and Graphical Methods)
Error Estimation/Re-sampling Techniques
Statistical Significance Testing
Issues in Data Set Selection and Evaluation Benchmarks Design

Nathalie Japkowicz

University of Ottawa

Dr. Nathalie Japkowicz is a Professor of Computer Science in the School of Electrical Engineering and Computer Science at the University of Ottawa. She is the immediate past president of the Canadian Artificial Intelligence Association and was program co-chair of Canadian AI 2009. She is the co-author of "Evaluating Learning Algorithms: A Classification Perspective", with Mohak Shah, Cambridge University Press (2011) and the co-editor of "Big Data Analysis; New Algorithms for a New Society", edited with Jerzy Stefanowski, Springer (2015).

Learning Bayesian Networks for Complex Relational Data

(Invited Tutorial)

Abstract. Many organizations maintain critical data in a relational database. The tutorial describes the statistical and algorithmic challenges of constructing models from such data, compared to the single-table data that is the traditional emphasis of machine learning. We extend the popular model of Bayesian networks to relational data, integrating probabilistic associations across all tables in the database. Extensive research has shown that such models, accounting for relationships between instances of the same entity (such as actors featured in the same movie) and between instances of different entities (such as ratings of movies by viewers), have greater predictive accuracy than models learned from a single table. We illustrate these challenges and their solutions in a real-life running example, the Internet Movie Database. The tutorial shows how Bayesian networks support several relational learning tasks, such as querying multi-relational frequencies, link-based classification, and relational anomaly detection.

The resulting relational Bayesian models (and the related Markov Logic Networks model) combine probability, logic, and learning; unifying these different areas is a long-standing goal of AI research. This tutorial addresses two AI audiences: (1) Researchers with a background in machine learning who wish to combine graphical models with first-order logic. This powerful combination has achieved state-of-the-art results on such AI tasks as information extraction, ontology matching, entity resolution, and relational query optimization. (2) Researchers with a background in knowledge representation and logic who wish to learn about techniques for learning first-order probabilistic rules/models from data.

Ted Kirkpatrick

Simon Fraser University

Ted Kirkpatrick is an Associate Professor of Computing Science at Simon Fraser University. With Oliver Schulte, he has co-authored articles on modeling relational data via Bayesian networks. Previously, he studied statistical design space exploration and interaction techniques using haptic interfaces.

Oliver Schulte

Simon Fraser University

Oliver Schulte is a Professor in the School of Computing Science at Simon Fraser University, Vancouver, Canada. He received his Ph.D. from Carnegie Mellon University in 1997. His current research focuses on machine learning for structured data, such as relational databases and event data. He has published papers in leading AI and machine learning venues on a variety of topics, including relational learning, learning Bayesian networks, game theory, and scientific discovery. While he has some nice awards, including the best paper award at Canadian AI 2010, his biggest claim to fame may be a draw against chess world champion Gary Kasparov.

Contents

Actions and Behaviours

Action Recognition by Pairwise Proximity Function Support Vector
Machines with Dynamic Time Warping Kernels   3
   Mohammad Ali Bagheri, Qigang Gao, and Sergio Escalera

An Agent-Based Architecture for Sensor Data Collection and Reasoning
in Smart Home Environments for Independent Living   15
   Thomas Reichherzer, Steven Satterfield, Joseph Belitsos,
   Janusz Chudzynski, and Lamar Watson

Biometric Authentication by Keystroke Dynamics for Remote Evaluation
with One-Class Classification   21
   Chuan Chang, Thierry Eude, and Luis Eduardo Obando Carbajal

Fishing Activity Detection from AIS Data Using Autoencoders   33
   Xiang Jiang, Daniel L. Silver, Baifan Hu, Erico N. de Souza,
   and Stan Matwin

Simulating the Bubble Net Hunting Behaviour of Humpback Whales:
The BNH-Whale Algorithm   40
   Stacey Hala, Howard J. Hamilton, and Paolo Domenici

Gaussian Neuron in Deep Belief Network for Sentiment Prediction   46
   Yong Jin, Donglei Du, and Harry Zhang

Grounding Social Interaction with Affective Intelligence   52
   Joshua D.A. Jung, Jesse Hoey, Jonathan H. Morgan, Tobias Schröder,
   and Ingo Wolf

The Mismeasure of Machines   58
   Eric Neufeld and Sonje Finnestad

Uncovering Hidden Sentiment in Meetings   64
   Gabriel Murray

Fuzzy Computational Model for Emotions Originated in Workplace Events . . .   73
   Ahmad Soleimani and Ziad Kobti

Audio and Visual Recognition

Tolerance-Based Approach to Audio Signal Classification   83
  Sheela Ramanna and Ashmeet Singh

A Novel Dataset for Real-Life Evaluation of Facial Expression Recognition
Methodologies   89
  Muhammad Hameed Siddiqi, Maqbool Ali, Muhammad Idris,
  Oresti Banos, Sungyoung Lee, and Hyunseung Choo

Visual Perception Similarities to Improve the Quality of User Cold Start
Recommendations   96
  Crícia Z. Felício, Claudianne M.M. de Almeida, Guilherme Alves,
  Fabíola S.F. Pereira, Klérisson V.R. Paixão, and Sandra de Amo

Object-Based Representation for Scene Classification   102
  Xuhui Luo and Jinhua Xu

Salient Object Detection in Noisy Images.   109
  Nitin Kumar, Maheep Singh, M.C. Govil, E.S. Pilli, and Ajay Jaiswal

An Approach to Improving Single Sample Face Recognition Using High
Confident Tracking Trajectories   115
  M. Ali Akber Dewan, Dan Qiao, Fuhua Lin, Dunwei Wen,
  and Kinshuk

Neural Network-POMDP-Based Traffic Sign Classification Under Weather
Conditions   122
  Shervin Shahryari and Cameron Hamilton

Natural Language Processing

Poetry Chronological Classification: Hafez.   131
  Arya Rahgozar and Diana Inkpen

f: Phrase Relatedness Function Using Overlapping Bi-gram Context   137
  Md. Rashadul Hasan Rakib, Aminul Islam, and Evangelos Milios

Harnessing Open Information Extraction for Entity Classification
in a French Corpus   150
  Fabrizio Gotti and Philippe Langlais

A Novel Genetic Algorithm for the Word Sense Disambiguation Problem . . .   162
  Wojdan Alsaeedan and Mohamed El Bachir Menai

Mining Biomedical Literature: An Open Source and Modular Approach   168
  Hayda Almeida, Ludovic Jean-Louis, and Marie-Jean Meurs

Word Normalization Using Phonetic Signatures 180
Vincent Jahjah, Richard Khoury, and Luc Lamontagne

Forecasting Canadian Elections Using Twitter...................... 186
Kenton White

Time-Sensitive Topic-Based Communities on Twitter 192
*Hossein Fani, Fattane Zarrinkalam, Ebrahim Bagheri,
and Weichang Du*

Reasoning and Learning

On Tree Structures Used by Simple Propagation.................... 207
*Anders L. Madsen, Cory J. Butz, Jhonatan S. Oliveira,
and André E. dos Santos*

A Simple Method for Testing Independencies in Bayesian Networks....... 213
*Cory J. Butz, André E. dos Santos, Jhonatan S. Oliveira,
and Christophe Gonzales*

Flexible Approximators for Approximating Fixpoint Theory............. 224
*Fangfang Liu, Yi Bi, Md. Solimul Chowdhury, Jia-Huai You,
and Zhiyong Feng*

Learning Statistically Significant Contrast Sets 237
Mohomed Shazan Mohomed Jabbar and Osmar R. Zaïane

A Connection Calculus for the Description Logic $\mathcal{ALC}$ 243
Fred Freitas and Jens Otten

Representation, Reasoning, and Learning for a Relational Influence
Diagram Applied to a Real-Time Geological Domain 257
Matthew Dirks, Andrew Csinger, Andrew Bamber, and David Poole

Nearly Counterfactual Revision................................ 263
Aaron Hunter

Improving Conversation Engagement Through Data-Driven Agent
Behavior Modification... 270
Michael Procter, Fuhua Lin, and Robert Heller

Active Recruitment Mechanisms for Heterogeneous Robot Teams
in Dangerous Environments 276
Geoff Nagy and John Anderson

Streams and Distributed Computing

Compression of General Bayesian Net CPTs 285
  Yang Xiang and Qian Jiang

Sampling Graphical Networks via Conditional Independence Coupling
of Markov Chains.. 298
  Guichong Li

Threaded Ensembles of Supervised and Unsupervised Neural Networks
for Stream Learning 304
  Yue Dong and Nathalie Japkowicz

A Density-Grid Based Clustering Algorithm on Data Stream Using
Resilient Distributed Datasets 316
  Yuan Zhang and Jiongmin Zhang

Distributed Gaussian Mixture Model Summarization Using the MapReduce
Framework... 323
  Arina Esmaeilpour, Elnaz Bigdeli, Fatemeh Cheraghchi, Bijan Raahemi,
  and Behrouz H. Far

Erratum to: Advances in Artificial Intelligence E1
  Richard Khoury and Christopher Drummond

Author Index 337

Actions and Behaviours

Action Recognition by Pairwise Proximity Function Support Vector Machines with Dynamic Time Warping Kernels

Mohammad Ali Bagheri[1,2(✉)], Qigang Gao[1], and Sergio Escalera[3,4]

[1] Faculty of Computer Science, Dalhousie University, Halifax, Canada
bagheri@cs.dal.ca, ma.bagheri@gmail.com
[2] Faculty of Engineering, University of Larestan, Lar, Iran
[3] Computer Vision Center, Campus UAB, Edifici O, 08193 Bellaterra, Spain
[4] Dept. Matemtica Aplicada i Anlisi, Universitat de Barcelona,
Gran Via de les Corts Catalanes 585, 08007 Barcelona, Spain

Abstract. In the context of human action recognition using skeleton data, the 3D trajectories of joint points may be considered as multi-dimensional time series. The traditional recognition technique in the literature is based on time series dis(similarity) measures (such as Dynamic Time Warping). For these general dis(similarity) measures, k-nearest neighbor algorithms are a natural choice. However, k-NN classifiers are known to be sensitive to noise and outliers. In this paper, a new class of Support Vector Machine that is applicable to trajectory classification, such as action recognition, is developed by incorporating an efficient time-series distances measure into the kernel function. More specifically, the derivative of Dynamic Time Warping (DTW) distance measure is employed as the SVM kernel. In addition, the pairwise proximity learning strategy is utilized in order to make use of non-positive semi-definite (PSD) kernels in the SVM formulation. The recognition results of the proposed technique on two action recognition datasets demonstrates the ourperformance of our methodology compared to the state-of-the-art methods. Remarkably, we obtained 89 % accuracy on the well-known MSRAction3D dataset using only 3D trajectories of body joints obtained by Kinect.

1 Introduction

Support Vector Machine (SVM) is one of the leading pattern classification techniques used in various vision application tasks, such as image and video recognition [29]. Given labeled training data of the form $\{(x_i, y_i)\}_{i=1}^m$, with $y_i \in \{-1, +1\}$[1], the standard form of SVM finds a hyperplane which best separates the data by minimizing a constrained optimization problem:

$$\tau(w, \xi) = \frac{1}{2}||w||^2 + C \sum_{i=1}^{m} \xi_i \tag{1}$$

---

[1] In our formulation, the input samples, x_i, are not restricted to be a subset of R^n and can be any set, e.g. set of images or videos.

© Springer International Publishing Switzerland 2016
R. Khoury and C. Drummond (Eds.): Canadian AI 2016, LNAI 9673, pp. 3–14, 2016.
DOI: 10.1007/978-3-319-34111-8_1

$$\text{subject to: } y_i((w.x_i) + b) + \xi_i \geq 1$$
$$\xi_i \geq 0$$

where ξ_i are slack variables and $C > 0$ is the tradeoff between a large margin and a small error penalty.

The cornerstone of SVM is that non-linear decision boundaries can be learnt using the so called 'kernel trick'. A *Kernel* is a function $\mathcal{K} : \mathcal{X} \times \mathcal{X} \mapsto \mathcal{R}$, such that for all x_i, $i = 1, \ldots, m$ yields to a symmetric positive semi-definite (PSD) matrix K, where $K_{ij} = \kappa(x_i, x_j)$. Indeed, the kernel function implicitly maps their inputs into high-dimensional *feature spaces*, $x \mapsto \Phi(x)$. Two common kernel functions are the Gaussian Kernel and linear kernel.

In the dual formulation, the SVM algorithm maximizes:

$$W(a) = \sum_{i=1}^{m} \alpha_i - \frac{1}{2} \sum_{ij} \alpha_i \alpha_j y_i y_j \kappa(x_i, x_j) \tag{2}$$

$$\text{subject to: } 0 \leq \alpha_i \leq C \text{ and } \sum_{\alpha_i y_i} = 0$$

The decision function is given by:

$$f(x) = sign\left(\sum_{i=1}^{m} y_i \alpha_i \kappa(x, x_i) + b \right) \tag{3}$$

where the threshold b is defined as:

$$b = y_i - \sum_{i=1}^{m} y_i \alpha_i \kappa(x_i, x_j) \tag{4}$$

In this paper, we aim to classify human actions by employing spatio-temporal information of skeleton joint points, i.e. the real positions of body joints over the time. More specifically, we use the 3D trajectories of dominant body joints obtained by the Kinect camera. These trajectories encode significant discriminative information and is sufficient for human beings to recognize different actions [7]. In addition, according to an influential computational model of human visual attention theory [21], visual attention leads to visual salient entities, which provide selective visual information to make human visual perception efficient and effective. Trajectories of skeleton joints are visual salient points of human body, and their movements in 4D space reflect motion semantics.

From the classification point of view, these trajectories may be considered as multi-dimensional time series. The traditional recognition technique in the literature is based on time series dis(similarity) measures (such as Dynamic Time Warping). For these general dis(similarity) measures, k-nearest neighbor algorithms are a natural choice. In practice, given two actions represented by two multi dimensional time series, a time series distance measure calculates the distance between two actions. To classify an unlabeled test action (sample), its distance to all training samples is calculated. Consequently, the nearest neighbor algorithm is employed for classification. Given a test action, we calculate its distance to all training actions, e.x. by using DTW, and the target of the closest sample is predicted as the target class.

In general, the k-NN classification algorithms work reasonably well; but are known to be sensitive to noise and outliers. Since SVMs often outperform k-NNs on many practical classification problems where a natural choice of PSD kernels exists, it is desirable to extend the applicability of kernel SVMs.

In our action classification problem, however, time series distances measures are generally non-PSD kernels and basic SVM formulations are not directly applicable. To include non-PSD kernels in SVM, several ad-hoc strategies have been proposed. The straightforward strategy is to simply overlook the fact that the kernel should be non-PSD. In this case, the existence of a Reproducing Kernel Hilbert Space is not guaranteed [18] and it is no longer clear what is going to be optimized.

Another strategy, which has been applied in our work, is based on *pairwise proximity function* SVM(ppfSVM) [5]. This strategy involves the construction of a set of inputs such that each sample is represented with its dis(similarity) to all other samples in the dataset. The basic SVM is then applied to the transformed data in the usual way. As a consequence, sparsity of the solution may be lost. The ppfSVM is related to the arbitrary kernel SVM, a special case of the generalized Support Vector Machines [13]. The name is due to the fact that no restrictions such as positive semi-definiteness, differentiability or continuity are put on the kernel function.

In this paper, we investigate the effectiveness of this strategy for human action classification when the pairwise similarities are based on time-series distances measures. More specifically, we demonstrate the effectiveness of the derivative of Dynamic Time Warping (DTW), as SVM kernel function. The experimental results on two benchmark datasets prove the outperformance of the proposed method compared to the state-of-the-art techniques.

The contributions of our work are as follows: (1) we propose a new class of Support Vector Machine (SVM) that is applicable to trajectory classification, such as action recognition; (2) we introduce the derivatives of Dynamic Time Warping distance measures as pairwise similarity measures for SVM kernel; (3) we demonstrate the validity of the proposed methodology for action/gesture classification.

The rest of the paper is organized as follows: Sect. 2 reviews the related work on action recognition, and briefly introduces Dynamic Time Warping. Section 3 presents our methodology for action recognition. Section 4 evaluates the proposed method and Sect. 5 concludes the paper.

2 Related Work

2.1 Action Recognition

The fast and reliable recognition of human actions from captured videos has been a goal of computer vision for decades. Robust action recognition has diverse applications including gaming, sign language interpretation, human-computer interaction (HCI), surveillance, and health care. Understanding gestures/actions from a real-time visual stream is a challenging task for current computer vision algorithms. Over the last decade, spatial-temporal (ST) volume-based holistic approaches and local ST feature representations have been reportedly achieved good performance on some action datasets, but they are still far from being able to express the effective visual information for efficient high-level interpretation.

Various representational methodologies have been proposed to recognize human actions/gestures. Based on extracted salient points or regions [9] from ST volume, several local ST descriptor methods, such as HOG/HOF [10] and extended SURF [3] have been widely used for human action recognition from RGB data. Inspired from the text mining area, the intermediate level feature descriptor for RGB videos, Bag-of-Word (BoW) [12,23], has been developed due to its semantic representation and robustness to noise. Recently, BoW-based methods have been extended to depth data.

Development of low-cost depth sensors with acceptable accuracy has greatly simplified the task of action recognition [19]. Most importantly, the recent release of the Microsoft Kinect camera and its evolving skeleton joints detection technique in late 2011 led to a substantial revolutionary effect in the field of Computer Vision and created a wide range of opportunities for demanding applications. Shotton et al. [19] proposed one of the greatest advances in the extraction of the human body pose from depth data, which is provided as a part of the Kinect platform. Their work enables us to recover 3D positions of skeleton joints in real time and with reasonable accuracy.

Since 2011, hundreds of studies are devoted to action analysis using depth information. In [28], visual features for activity recognition are computed based on the spatial and temporal differences among detected joints. This feature set contains information about static posture, motion, and offset. Then, Naive Bayes Nearest Neighbor method was applied for the classification task. Alternatively, a histogram of 3-D joint locations (HOJ3-D) for body posture representation is proposed in [27]. In this representation, the 3D space is partitioned into bins using a spherical coordinate system, and the HOJ3-D histogram is constructed by casting joints into certain bins. After applying linear discriminant analysis (LDA) for dimensionality reduction, HOJ3-D vectors are clustered into k posture visual words. The temporal behaviour of these visual words is coded by discrete HMMs. Reyes et al. [16] used 15 joints from Primesense API to represent a human model. Dynamic Time Warping (DTW) with weighted joints is used to achieve real-time action recognition. Sung et al. [20] proposed a 459-element feature vector from various body joints for each frame, and then a two-layered Maximum Entropy Markov Model (MEMM) was applied to recognize single person activities. Despite active research for action/gesture recognition, none of the previous skeleton-based approaches considers a multiple classifier system philosophy. In [6], Bag-of-Visual-and-Depth-Words defined containing a vocabulary from RGB and depth sequences. This novel representation was also used to perform multi-modal action recognition. In [1], the authors proposed an ensemble

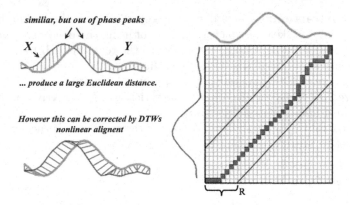

Fig. 1. Top left: two time series which are similar but out of phase produce a large Euclidean distance. Bottom left: this can be corrected by DTWs nonlinear alignment. Right: to align the signals we construct a warping matrix, and search for the optimal warping path

of five action learning techniques, each performing the recognition task from a different perspective and combined the outputs of these classifiers based on the Dempster-Shafer combination theory.

2.2 Dynamic Time Warping

Dynamic Time Warping (DTW) is a well-known algorithm which aims to compare and align two temporal sequences, taking into account that sequences may vary in length (time) [16]. DTW employs the dynamic programming technique to find the minimal distance between two time series, where sequences are warped by stretching or shrinking the time dimension. Although it was originally developed for speech recognition [17], it has also been employed in many other areas like handwriting recognition, econometrics, and action recognition.

An alignment between two time series can be represented by a warping path which minimizes the cumulative distance, shown in Fig. 1. The DTW distance between time series x and y of length n and m will be recursively defined as:

$$DTW(i,j) = d(i,j) + min \begin{cases} DTW(i, j-1) \\ DTW(i-1, j) \\ DTW(i-1, j-1) \end{cases}$$

Here, $d(i,j)$ is the square Euclidean distance of x_i and y_j.

3   The Proposed Algorithm

The proposed algorithm works as follows:

1. **Feature extraction:** Given a depth image, 20 joints of the human body can be tracked by the skeleton tracker. Instead of using the positions of joints, we employ the relative position of each joint to the torso at each frame, as more discriminative and intuitive 3D joint features.
2. **Compute non-PSD kernels:** we compute the pairwise distance of each normalized 3D trajectory to other trajectories, using the derivative of DTW, as described in following subsections.
3. **Classification:** we train the ppfSVM using the computed kernel and evaluate the model on unseen test samples.

3.1   Kernels from Pairwise Data

According to [5], it is assumed that instead of a standard kernel function, all that is available is a proximity function, $P : \mathcal{X} \times \mathcal{X} \mapsto R$. No restrictions are placed on the function P, not symmetry nor even continuity. The mapping $\Phi(x)$ is defined by:

$$\Phi(x) : x \mapsto (P(x, x_1), P(x, x_2), \dots, P(x, x_m)^T \tag{5}$$

where $x_i, i = 1, \dots, m$ are the examples in dataset. Here, we represent each sample x_i by $x_i = \Phi_m(x_i)$ i.e. an m-dimensional vector containing proximities to all other samples in the dataset. Let P denote the $m \times m$ matrix with entries $P(x_i, x_j), i, j = 1, \dots, m$. Using the linear kernel on this data representation, the resulting kernel matrix becomes $K = PP^T$. In this case the decision rule (3) simplifies to

$$f(x) = sign\left(\sum_{i=1}^{m} y_i \alpha_i P \Phi_m(x) + b \right) \tag{6}$$

All elements of $\Phi_m(x_i)$ must be computed when classifying a point x.

3.2   Kernel Using Derivative of Dynamic Time Warping Distance

Despite the success of time series dis(similarity) measures they may fail in some situations. For example, since the DTW algorithm aims to explain variability in the Y-axis by warping the X-axis, it may results in unintuitive alignments where a single point on one sequence maps onto a large subsection of the other sequence; which is referred to as *"singularity"* in the related literature [8]. Also, they may fail to find obvious, natural alignments of two time series simply because a feature (i.e. peak, valley, inflection point, plateau etc.) in one series is slightly higher or lower than its corresponding feature in the other time series.

To deal with such problems, the derivatives version of DTW are employed in this work in order to consider the higher level features. This modified version is named *Derivative DTW (DDTW)* as defined as following:

$$DDTW(x, y) = DTW(\nabla x, \nabla y) \tag{7}$$

where ∇x and ∇y are the first discrete derivatives of x and y.

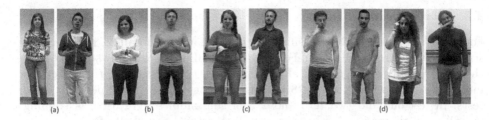

Fig. 2. Some example gestures in the Chaleran dataset are very easy to be confused, even from human visual perception. (a) *Che vuoi* vs. *Che due palle*. For the *Che vuoi* gesture, both hands are in front of the chest area; where for *Che due palle* gesture they are near the waist region. (b) *Vanno d'accordo* vs. *Cos hai combinato*: both hand positions are very close and with the same motion directions; (c) both gestures, *Si sono messid'accordo* and *non ce ne piu*, require hand rotations; (d) four gestures, *Furbo, seipazzo, buonissimo*, and *cosatifarei* are required with the finger pointing to the head area, which cannot be easily determined, even with human eyes.

4 Experimental Evaluation

Here, we present the experimental details of evaluation, including the datasets used, settings of the experiments, as well as the obtained results. The codes were implemented in C/C++ with an interface in Matlab and is available upon request.

4.1 Datasets

We evaluated our framework on two publicly available datasets: the Multi-modal Gesture Recognition Challenge 2013 (Chalearn) and MSR Action3D.

Chalearn Dataset: This dataset is a newly released large video database of 13,858 gestures from a lexicon of 20 Italian gesture categories recorded with a Kinect camera, including audio, skeletal model, user mask, RGB and depth images [4]. It contains image sequences capturing 27 subjects performing natural communicative gestures and speaking in fluent Italian, and is divided into development, validation and test parts. We conducted our experiments on the depth images of development and validation samples which contains 11,116 gestures across over 680 depth sequences. Each sequence lasts between 1 and 2 min and contains between 8 and 20 gesture samples, around 1,800 frames. Some examples of RGB images are shown in Fig. 2.

MSRAction3D Dataset: This dataset [11] is a well-known benchmark dataset for 3D action recognition. This dataset contains 20 actions, including *high arm wave, horizontal arm wave, hammer, hand catch, forward punch, high throw, draw x, draw tick, draw circle, hand clap, two hand wave, side-boxing, bend, forward kick, side kick, jogging, tennis swing, tennis serve, golf swing, pick up & throw*. Each action was performed 2 or 3 times by each subject. Skeleton joint data of each frame is available having a variety of motions related to arms, legs, torso, and their combinations. In total, there are 567 depth map sequences with a resolution of 320 × 240.

4.2 Classification Results

For Chalearn dataset, the classification performance is obtained by means of stratified 5-fold cross-validation. For MSR Action3D dataset, many studies follow the experimental setting of Li et al. [11], such that they first divide the 20 actions into three subsets, each having 8 actions. For each subset, they perform three tests. In test one and two, 1/3 and 2/3 of the samples were used as training samples and the rest as testing samples. In the third test, half of the subjects are used as training and the rest subjects as testing. The experimental results on the first two tests are generally very promising, mainly more than 90 % accuracy. On the third test, however, the recognition performance dramatically decreases. It shows that many of these methods do not have good generalization ability when a different subject is performing the action, even in the same environmental settings. In order to have more reliable results, we followed the same experimental setup of [15, 25]. In this setting, actors 1,3,5,7, and 9 are used for training and the rest for testing.

The summaries of the results are reported in Table 1 for Chalearn and MSRAction3D datasets. In these tables, accuracies of traditional k-NN-based techniques using DDTW distance measures along with the corresponding accuracies using ppfSVMs are reported. It is important to note the outperformance of the results in comparison with the traditional kNN-based classifiers. The result are quite promising, considering the facts that the skeleton tracker sometimes fails and the tracked joint positions are quite noisy. In addition, the confusion

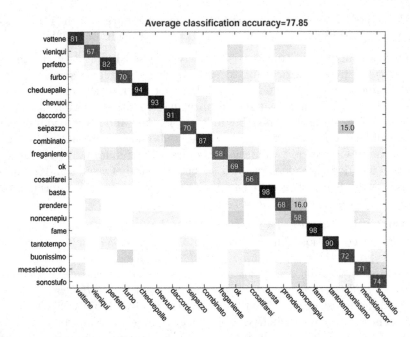

Fig. 3. Confusion matrices of the proposed technique on the Chalearn dataset.

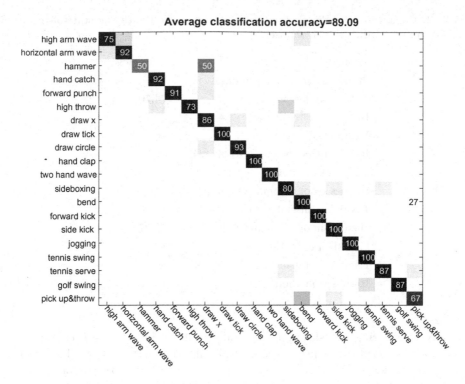

Fig. 4. Confusion matrices of the proposed technique on the MSRAction3D dataset.

Table 1. Classification accuracy of different learners on the Chalearn and MSRAction3D datasets.

	MSRAction3D dataset	Chalearn dataset
K-NN	80.12	68.90
ppfSVM	89.09	77.85

matrix of the proposed classification algorithm for these datasets are demonstrated in Figs. 3 and 4. It is important to note the outperformance of the results in comparison with the traditional kNN-based classifiers. The result are quite promising, considering the facts that the skeleton tracker sometimes fails and the tracked joint positions are quite noisy.

We then compare our classification results on MSRAction3D dataset with state-of-the-art methods. Table 2 shows the accuracy of our method and the rival methods on this dataset based on the cross-subject test setting [11]. Most studies use depth data in addition to skeleton joint information. However, processing sequences of depth images is much more computationally intensive.

The results provided in Table 2 along with the confusion matrix depicted in Figs. 3 and 4 demonstrate the superiority of the proposed methodology.

Table 2. Comparing classification accuracy of our method with the state-of-the-art methods on the MSRAction3D dataset.

Method	Accuracy
Studies employed depth data	
Action Graph [11]	74.70
HON4D [15]	85.85
Vieira et al. [22]	78.20
Random Occupancy Patterns [24]	86.50
DMM-LBP-FF [2]	87.90
Studies employed only skeleton data	
Actionlet Ensemble [26]	88.20
Histogram of 3D Joint [27]	78.97
GB-RBM & HMM [14]	80.20
Ensemble classification [1]	84.85
Proposed method	89.09

By only considering the skeleton data, our results achieved the accuracies of many works based on depth data Considering the fact that we have only employed the skeleton data, not depth sequences, the results are promising. The confusion matrix also reveals that almost all classes, except the "Hammer" class have been classified very well. This is due to fact that skeleton tracker sometimes fails and the tracked joint positions are quite noisy.

5   Conclusion

In this paper, we tackled the problem of human action classification using the 3D trajectories of body joint positions over the time. To do that, we utilized the derivatives of two time series distance measures, including Dynamic Time Warping and Longest Common subsequences. However, instead of employing these general measures as a distance measure for k-NN, we transformed these measures using the pairwise proximity function in order to be used for powerful SVM classification algorithm. Comparing the recognition results of the proposed methods with state-of-the-art techniques on two action recognition datasets, showed significant performance improvements. Remarkably, we obtained 89 % accuracy on the well-known MSRAction3D dataset using only 3D trajectories of body joints obtained by Kinect.

References

1. Bagheri, M.A., Hu, G., Gao, Q., Escalera, S.: A framework of multi-classifier fusion for human action recognition. In: 2014 22nd International Conference on Pattern Recognition (ICPR), pp. 1260–1265. IEEE (2014)
2. Chen, C., Jafari, R., Kehtarnavaz, N.: Action recognition from depth sequences using depth motion maps-based local binary patterns. In: WACV, pp. 1092–1099 (2015)
3. Dollar, P., Rabaud, V., Cottrell, G., Belongie, S.: Behavior recognition via sparse spatio-temporal features. In: IEEE Workshop on Visual Surveillance and Performance Evaluation of Tracking and Surveillance, pp. 65–72. IEEE (2005)
4. Escalera, S., Gonzlez, J., Bar X., Reyes, M., Lopes, O., Guyon, I., Athistos, V., Escalante, H.: Multi-modal gesture recognition challenge 2013: dataset and results. In: ICMI (2013)
5. Graepel, T., Herbrich, R., Bollmann-Sdorra, P., Obermayer, K.: Classification on pairwise proximity data. In: Advances in Neural Information Processing Systems, pp. 438–444 (1999)
6. Hernndez-Vela, A., Bautista, M.A., Perez-Sala, X., Ponce, V., Bar X., Pujol, O., Angulo, C., Escalera, S.: BoVDW: bag-of-visual-and-depth-words for gesture recognition. In: ICPR, pp. 449–452. IEEE (2012)
7. Johansson, G.: Visual perception of biological motion and a model for its analysis. Percept. Psychophy. **14**(2), 201–211 (1973)
8. Keogh, E.J., Pazzani, M.J.: Derivative dynamic time warping. SIAM (2001)
9. Laptev, I.: On space-time interest points. IJCV **64**(2–3), 107–123 (2005)
10. Laptev, I., Marszalek, M., Schmid, C., Rozenfeld, B.: Learning realistic human actions from movies. In: IEEE Conference on Computer Vision and Pattern Recognition, pp. 1–8. IEEE (2008)
11. Li, W., Zhang, Z., Liu, Z.: Action recognition based on a bag of 3D points. In: CVPR Workshop (CVPRW), pp. 9–14. IEEE (2010)
12. Liu, J., Kuipers, B., Savarese, S.: Recognizing human actions by attributes. In: IEEE Conference on Computer Vision and Pattern Recognition, pp. 3337–3344. IEEE (2011)
13. Mangasarian, O.L.: Generalized support vector machines. In: Advances in Neural Information Processing Systems, pp. 135–146 (1999)
14. Nie, S., Ji, Q.: Capturing global and local dynamics for human action recognition. In: 2014 22nd International Conference on Pattern Recognition (ICPR), pp. 1946–1951. IEEE (2014)
15. Oreifej, O., Liu, Z., Redmond, W.: HON4D:: histogram of oriented 4D normals for activity recognition from depth sequences. In: IEEE Conference on Computer Vision and Pattern Recognition (2013)
16. Reyes, M., Dominguez, G., Escalera, S.: Feature weighting in dynamic timewarping for gesture recognition in depth data. In: CVPR Workshops (CVPRW), pp. 1182–1188. IEEE (2011)
17. Sakoe, H., Chiba, S.: Dynamic programming algorithm optimization for spoken word recognition. IEEE Trans. Acoust. Speech Sig. Process. **26**(1), 43–49 (1978)
18. Schölkopf, B., Smola, A.J.: Learning with Kernels: Support Vector Machines, Regularization, Optimization, and Beyond. MIT Press, Cambridge (2002)
19. Shotton, J., Fitzgibbon, A., Cook, M., Sharp, T., Finocchio, M., Moore, R., Kipman, A., Blake, A.: Real-time human pose recognition in parts from single depth images. In: IEEE Conference on Computer Vision and Pattern Recognition (2011)

20. Sung, J., Ponce, C., Selman, B., Saxena, A.: Unstructured human activity detection from RGBD images. In: ICRA, pp. 842–849. IEEE (2012)
21. Treisman, A., Schmidt, H.: Illusory conjunctions in the perception of objects. Cogn. Psychol. **14**(1), 107–141 (1982)
22. Vieira, A.W., Nascimento, E.R., Oliveira, G.L., Liu, Z., Campos, M.F.M.: STOP: space-time occupancy patterns for 3D action recognition from depth map sequences. In: Alvarez, L., Mejail, M., Gomez, L., Jacobo, J. (eds.) CIARP 2012. LNCS, vol. 7441, pp. 252–259. Springer, Heidelberg (2012)
23. Wang, H., Klaser, A., Schmid, C., Liu, C.L.: Action recognition by dense trajectories. In: IEEE Conference on Computer Vision and Pattern Recognition, pp. 3169–3176. IEEE (2011)
24. Wang, J., Liu, Z., Chorowski, J., Chen, Z., Wu, Y.: Robust 3D action recognition with random occupancy patterns. In: Fitzgibbon, A., Lazebnik, S., Perona, P., Sato, Y., Schmid, C. (eds.) ECCV 2012, Part II. LNCS, vol. 7573, pp. 872–885. Springer, Heidelberg (2012)
25. Wang, J., Liu, Z., Wu, Y., Yuan, J.: Mining actionlet ensemble for action recognition with depth cameras. In: IEEE Conference on Computer Vision and Pattern Recognition, pp. 1290–1297. IEEE (2012)
26. Wang, J., Liu, Z., Wu, Y., Yuan, J.: Learning actionlet ensemble for 3D human action recognition. PAMI **36**(5), 914–927 (2014)
27. Xia, L., Chen, C.C., Aggarwal, J.: View invariant human action recognition using histograms of 3D joints. In: CVPR Workshops (CVPRW), pp. 20–27. IEEE (2012)
28. Yang, X., Tian, Y.: Eigenjoints-based action recognition using naive-bayes-nearest-neighbor. In: CVPR Workshops (CVPRW), pp. 14–19. IEEE (2012)
29. Yang, X., Tian, Y.L.: Action recognition using super sparse coding vector with spatio-temporal awareness. In: Fleet, D., Pajdla, T., Schiele, B., Tuytelaars, T. (eds.) ECCV 2014, Part II. LNCS, vol. 8690, pp. 727–741. Springer, Heidelberg (2014)

An Agent-Based Architecture for Sensor Data Collection and Reasoning in Smart Home Environments for Independent Living

Thomas Reichherzer$^{(\boxtimes)}$, Steven Satterfield, Joseph Belitsos, Janusz Chudzynski, and Lamar Watson

Department of Computer Science, University of West Florida, 11000 University Pkwy, Pensacola, FL 32514, USA
{treichherzer,jchudzynski}@uwf.edu,
{sms23,jrb68,lw28}@students.uwf.edu

Abstract. There has been a tremendous growth in new sensor technology and wireless, handheld computing devices that give rise to new opportunities for smart home applications. With advances in the technology, reduced costs of operation, and the ubiquity of WiFi and cellular networks, the reach and value of smart home technology is growing giving rise to applications of smart home systems to support independent living of the elderly. Such systems must collect and analyze data in the home to recognize unusual activities and alert care givers and/or family members in emergency situations. To address this challenge we developed a flexible and scalable multi-agent system for sensor data collection, integration, processing, and alert management. A prototype system for activity recognition has been built and preliminary tests demonstrate that activities can be successfully recognized based on data captured in the home. As a next step, the sensor network will be deployed in an inpatient residence facility to collect real-world data for evaluating the system's performance and to develop new applications.

Keywords: Agent-based architecture · Sensor network · Activity recognition · Case-based reasoning

1 Introduction

Smart devices are increasingly present in our daily lives and they are projected to be progressively in demand over the next 20 years for applications of healthcare driven in part by the need to improve healthcare services and products for an aging population. Projections show that as the baby-boomer generation reaches age 65 at the rate of 10,000 per day, nearly one in every five Americans will be over the age of 65 by 2029 [1]. Increasing healthcare costs and a growing shortage of professional caregivers as well as a strong desire of elderly people to age in their home are strong predictors of the coming market for smart home technology for elder care [2, 3].

If smart home technology is to gain a wider acceptance for applications in elder care, it must be capable of monitoring the resident around the clock, assisting residents with their daily tasks and alerting family members and caregivers in emergency situations.

R. Khoury and C. Drummond (Eds.): Canadian AI 2016, LNAI 9673, pp. 15–20, 2016.
DOI: 10.1007/978-3-319-34111-8_2

To provide such capabilities, the technology must acquire contextual information from sensors in the home and reason without intervention or maintenance to make decisions and provide support. The technology must be extensible allowing newly available sensors to be easily integrated into the smart home while not interfering with other, existing components of the system. Furthermore, the system's reasoning capabilities must be adaptable using machine-learning techniques to be able to change to subtle variations in the behavior patterns of the resident and new emerging needs in the home.

This paper describes the current status and recent achievements in building a Smart Independent Living for Elders (SMILE) home focusing on the sensor technology, system architecture, and revised machine-learning methods, specifically case-based reasoning, based on prior work [4] to improve activity recognition and adaptation capabilities. The objective for the SMILE project is to use off-the-shelf sensors and small computing devices for building wireless sensor networks and applying novel computing algorithms for data processing and analysis to provide support for SMILE residents in the form of reminders or suggestions to complete activities.

The paper is organized as follows: the second and next section describes the sensor network and the agent-based architecture for collecting and analyzing sensor data, providing reasoning and learning support, and generating responses. The third section describes the agents that have been implemented for the prototype system along with a short description of the reasoning algorithm used for activity recognition. The fourth section describes recent experiments and their results to evaluate the performance of the applied methods. The fifth and final section discusses future experiments and directions for the SMILE project.

2   The SMILE Home Multi-agent Architecture

An important design criteria for a smart home system's architecture is its ability to incorporate new technology into the home to meet the needs of the residents without causing disruption to services already provided by the home. Any bug fixes or upgrades to software that improve system reliability and performance must not be disruptive to the system's data collection, processing, or response services. Thus, the SMILE home was designed to be comprised of independent, distributed components modeled as software agents that use multiple databases to record and exchange messages. Each SMILE device, whether a sensor, a widget for visualization in a browser, or an actuator, is managed by a software agent that acts as a wrapper to the technology providing an interface for data collection and control of the device to generate a response of the home.

There are three different types of software agents operating within the SMILE home, each integrated within a three-layer system architecture comprised of a sensor, a middleware, and an application layer. Sensor agents operate at the sensor layer to collect raw sensor data from programmable sensors in the home and store them in a sensor database. They record environmental changes such as the light and temperature in a room, the movement of the resident or things, or the status of appliances. Sensor agents may run on small computing devices permanently installed in the home or on mobile devices carried by a resident.

Middleware layer agents process captured sensor data transforming and combining them to describe events in the home and infer a resident's activities. They can either subscribe to specific events triggered by the database such as the addition of certain environmental data points or poll information of interest from the databases. Agents at the middleware layer use the system's databases for information sharing which enables the agents to respond to specific types of information they need to process to produce new information for other agents to act upon. There is no centralized control system needed to coordinate information and work flows. Middleware layer agents are run by an application server that also executes multiple Web services to provide secured, controlled access to information stored in the system's databases to application layer agents including interface agents and all actuators.

The application layer agents provide control services for devices in the home such as turning off and on the light, offer suggestions or warnings to the residents to support activities or prevent accidents, and provide information about the resident's well being to care takers and family members. The agents may execute on handheld devices or on small computing devices that drive flat screen displays installed in the home to provide opportunities for residents to interact with the SMILE home by accessing recommendations that agents generate or giving feedback to the system.

3   Data Collection, Processing, and Reasoning

As part of the SMILE project a wireless sensor network was built from multiple, low-cost sensors wired to Arduino boards, Raspberry Pi computing devices, and the Nordic nRF24L01+ wireless Radio Frequency (RF24) transceiver modules [5] that connect to the Arduino boards and Raspberry Pis. The RF24 transceivers form together a mesh network for sending data wirelessly from the Arduino boards to the Pis as the sensor base station. Each Arduino board executes a Sensor Data Collector (SDC) software for collecting data from attached sensors; the Raspberry Pi executes the Sensor Data Distribution (SDD) software that receives data packets from one or several SDCs through the RF24 mesh network. The SDD passes received data to sensor agent, also executing on the PI, via internal UDP communication. Figure 1 illustrates the devices and the process for collecting sensor data.

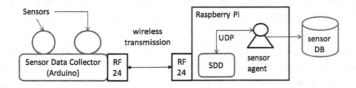

Fig. 1.   Collecting sensor data in the home.

The home's sensor agents execute on a Raspberry Pi or an Android mobile device. Each agent is programmed to collect data from a specific sensor that is given a unique sensor ID. Multiple agents can execute on the same Pi to collect data from different sensors that are within the Pi's reach of the wireless network. Currently, the home provides sensor agents for collecting environmental data such as ambient temperature and light, pressure on sofa chair cushions or a bed to detect residents sitting or laying on them, stove top temperatures and infrared sensors to detect when a burner is on and in use, water flow of a sink faucet, and water level in a kitchen sink. These sensors have been chosen to collect sufficient data to infer normal activities such as getting up in the morning and using the stove and the kitchen sink to prepare a meal.

The information and aggregation agents poll the sensor database at close intervals to transform raw sensor data into meaningful descriptions of environmental events as Resource Description Framework (RDF) triples. The agents use an ontology that models entities and entity types in the home such as rooms, appliances, environmental conditions as well as other things in the home and store produced triples in a triple store database. The activity recognition agent uses case-based reasoning methods [6] to infer activities using data from the triple store and a knowledge base that stores descriptions of activities in the home as a set of individual cases. The case representation for each case includes (1) a textual description and a classification of the case, (2) the origin of the case, which can be either adapted or designed, (3) a history of the case, (4) a description of the problem space including temporal and spatial constraints that must be met for a case to match, and (5) a solution space including the recognized activity and risks the activity may pose on the resident or the home. A case's problem space is modeled using a set of weighted features describing events that are relevant to an activity. The weight associated with each feature, a proportional value among all features in the problem description, indicates the strength of relevance. It is assigned by the knowledge engineer but future work will explore techniques for learning them from training data. A history field records how a case has been previously used for recognition of successfully and unsuccessfully completed activities or whether it had been used for generating new cases using adaptation techniques.

To recognize activities, the agent polls the triple store at regular intervals for new events and attempts to map those events onto the features of cases in the case base. This comparison of the events representing the current environmental context with the existing cases in the case base is performed in a two-stage process: an initial surface-level comparison of the events occurring in the home with the RDF triples representing the known patterns in activities so that a set of possible activities can be isolated, followed by a deeper examination of the spatio-temporal aspects of events in the home. In the initial comparison stage, the agent (1) selects new cases from the case base whose features match any of the newly observed events to compile a list of candidate cases, (2) checks if candidate cases match additional features in the cases' problem space descriptions to reach a minimum threshold for deeper examination of the cases, and (3) checks if features no longer match in candidate cases due to changes in events. In a second stage, the agent examines marked cases more closely to find if the temporal and spatial constraints involving case feature match observed events.

During this stage the agent forms expectations on how to complete an activity successfully by examining features not yet matching in candidate cases but expected to match next. If an event occurs that deviates from what is expected by a case, the agent records an expectation failure. It then decides whether cases may be adapted to model new activities not yet recognized to match observed events or whether the observed events truly represent a failed execution of an activity and warrants intervention by the home. The agent uses annotations associated with constraints to distinguish between adaptable cases and cases representing a serious violation of an activity. Finally, cases representing confirmed activities may be demoted to inactive cases when changes in observed events no longer match features or constraints in the cases.

The interface agents are designed to provide information about the health and well being of the residents and to generate suggestive actions to the residents of the home. Currently, the system provides an information visualization agent operating inside a Web-based dashboard that visualizes the current situation in the home such as temperature and light of individual rooms, status of appliances, and resident location. The agent executes queries on regular intervals to poll environmental information from the sensor and system databases and update graphical widgets integrated into the dashboard. The dashboard makes such information available to trusted individuals. It will be used in upcoming experiments to study patterns of activities in the home.

4 Experiments and Results

A prototype SMILE home has been built and deployed in a test lab environment for initial experimentations. In total, eight sensors (light, temperature, infrared, pressure, water-level sensors), two Raspberry Pis for collecting and distributing sensor data, and ten different iBeacons for indoor localization have been installed. In addition database, application, and Web servers have been deployed on physical servers of the Computer Science department's data center to run middle-layer services. A basic dashboard with widgets for visualization has been implemented and deployed.

To evaluate the introduced algorithm for activity recognition, an experiment was conducted involving several activities in the home. However, as individual components of the system are not yet fully integrated, a synthesized data set of observed events was created and used for the experiment involving five different activities of daily living. The data set was created by a student capturing real-world events that might occur in the home as a result of a series of envisioned activities planned over the course of a day, recording the events as time-stamped RDF triples in the triple store with annotations of the original activity that generated them for evaluation purposes. To make the data set more realistic, noise was added in the form of randomly inserted events describing movements of the resident and objects in the home that are not directly related to the envisioned activities. Finally, a case-base was manually created modeling a total of nine different activities in the home (cooking, managing finances, reading for study, reading for leisure, cleaning the library, cleaning the kitchen, washing dishes, watching television, going to bed) including the five activities that were envisioned by the student.

The algorithm for the activity agent was then executed using the triple store and the knowledge base as described above. Intermediate results of selected and matching cases were recorded. The results were that all nine cases as modeled by the case base were at some point selected as candidate cases of matching activities, six of them were closely examined, and five were recognized as activities matching exactly the activities that were performed during the course of the day. However, the experiment uncovered a weakness in one of the cases, "cooking on left front burner". The case did not include a constraint to adequately recognize an expectation failure when the resident leaves the burner on and walks away from the kitchen. Future work will examine methods to build and evaluate cases for completeness before they are added to the case base.

5 Conclusions and Future Work

This paper describes ongoing work of the SMILE project that aims to build an intelligent system inferring activities and providing suggestions to the residents in the home. In collaboration with the Center on Aging at The University of West Florida and Covenant Hospice, the prototype system will be deployed in the near future to an inpatient residence facility. It will be used in a number of already approved human-subject studies in which residents of the facility perform daily activities to record real-world data. The data will be used to (1) evaluate architectural bottlenecks and the accuracy of inferred activities, (2) develop a process to build a case base with critical cases not yet sufficiently covered, and (3) study issues of trust and security in the system. The ultimate goal will be to develop new applications for SMILE residents, care-takers, and family members to support independent living of at-risk older adults.

References

1. Cohn, D., Taylor, P.: Baby Boomers Approach 65 – Glumly. Pew Research Center, December 2010. http://pewresearch.org/pubs/1834/baby-boomers-old-age-downbeat-pessimism
2. Harkleroad, J.: Boomers' desire to 'age in place' driving new home trends. Silicon Valley/San Jose Bus. J., 26 February 2006. http://www.bizjournals.com/sanjose/stories/2006/02/27/focus3.html
3. El-Basioni, B.M.M., El-Kader, S.M.A., Eissa, H.S.: Designing a local path repair algorithm for directed diffusion protocol. Egypt. Inf. J. **13**(3), 155–169 (2012)
4. Satterfield, S., Reichherzer, T., Coffey, J., El-Sheikh, E.: Application of structural case-based reasoning to activity recognition in smart home environments. In: 11th International Conference on Machine Learning and Applications, Vol. 1, pp. 1–6. IEEE (2012). doi:10.1109/ICMLA.2012.10
5. Nordic Semiconductor, nRF24L01+ Single Chip 2.4 GHz Transceiver Product Specification. Nordic Semiconductor (2007). http://www.nordicsemi.com/
6. Aamodt, A., Plaza, E.: Case-based reasoning: foundational issues, methodological variations, and system approaches. AI Commun. **7**(1), 39–59 (1994)

Biometric Authentication by Keystroke Dynamics for Remote Evaluation with One-Class Classification

Chuan Chang[✉], Thierry Eude, and Luis Eduardo Obando Carbajal

Department of Computer Science and Software Engineering,
Laval University, Quebec city, Canada
{chuan.chang.1,luis-eduardo.obando-carbajal.1}@ulaval.ca,
cchang1002@gmail.com, thierry.eude@ift.ulaval.ca
http://www.ift.ulaval.ca/~eude/index.htm

Abstract. One-Class SVM is an unsupervised algorithm that learns a decision function from only one class for novelty detection: classifying new data as similar (inlier) or different (outlier) to the training set. In this article, we have applied the One-Class SVM to Keystroke Dynamics pattern recognition for user authentication in a remote evaluation system at Laval University. Since all of their students have a short and unique identifier at Laval University, this particular static text is used as the Keystroke Dynamics input for a user to build our own dataset. Then, we were able to identify weaknesses of such a system by evaluating the recognition accuracy depending on the number of signatures and as a function of their number of characters. Finally, we were able to show some correlations between the dispersion and mode of distributions of features characterizing the signatures and the recognition rate.

Keywords: Keystroke dynamics · Biometrics · Person authentication · Machine learning · One-class classification · Support vector machine

1 Introduction

Online education has been growing tremendously over the past years as technology has been integrated into education and training. However, one of the concerns of remote learning is the assessment that requires a process of identification to guarantee learning results.

The authentication is the process of determining whether someone or something is who or what it claims to be [1]. An authentication system can fail in one of two ways: either an authorized user is rejected or an illegal user is incorrectly granted access to the system. The biometric authentication is an automatic method of verifying the identity of an individual based on the measurement of his or her unique physiological traits or behavioral characteristics [4]. Since biometrics cannot be lost, stolen, or overheard like keys or passwords, it is an excellent candidate for identity verification [6].

© Springer International Publishing Switzerland 2016
R. Khoury and C. Drummond (Eds.): Canadian AI 2016, LNAI 9673, pp. 21–32, 2016.
DOI: 10.1007/978-3-319-34111-8_3

The biometrics can be generally divided into two categories: physiological biometrics, which are biological/chemical traits that are innate or naturally developed (face, palm, iris, etc.), and behavioral biometrics, which are mannerisms or traits that are learned or acquired (voice, handwriting, signature, keystroke dynamics, etc.) [4].

Keystroke dynamics is a behavioral biometrics based on the manner and rhythm in which each individual types [4]. It possesses multiple advantages: uniqueness, low implementation and deployment cost, transparency and non-invasiveness [11]. As a behavioral biometrics, keystroke dynamics can be used to identify students for remote evaluations. In order to expect lower error rate, behaviors should be as natural and consistent as possible.

The organization of this paper is as follows. The next section introduces the process steps used for individual identification. After the explanation of the data collection in Sect. 3, Sect. 4 presents the proposed methodology. The experimental results are shown in Sects. 5 and 6 concludes the paper.

2   Keystroke Dynamics Identification

2.1   Input Data

Keystroke dynamics can be applied as either static or dynamic text input. The first one requires the user to type a pre-defined text string. The other allows the user to type any text in an unconstrained way and involves continuous or periodic monitoring of his/her keystroke behavior. During the last three decades, the state of the art of keystroke dynamics says that static text input provides much more acceptable error rates and is significantly easier to be implemented [2]. Robinson et al. have obtained good results using only first names of an average length of 6.4 characters [8]. Bleha et al. used users' first and last names to get slightly better results. In this case, the string length doubled (the length of a full name varies from 11 to 17 characters) [2].

Otherwise, to expect lower error rate, behaviors should be as natural and consistent as possible. Regarding the case of Laval University, the use of the IDUL[1] seems appropriated. All members of Laval University get an unique identifier (IDUL) at their registration. The IDUL is composed of five letters which were generated from the member's first name and last name, followed or not by up to three digits. The IDUL is not confidential information, while the password created by the member should always be kept as a secret. The IDUL and the password together give the member access to his/her personal data and to necessary computer systems to the realization of online activities in Laval University (example: online self-service system for studies, email, online library, the course portal, etc.). In order to provide an identification solution that integrates easily with the existing system, we chose the IDUL to identify the user, which is a very common static string for each member and can be entered regularly and naturally.

---

[1] IDentifiant Université Laval (IDUL).

Furthermore, as shown in Fig. 1, the authentication process by keystroke dynamics can be logically divided into two phases: enrollment phase and verification phase, respectively called training phase and test phase in the domain of classification. During the enrollment/training phase, user's biometric data are acquired, processed and stored as a pattern in a database, for future use by the system. During the subsequent verification/test phase, new user biometric data are acquired and processed. Therefore, the authentication decision will be based on the output of the matching process of the newly presented biometric with the pre-stored reference patterns [4,7].

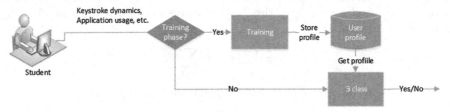

Fig. 1. The two phases for authentication process by keystroke dynamics (training phase and test phase).

2.2 Feature Extraction

The instants when each key is pressed and released are stored as basic data of keystroke dynamics. Two feature vectors based on the data are then generated as input for the classification algorithm (Fig. 2):

- Dwell time: interval between the instants in which a key is pressed and released. It represents the time that the key keeps being pressed.
- Flight time: interval between the instants in which a key is released and the next is pressed. Sometimes it can be negative.

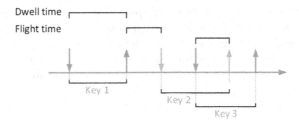

Fig. 2. Feature vectors in keystroke dynamics: the down and up arrows represent respectively the instants of pressing and releasing of each key.

A high resolution time API in JavaScript was used to improve the captured time resolution up to 1 ms. Normally, higher resolution of the captured data results in better classification accuracy.

2.3   Classification with SVM

A classification task usually involves separating data into training and testing sets. Each instance in the training set contains one target value (i.e. the class labels) and several attributes (i.e. features of the keystroke durations).

A SVM (Support Vector Machine) constructs a hyperplane in high or infinite dimensional space for classification [9]. The goal of SVM is to produce a model based on the training data which predicts the target values of the test data, given only the test data attributes. It's normally used for Binary-Class classification. For example, given a training set of instance-label pairs $(x_i, y_i), i = 1, ..., l$ where $x_i \in R^n$ and $y \in \{1, -1\}^l$ the SVM requires the solution of the following optimization problem:

$$\min_{W,b,\xi} \frac{1}{2} W^T W + C \sum_{i=1}^{l} \xi_i$$

$$\text{subject to} \qquad y_i(W^T \phi(x_i) + b) \geq 1 - \xi_i,$$

$$\xi_i \geq 0.$$

In which vector x_i are mapped into a higher or even infinite dimensional space by the function ϕ. SVM finds a linear separating hyperplane with the maximal margin in this higher dimensional space. $C > 0$ is the penalty parameter of the error term. Furthermore, $K(x_i, x_j) \equiv \phi(x_i)^T \phi(x_j)$ is called the kernel function. With the kernel trick, the computational efficiency has been greatly improved and the SVM has been extremely successful as a non-linear classifier. Four basic kernels often used are: linear, polynomial, radial basis function (RBF) and sigmoid [5].

Then we search for an appropriate kernel function and the best values of hyper-parameters to achieve a good separation by the hyperplane in feature space that has the largest distance to the nearest training data point of any class. As a result, the separating function is able to create complex boundaries in original space. Thanks to its superior performance in handling large dimensional data without increasing the system complexity, the SVM has been extensively used in a wide variety of applications. For example, it was adopted for distinguishing imposter patterns by creating a margin that separates normal patterns from imposters by many studies in the state of the art of Keystroke Dynamics. SVM has a competitive advantage regarding performance compared to Neural Network [11]. The LibSVM is apparently a good and simple tool available for classification for Keystroke Dynamics, as an open source and widely used machine learning library.

3   Data Collection

3.1   Positive Data

Due to the particularity of IDUL and the difference between each IDUL, we could not reuse any existing dataset. Instead, we built our own dataset of IDULs with the students who took one remote course in the same class at Laval University.

At this stage, the collection of Keystroke Dynamics data will be done at every access to educational materials. At each connection for visiting the page, we ask the student to input his/her IDUL in order to build his/her keystroke signature. In this way, the collection of signatures is not affected by the ripple effect that appears when the same character string is input several times consecutively. The input strings (IDUL) and their temporal features of keystroke (Dwell time and Flight time) will be calculated and stored in secure data files to which only the researchers have access. A validation process is used to insure the quality of input data. Afterward, the student will be asked if he/she accepts or not to store his/her data for future use in other research. By clicking on a "yes" or "no" button, the response will be stored. Then it will automatically redirect to the page of teaching materials. However, at the first connection of the page, a student could click the link "Do not participate in this research" to refuse the participation. In this case, the requiring of input will no longer be shown to the student, and he/she will be redirected straight to the page of teaching materials every time he/she visits the page.

Since all keystroke signatures were input by the students on their own, this Keystroke Dynamics dataset is called positive data in this research.

3.2   Negative Data

In order to test the performance of the SVM classifier (measured by the rate of successful classification of positive and negative signatures), we must collect negative data which is in fact false students' keystroke signatures. These false signatures simulate the inputs that would come from impostors trying to log in with students' IDULs.

In order to get as close as possible to the collection condition of positive data, we perform the collection of negative data on several sessions for several days, avoiding to train users on the keystroke signatures. We first asked 20 people to create our negative data. They had to type the IDULs of the students who participated earlier to the positive data collection at the first stage of the study. When it is possible, we searched participants with similar characteristics (age or keystroke speed) to our initial students (positive data), for having representative data.

To collect all of this data, we created and implemented webpages where the participants could enter the false keystrokes signatures. The participants were asked to access the webpage where IDULs were listed. Then they had to retype these different IDULs coming from our dataset of positive data. The IDULs and their temporal features of keystroke (dwell time and flight time) were extracted and stored in secure data files as it was done for the positive data collection.

In total, we had three different webpages with 20 IDULs that had to be retyped each time. The participant was given the indication to access the three webpages at three different moments during the day or on different days. We asked them to do so as a way to repeat the action of the original students who connected to the log in page at different times during the day. Assuming that the moment during the day could affect the behavior of a student's typing, we wanted to collect the negative data as representative as possible.

4   Methodology

4.1   One-Class classification

The target application is the identification of a student wishing to take an exam. It aims to determine, from their keystroke, if he/she is the right student or an impostor who wants to take the exam. In fact, it's not always possible in practical condition to obtain sufficient negative data created by imposters for building models. Thus, according to the proposed application, it relates to solve a One-Class classification problem. As a matter of fact, for one-class SVM, it's unnecessary for its training set to contain data patterns of impostors. The negative keystroke signatures were only collected for testing the performance of SVM classifier. After learning the authorized user's data patterns, One-Class SVM can find the domain where his patterns probably live and impostors can be detected as outliers because their patterns are outside the domain, which is also called novelty detection.

Solving the One-Class SVM optimization problem is equivalent to solving the dual quadratic programming problem:

$$\min_{\alpha} \frac{1}{2} \sum_{ij} \alpha_i \alpha_j K(x_i, x_j)$$

subject to the constraints $0 \leq \alpha_i \leq \frac{1}{\nu l}$ and $\sum_i \alpha_i = 1$, where α_i is a lagrange multiplier. ν is a parameter that controls the trade-off between maximizing the distance of the hyperplane from the origin and the number of data points contained by the hyperplane, l is the number of points in the training dataset.

4.2   RBF Kernel and Tuning of Hyper-Parameters

As we mentioned before, the One-Class SVM first requires the choice of a kernel and certain hyper-parameters to define a frontier. Since the distributions of both Dwell time and Flight time of keystrokes are continuous, the RBF kernel is in general a reasonable good choice to deal with non-linear, multi-mode distributions [9,12], with

$$K(x_i, x_j) = exp(-\gamma \parallel x_i - x_j \parallel^2), \gamma > 0$$

where γ is the only one hyper-parameter of RBF kernel, which makes it simpler than other kernels.

The hyper-parameter γ of RBF kernel plays a key role in classifiers. The effect of γ on the tightness of decision boundaries is monotonic. With increasing of γ, the RBF kernel will be more narrow, the number of support vectors increases and the decision boundaries become tighter. Meanwhile the False Negative rate increases and the False Positive rate decreases. If γ is very large, all training vectors will end up as support vectors, which is called Overfitting and is clearly not desirable. Otherwise if γ is very small, the frontier will be too loose and

Underfitting appears. Thus we could get a weak accuracy in the training phase, and it will also decrease the accuracy in the testing phase.

On the other hand, the parameter $\nu \in [0, 1]$ is used to tune the upper bound on the fraction of outliers outside the decision boundaries and the lower bound on the fraction of support vectors in the training dataset [10]. Therefore, it is still possible for One-Class SVM to contain outliers in the training data, since the training data may be polluted by noises caused by the measurement.

However, there is no exact formula or algorithm to set its two main hyper-parameters γ and ν for tuning the frontier. We recommend a "grid-search" on γ and ν using cross-validation. Various pairs of (γ, ν) values are tried and the one with the best cross-validation accuracy is picked. We found trying exponentially growing sequences of γ and ν is a practical way to identify good parameters (for example, $\gamma = 2^{-40}, 2^{-39}, ..., 2^{-13}$, $\nu = 2^{-10}, 2^{-9.9}, ..., 2^{-6}$).

4.3   Experimental Protocol

In order to reduce the noises existing in the measurement of dataset, a pre-processing of data is necessary. According to Joyce and Grupta [3], all keystroke timing data greater than three standard deviations (σ) above the mean reference signature are discarded.

Then, as shown in Fig. 3, we divide the positive data into two parts (4/5 + 1/5). The signatures in the second part (1/5) are selected randomly from the whole positive data. The first part (4/5, called Data 1) is used for training the model in One-Class SVM. The second part is mixed with the negative data (together called Data 2) for testing the obtained classifier model in Binary-Class SVM. Here, we use as many negative signatures as positive ones for each IDUL. In order to avoid the influence of random selection of data to the classification process, we repeat the positive data separation to further carry out the experiment for 3 times.

During the training phase, we use Data 1 and RBF kernel to train the classifier model for each IDUL. The best value pair of hyper-parameters (γ, ν) is searched to find the best model. In this phase, a 5-fold Cross-Validation[2] is used to relieve the Overfitting problem. Then a grid search method is used for searching the best value pair (γ, ν) for each IDUL with LibSVM. This could be regarded as tuning the frontier of classification by trying out all combinations of γ and ν value in a certain range to see the results. At last, we chose the value pair (γ, ν) with the best performance measurement achieved. It is worthwhile

---

[2] k-fold Cross-Validation: the training set is split into k smaller sets. For each of the k "folds", a model is trained using $k - 1$ of the folds as training data, and the resulting model is tested on the remaining part of the data to compute the accuracy. The accuracy reported by k-fold cross-validation is then the average of the values computed in the loop.

to note that the performance measurement of LibSVM "Accuracy[3]" becomes "Recall[4]" in our One-Class training phase.

During the test phase, the obtained classifier model is tested in Binary-Class SVM with Data 2 which has never been learnt by the model. The best performance measurement ("Accuracy" in our Binary-Class test phase) achieved is stored as the final successful classification result.

As mentioned above, in order to avoid the variations due to the random selection of data, we carried out the experiment for 3 times and took the average of the 3 obtained "Accuracy" values for each IDUL as the final performance measurements.

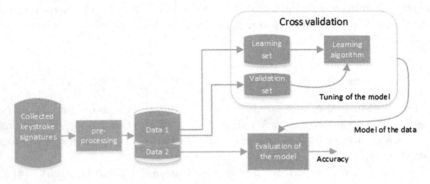

Fig. 3. Learning and model's evaluation process.

5   Results

The collection of positive data took one Canadian semester (15 weeks). After eliminating all invalid data, we finally had 1387 keystroke signatures of 36 students in total. The quantity of keystroke signatures for each student varies from 194 to 12 in the dataset. The collection of negative data took two weeks. After removing all invalid data, we finally had a total of 2247 keystroke signatures associated to a total of 36 students.

First, we searched the tendency of Recall obtained for each IDUL in the training phase. As explained previously, we tried 3 random selections in the positive data to divide it into two parts before the training phase. We have obtained 3 groups of best Recall and hyper-parameters value pair (γ, ν). If we draw all the Recall values obtained in the 3 experiments for each IDUL in terms of their signature numbers in descending order, we will find that only a few signatures showed wide variation of Recall. For example, as shown in Fig. 4, wide variation of Recall was achieved in the case of IDUL #29. This means

---

[3] Accuracy: $\frac{(TP+TN)}{TP+FP+TN+FN}$, Where TP, TN, FP, FN denotes respectively True Positive, True Negative, False Positive and False Negative.

[4] Recall: $\frac{TP}{(TP+FN)}$ (also called True Positive Rate) measures the proportion of positives that are correctly identified.

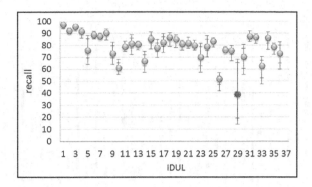

Fig. 4. Best Recall for each IDUL when they are sorted in terms of their signature numbers in descending order.

that only a few positive signatures were able to be recognized with a best model which could be obtained in training phase, while a lot of positive signatures were rejected.

Then, we checked the performance of the models obtained in the training phase. We used the collected negative data together with the three selections of positive data. For each selection of positive data, we add the same number of negative data to obtain the Data 2. The performance measurement in the test phase is given by the recognition rate "Accuracy" which varies from 98 % to 55 % with an average value of 85 % and a standard deviation of 11.2. The average Accuracy obtained in 3 experiments for each IDUL was sorted in terms of the number of its signatures in descending order. We can see the trend by means of the grey line on the top of the Fig. 5. It points out that the Accuracy tends to decrease slightly with the decreasing of the number of signatures (from 197 to 12). However, it appears that the number of signatures has no influence on the False Positive number (green trend line at the bottom of the figure). For some signatures, though the Recall value was good for the model training (see Fig. 4), we found that the performance can be poor with low Accuracy value (Fig. 5). For instance, the IDUL 17 showed a quite high Recall value in the training phase, while the obtained Accuracy was quite low in the testing phase. This may be principally caused by the high False Positive rate (see the green column lines in the Fig. 5).

Besides, great performance difference between some IDULs cannot be ascribed to different quantities of signature for each IDUL as observed above. A statistical study of correlation with distribution characteristics allows us to point out that different behaviours between persons have an influence on the recognition. For example, we can compare the distributions of two IDULs which present different performances. As we see in Fig. 6a, the IDUL #7, which has mono-modal and concentrated distribution within a single keystoke, gets a very good recognition rate. Whereas, if the distribution appears multi-modal and dispersed for a single time feature as shown in Fig. 6b in the case of IDUL #17,

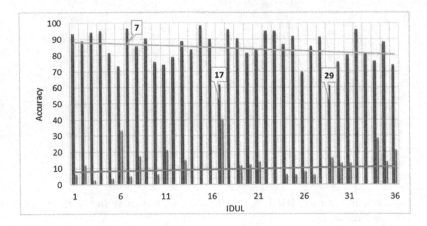

Fig. 5. Evaluation of the performance by the Accuracy measurement. The Accuracy tends to decrease slightly with the decreasing of the number of signatures (grey line). But the number of signatures seems to have no influence on the False Positive number (green line). (Color figure online)

worse recognition rates are found. So we observed a correlation with distribution characteristics, which can be used in the recognition process and can explain why the recognition rates of some IDULs are not satisfied.

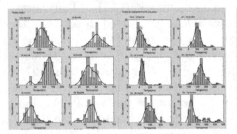

(a) Better Accuracy thanks to mono-modal and concentrated distribution within a single keystoke shown in the case of IDUL #7.

(b) Worse Accuracy caused by multi-modal and dispersed distribution within a single keystoke shown in the case of IDUL #17.

Fig. 6. Distribution of the time features: dwell time in blue, flight time in orange. (Color figure online)

The Fig. 7 shows the influence of the IDUL length (quantity of characters included in each IDUL) on the performance, by displaying the Accuracy as well as False Positive rate in terms of IDUL length in descending order. We can observe that the Accuracy decreases slightly but the False Positive rate increases rapidly with the decrease of IDUL length. Therefore the IDUL length imposed by the registration system is also an important factor to consider.

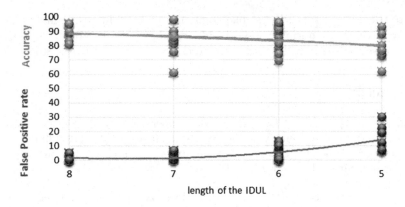

Fig. 7. Evaluation of the performance: Accuracy decreases slightly but the False Positive rate increases rapidly with the decrease of IDUL length.

6 Conclusion

In this article, we study how to identify students for the authentication in a remote evaluation system, using their keystroke signatures of the unique identifiers which are given at their University registration. We established a database of keystroke signatures on the basis of an existing online learning system at Laval University. Then, we used this database to demonstrate that the main difficulty to identify a student who wants to take an exam is not necessarily from the number of signatures which is necessary to collect. Instead, it is more important to obtain homogeneous signatures without too much disparity.

We applied the One-Class SVM which is an unsupervised algorithm that learns a decision function for novelty detection. The RBF kernel was used, and the grid search method was applied for tuning the frontier of its two main hyper-parameters γ and ν. Thus, we determined the best pairs of hyper-parameters for each set of signature associated with the IDULs. After that, we evaluated the performance of models found in terms of accuracy. We were able to show that the authentication system could be effective with satisfactory results after only a few weeks of collection once we had obtained just 12 signatures. Besides, the error rates decrease as the number of signatures increases. We also observed the distributions of the time features in the keystroke signatures. However, we found that the signatures which have multimodal distributions could have poor performance in recognition. Finally, we were able to find that the IDUL length has more influence on the False Positive rate than the quantity of collected signatures.

Future work aims at finding a better method for selecting hyper-parameters of SVM than the one based on Recall by grid search. Furthermore, other kernels could be studied for resolving the problem of multi-modal distributions. Combining kernels is one possible way to combine multiple information sources and allow to learn different modes separately. At last, we could add an additional factor to the data vector of Keystroke Dynamics by including the Shift and Cap-sLock key, which are normally used to identify how a user types in upper case or lower case letters. It will help to reduce the error rates of the classifier.

References

1. Banerjee, S.P., Woodard, D.L.: Biometric authentication and identification using keystroke dynamics: a survey. J. Pattern Recognit. Res. **7**(1), 116–139 (2012)
2. Bleha, S., Slivinsky, C., Hussien, B.: Computer-access security systems using key-stroke dynamics. IEEE Trans. Pattern Anal. Mach. Intell. **12**(12), 1217–1222 (1990)
3. Joyce, R., Gupta, G.: Identity authentication based on keystroke latencies. Commun. ACM **33**(2), 168–176 (1990)
4. Karnan, M., Akila, M., Krishnaraj, N.: Biometric personal authentication using keystroke dynamics: a review. Appl. Soft Comput. **11**(2), 1565–1573 (2011)
5. Lin, C.J., Hsu, C.W., Chang, C.C.: A practical guide to support vector classification. National Taiwan University (2003). www.csie.ntu.edu.tw/cjlin/papers/guide/guide.pdf
6. Monrose, F., Rubin, A.D.: Keystroke dynamics as a biometric for authentication. Future Gener. Comput. Syst. **16**(4), 351–359 (2000)
7. Pisani, P.H., Lorena, A.C.: A systematic review on keystroke dynamics. J. Braz. Comput. Soc. **19**(4), 573–587 (2013)
8. Robinson, J., Liang, V.M., Chambers, J., MacKenzie, C.L., et al.: Computer user verification using login string keystroke dynamics. IEEE Trans. Syst. Man Cybern. Part A: Syst. Hum. **28**(2), 236–241 (1998)
9. Sang, Y., Shen, H., Fan, P.: Novel impostors detection in keystroke dynamics by support vector machine. In: Liew, K.-M., Shen, H., See, S., Cai, W., Fan, P., Horiguchi, S. (eds.) Parallel and Distributed Computing: Applications and Technologies. LNCS, vol. 3320, pp. 666–669. Springer, Heidelberg (2005)
10. Schölkopf, B., Platt, J.C., Shawe-Taylor, J., Smola, A.J., Williamson, R.C.: Estimating the support of a high-dimensional distribution. Neural Comput. **13**(7), 1443–1471 (2001)
11. Teh, P.S., Teoh, A.B.J., Yue, S.: A survey of keystroke dynamics biometrics. Sci. World J. **2013**, 24 (2013). doi:10.1155/2013/408280. Article ID: 408280
12. Yu, E., Cho, S.: Novelty detection approach for keystroke dynamics identity verification. In: Liu, J., Cheung, Y.-M., Yin, H. (eds.) Intelligent Data Engineering and Automated Learning. LNCS, vol. 2690, pp. 1016–1023. Springer, Heidelberg (2003)

Fishing Activity Detection
from AIS Data Using Autoencoders

Xiang Jiang[1], Daniel L. Silver[2], Baifan Hu[1], Erico N. de Souza[1],
and Stan Matwin[1(✉)]

[1] Dalhousie University, Halifax, Canada
{xiang.jiang,baifanhu,erico.souza}@dal.ca, stan@cs.dal.ca
[2] Acadia University, Wolfville, NS, Canada
danny.silver@acadiau.ca

1 Introduction

Marine life has significant impact on our planet, providing food, oxygen and
biodiversity. But, 90 percent of the large fish are gone primarily because of
overfishing, according to the 2010 Census of Marine Life. Thus it is desirable
to detect fishing activities in the ocean. Satellite AIS (Automatic Identification
System) is a vessel identification system that monitors the position of ships
worldwide for collision avoidance, allowing us to track vessels on the ocean. AIS
equipment is required to be fitted aboard international voyaging ships that are
300 tons or above, and all passenger ships. This makes AIS data an ideal source to
detect ship trajectories and fishing activities. This work follows in the footsteps
of [5], which was the first use of Machine Learning approaches for fishing activity
detection. In this paper, we present a new approach to detect fishing activities
using autoencoders with AIS data. This work is the first attempt to use deep
learning approaches to automatically find features in AIS data to detect fishing
activities. The performance of autoencoders is compared with support vector
machines and Random Forests.

2 Background and Related Work

A *ship trajectory* can be defined as the curved path along which a vessel moves.
In the context of AIS data, a ship trajectory consists of a vector of discrete
data points in chronological order of ship movements, $\mathbf{T} = [\mathbf{t}_1, \mathbf{t}_2, \ldots, \mathbf{t}_N]$. Each
data point $\mathbf{t}_i$ includes information about longitude, latitude, speed over ground,
and course over ground. The data arrives at irregular time intervals between
consecutive data points. The task of *fishing activity detection* can be defined
as: given a ship trajectory $\mathbf{T}$, predict a label $\mathbf{y}_i$ for each data point $\mathbf{t}_i$ where
$\mathbf{y}_i \in \{Fishing, NonFishing\}$. There are at least three perspectives that can
be taken on the task of fishing activity detection: structured prediction, time
series classification and image recognition. From the perspective of *structured
prediction*, fishing activity detection can be viewed as a problem of sequence
labeling in which the input is a sequence of ship trajectories and the output is

© Springer International Publishing Switzerland 2016
R. Khoury and C. Drummond (Eds.): Canadian AI 2016, LNAI 9673, pp. 33–39, 2016.
DOI: 10.1007/978-3-319-34111-8_4

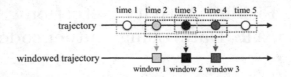

Fig. 1. An example of a sliding window algorithm.

a sequence of labels. From the perspective of *time series classification*, fishing activity detection can be seen as a problem of time series sequence mapping in which the similarity of two sequences are measured. From the perspective of *image recognition*, fishing activity detection can be seen as a problem of image segmentation in which a trajectory is segmented into fishing and non-fishing segments.

Different types of vessels follow different pattens of movement. For longliners, a pattern recognition approach named Lavielle's algorithm has been applied [5]. In contrast, this paper presents a deep learning approach that automatically finds features in the AIS data to detect fishing activities. An autoencoder is an artificial neural network that learns to reproduce its input data with backpropagation. Autoencoder always consists of two parts, encoder and decoder. An encoder converts the input into feature representation, and a decoder converts the feature representation back into a reconstructed input. A particular approach for developing autoencoders is Restricted Boltzmann Machines [1].

3 Theory and Approach

We focus on the problem of detecting if a longliner vessel is in the state of fishing vs. non-fishing. The following describes our approaches and methods. For details please see [3].

3.1 A Sliding Window Approach

We divide a ship trajectory into a sequence of discrete window segments (see Fig. 1). We hope that a window can provide useful context information to aid the classification for a particular data point. While training a model, the label of a window is given by the label of the middle data point from this window. When making predictions, we assign the label of the window centered at that particular data point as its label.

3.2 Data Preparation

Undersampling from Time Intervals. AIS data have different levels of granularity - the time intervals between consecutive AIS data points are irregular. This complicates the classification problem. To reduce the variations and noise introduced from different granularity, we undersample the data **T** based on time

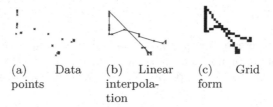

(a) Data points   (b) Linear interpolation   (c) Grid form

Fig. 2. Transforming data points from a sliding window to a matrix.

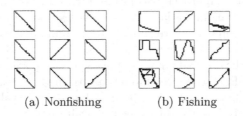

(a) Nonfishing   (b) Fishing

Fig. 3. Examples of non-fishing and fishing activities.

intervals: if a set of consecutive data points are within a certain range of time, say 100 s, one sample is selected from this set. This process aims at producing a more evenly distributed data vector $\mathbf{T_e}$ of time intervals.

Relative Longitudes and Latitudes. If we use absolute longitudes and latitudes, a model that is trained in one region of the ocean will not work in other regions. Thus, we transform absolute longitudes and latitudes of each time window to a relative form $\mathbf{T_r}$ by subtracting the starting position of the window.

Linear Interpolation and Matrix Representation. Because AIS data are samples from a trajectory, we use linear interpolation on $\mathbf{T_r}$ to recover the trajectory. We further convert the interpolated trajectory to an image matrix. Data is normalized in the process of transformation. Figure 2 shows an example of transforming data points from a sliding window to a matrix representation. Figure 3 shows fishing and non-fishing windows when the window size is 20.

Model Development. After the data are transformed to image matrices according to the above procedures, what we typically find is that the data is imbalanced in favour of the majority class. So we undersample the matrices based on the labels. The training data $\mathbf{T_r}$ is undersampled to create $\mathbf{T_r'}$ to balance the two classes. We use Restricted Boltzmann Machines to develop autoencoders using $\mathbf{T_r'}$, then use backpropagation to fine-tune a standard feedforward neural network built on top of the RBM using $\mathbf{T_r}$. This leads to better results than finetuning on $\mathbf{T_r'}$. This is because, during unsupervised learning, the features developed are equally distributed over both classes. We propose that the RBM autoencoder will discover features in the sliding window matrices that characterize the differences between fishing and non-fishing activities.

4   Experiments

The objective is to create a supervised classifier that can predict if a window of AIS data is fishing or not. The ten longliners studied in this paper have an average track size of 46427 data points, 77.96 % of which are fishing.

4.1   Methods

The performance of our approach is compared with SVM and Random Forests. The autoencoder has an input layer of 400 nodes and a hidden layer of 100 nodes. The same preprocessing techniques are applied to all approaches. We optimized SVM by selecting different kernel functions and optimized Random Forests by selecting the optimal number of decision trees. The models are trained and tested 10 times. Each time, the models are trained on 9 ships and tested on the remaining ship. Because the data is imbalanced, we evaluate the methods using F1 score. We also discuss quantiles as an alternative to parametric evaluations especially when distribution can not be simply represented as mean and standard deviation.

We use paired samples t-test and Wilcoxon Signed-Rank Test to investigate whether one method is consistently better than other methods. Paired samples t-test is a parametric statistical hypothesis test under the assumption that the differences between pairs are normally distributed. We use Shapiro-Wilk Test [4] to test if the normality assumption holds, if not, we use Wilcoxon Signed-Rank Test to compare different methods [2].

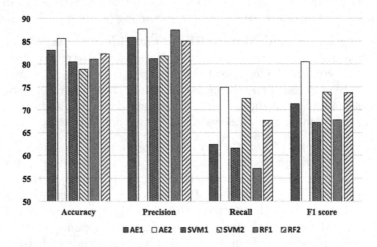

Fig. 4. Performance of autoencoders, SVM and Random Forests. AE1, SVM1 and RF1 are not undersampled from time intervals; AE2, SVM2 and RF2 are undersampled from time intervals.

4.2   Results

Figure 4 and Table 1 shows the classification accuracy, precision, recall and F1 score of models developed using autoencoders, SVM and Random Forests.

Table 1. Performance of autoencoders, SVM and Random Forests when data are undersampled from time intervals. Fishing activity is labeled as positive class.

	Accuracy			Precision			Recall			F1 score		
	AE	SVM	RF	AE	SVM	RF	AE	SVM	RF	AE	SVM	RF
Mean	85.6	78.9	82.2	87.6	81.8	85.1	74.8	72.4	67.6	80.5	73.8	73.8
SD	4.4	6.2	8.4	3.1	9.2	8.2	8.8	27.0	19.2	5.8	17.6	14.2
1 quantile	84.2	75.9	82.6	85.3	73.6	81.7	67.9	40.7	50.9	77.5	54.4	62.5
2 quantile	86.8	80.4	84.8	87.1	82.0	88.2	77.7	75.9	73.8	82.4	80.8	79.3
3 quantile	88.0	83.0	85.7	90.4	87.4	90.9	81.3	97.3	84.6	84.4	88.8	83.6

4.3   Discussion

In this section, we analyze the impact of undersampling from time intervals, and compare the performance of autoencoders with SVM and Random Forests.

Table 2. P-values for undersampling from time intervals.

	T-test		Shapiro-Wilk		Wilcoxon Signed-Rank	
	Accuracy	F1 score	Accuracy	F1 score	Accuracy	F1 score
AE	**0.010**	**0.001**	0.080	0.712	0.007	0.007
SVM	**0.006**	0.103	0.201	**0.002**	0.012	**0.093**
RF	**0.366**	**0.029**	0.115	0.699	0.173	0.047

To analyze the impact of undersampling from time intervals, we perform a hypothesis test on the classification accuracy and F1 score with and without undersampling. Table 2 shows the p-values of paired samples t-test[1] and its corresponding p-values for Shapiro-Wilk normality test[2]. In terms of F1 score: we use paired samples t-test for autoencoders and random forests, and use Wilcoxon Signed-Rank for SVM. On significance level 0.05, we find undersampling from time intervals can improve the F1 score of autoencoders, but does not necessary improve the F1 score of SVM and Random Forests.

To compare the performance of autoencoders with SVM and Random Forests, we perform hypothesis test on classification accuracies from undersampled data.

---

[1] The null hypothesis for paired samples t-test is that the mean difference between paired observations is zero. On significance level 0.05, the difference is statistically significant when p-value is less than 0.05.

[2] The null hypothesis for Shapiro-Wilk test is that the population is normally distributed. On significance level 0.05, if the p-value is greater than 0.05, the null hypothesis that the data came from a normally distributed population cannot be rejected.

Table 3. P-values for undersampling from time intervals using paired samples t-test, Shapiro-Wilk normality test and Wilcoxon Signed-Rank test.

	T-test		Shapiro-Wilk		Wilcoxon Signed-Rank	
	Accuracy	F1 score	Accuracy	F1 score	Accuracy	F1 score
AE vs. SVM	**0.0001**	**0.205**	0.9692	0.879	0.005	0.285
AE vs. RF	**0.0629**	0.062	**0.0005**	**0.006**	0.008	**0.013**

Table 3 shows the p-values of paired samples t-test, Shapiro-Wilk normality test and Wilcoxon Signed-Rank test of classification accuracy and F1 score.

In terms of F1 score, from Shapiro-Wilk test, we find the differences of the pairs when comparing autoencoders with SVM follow normal distribution, but the differences of the pairs when comparing autoencoders with Random Forests do not follow normal distribution. Thus, we should use paired samples t-test when comparing autoencoders with SVM, and use Wilcoxon Signed-Rank test when comparing autoencoders and random forest. On significance level 0.05, because the p-value of t-test between autoencoders and SVM is 0.205 which is bigger than 0.05, we suggest the difference between F1 score of autoencoders and SVM is not statistical significant. On significance level 0.05, because the p-value of t-test between autoencoders and Random Forests is 0.013 which is less than 0.05, we suggest autoencoders have better F1 score compared with Random Forests.

In terms of quantiles, we find the first and second quantiles of autoencoders are better than SVM and Random Forests both on classification accuracy and F1 score. This suggests that in the cases when the model is not performing well, autoencoders can generally have better lower bound of classification accuracy and F1 score compared to SVM and Random Forests. We also find autoencoders have a smaller standard deviation compared to SVM and Random Forests, meaning the model is more stable.

5   Conclusions and Future Work

This paper presented early results from applying Deep Learning tools on a specific task, related to AIS data. The volume and importance of AIS data is expected to increase sharply in the next few years, so it is relevant to develop effective and efficient tools for Machine Learning on this type of data. Here we have focused on the detection of fishing vs. non-fishing activities in AIS trajectories using autoencoders. We find autoencoders can perform at least as well as and sometimes better than SVM and Random Forests on this classification problem. As for future work, we consider using recurrent neural networks to take advantage of temporal information, and using convolutional networks to learn better spatial representation. Finally, we will investigate other problems in which AIS data might be useful, such as gear classification and ship category classification.

References

1. Hinton, G.E., Salakhutdinov, R.R.: Reducing the dimensionality of data with neural networks. Science **313**(5786), 504–507 (2006)
2. Japkowicz, N., Shah, M.: Evaluating Learning Algorithms: A Classification Perspective. Cambridge University Press, Cambridge (2011)
3. Jiang, X., Silver, D.L., Hu, B., Souza, E.N.d., Matwin, S.: Fishing activity detection from AIS data using autoencoders. Technical report, Dalhousie University (2016)
4. Shapiro, S.S., Wilk, M.B.: An analysis of variance test for normality (complete samples). Biometrika **52**(3/4), 591–611 (1965). http://www.jstor.org/stable/2333709
5. de Souza, E.N., Kristina Boerder, B.W., Matwin, S.: Improving fishing pattern detection from AIS using data mining and machine learning (to appear)

Simulating the Bubble Net Hunting Behaviour of Humpback Whales: The BNH-Whale Algorithm

Stacey Hala[1], Howard J. Hamilton[1(✉)], and Paolo Domenici[2]

[1] Department of Computer Science, University of Regina, Regina, Canada
stacey_hala@hotmail.com, howard.hamilton@uregina.ca
[2] C.N.R. - IAMC - Unitá Operativa di Oristano, Oristano, Italy
paolo.domenici@cnr.it

Abstract. We describe the BNH-Whale algorithm, which simulates a method of catching schooling fish employed by humpback whales called bubble net hunting. When this method is used, the fish are herded into a net of bubbles created by one or more whales. Modified flocking algorithms are used to guide whale and fish behaviour, with a relatively complex algorithm being used by the whales to trap and consume fish.

1 Introduction

We describe the BNH-Whale algorithm, which simulates and visualizes the bubble net hunting behaviour of humpback whales (*Megaptera novaeangliae*) in real-time by using a wide variety of information about the characteristics of these whales and their prey. Briefly, *bubble net hunting* (BNH) refers to a hunting strategy employed by humpback whales, in which one or more whales swim around a school of fish and release bubbles (which the fish avoid) while other whales herd the fish up into the area enclosed by the "net" of bubbles [1]. Here we describe only the hypothesized models of decision making and action used by whales, although we have also devised models for fish and the environment. This work is of interest to Artificial Intelligence researchers as a case study that simulates a non-human intelligence. We propose a specific form of decision making that the whales could be using to guide their behaviour. The observed success of the simulated whales in capturing their simulated prey provides evidence that this form of decision making is actually employed.

The BNH-Whale algorithm is intended to aid biologists in evaluating their theories about BNH. Simulating BNH is valuable because observation in the field is expensive, often has poor visibility, and can potentially disrupt normal hunting behaviour. In particular, a simulation may provide insight into events occurring inside the bubble net, which are difficult to observe in nature.

2 Background

Relevant background concerns whale characteristics and behaviour reported by biologists and representations and algorithms developed for simulating and animating large numbers of thinking creatures in real time.

© Springer International Publishing Switzerland 2016
R. Khoury and C. Drummond (Eds.): Canadian AI 2016, LNAI 9673, pp. 40–45, 2016.
DOI: 10.1007/978-3-319-34111-8_5

Biological Background: Humpback whales are a *rare enemy* to their target fish, i.e., these whales are not the primary predators of the fish [1]. The predator response of the fish is effective against predators targeting a single fish, but the whales take advantage of this response for bulk feeding. The maneuverability of slower, smaller prey is far higher than that of the faster, larger predators. At certain predator-prey length ratios, when prey is around 10 to 100 times smaller, higher maneuverability allows the small prey to avoid being caught by a direct whole body attack while still being fast enough to avoid filter feeding [2]. Some schooling fish, such as herring, fall into such a ratio for humpback whales. It is at these predator-prey length ratios where alternative methods such as BNH are employed to allow the predator to capture the prey.

Various strategies are used by whales to hunt individually and in groups called *pods*. Individual whales look for natural clumps of fish and lunge to feed [3]. One hunting strategy used by pods is *corralling*, whereby whales herd fish into better positions for consumption [3]. BNH is a form of coralling; it motivates fish to move into a compact formation, which permits easier feeding [4].

Detailed observations suggest BNH requires at least two whales working together [1]. Pod sizes for BNH range from two to twenty whales, but the mean size is 7.4 whales [1]. It is speculated that whales communicate during feeding [1]. Each whale in the pod has a role. At least one whale, here called a *bubble blower*, stays close to the surface to form a cylindrical net of bubbles, called a *bubble net*. The other whales, here called *prey herders*, herd the school of fish towards the net and circle the school. Since many fish in coastal waters tend to stay closer to the sea floor than to the surface, the BNH strategy is apparently more effective when the whales herd the school towards the bubble net from below. This net aids in trapping the fish by motivating them to form a compact sphere called a *bait ball*, a last ditch defensive maneuver to avoid predators [5].

When the net is complete with the fish trapped inside, the whales perform a vertical lunges towards the center. In some instances of BNH, the whales take turns lunging; in others, they all lunge at the same time [3]. The choice between these techniques likely depends on school size, pod size, and circling radius [4].

AI Decision Making: To provide an accurate model of the decision making behaviour of the whales and fish, decision making should be simulated in each organism. However, faster execution is produced if single decisions affect multiple whales and fish. Making wise choices between such local and global decisions is necessary to achieve an effective balance between the accuracy of the simulation and its speed [7]. We use local decision making for the whales, since they are relatively few in number.

Previous simulations of whale feeding behaviour have focused on lunge feeding [8,9]. These particular simulations lack animation and focus on the forces involved with lunging. The mechanics of the lunge are used to calculate energy expenditure and the effectiveness of lunge feeding. An earlier simulation program uses data collected from tags attached to whales to visualize their movement [10]. It does not generate the motion of the whale, but it does give insight into the dynamics of the whale's maneuvers.

Table 1. Meanings of symbols used in equations.

α	Some vector value	h_i	A humpback whale
β	Depth function	$\hat{\jmath}$	Unit vector pointing up
c	Some constant	λ	Height of school, bottom to top
c_{fh}	Distance between a fish and a whale:	m	Mass
c_{hh}	Distance between two whales	P	A preference
d	Distance function	p	Position
δ	Horizontal span of school	r	Radius
E_i	Fish eaten by whale h_i	Δt	Time interval
F_T	The target school	u	Unit vector function
γ	Hunting location	v	Velocity
H	A set of humpback whales	x, y, z	Location or vector components

A simulation for BNH behaviour needs to model artificial fish as well as whales. The most widely known algorithm for controlling groups of objects is flocking [11]. This algorithm approximates the schooling behaviour of fish, and has been used to simulate schooling behaviour in other studies [12].

Fish and their environment have been modelled by previous researchers with varying degrees of fidelity, including detailed models of the muscles used by fish for movement, their senses, their learning algorithms, and the physics of the water involved [13]. However, our simulation requires 50,000+ entities and uses simpler models in order to run at real-time speeds.

3   The BNH-Whale Algorithm

We now describe the BNH-Whale algorithm, which is used by the simulated whales when hunting fish using the BNH method. Symbols used in equations are defined in the Table 1. The algorithm, which is shown in Fig. 1, is executed for each whale at each time step. When animation is enabled, each time step results in a frame of the animation being displayed. In the whale behaviour model, communication with and perception of other whales are handled separately. Communication is permitted among all whales, but perception is restricted to only the neighbours of a whale. It is assumed that the whales use communication to decide when to start blowing bubbles and whose turn it is to lunge. Perception refers to checking whether it is time to lunge: the whales begin to lunge when the center of the target school is above all of the whales or sufficiently close to the surface. When determining which whale is to lunge, the simulated whales communicate a level of hunger, and a hierarchy settles whose turn it is when multiple whales are equally hungry.

To identify a target school for the whales, we use a single global calculation (since the whales in a pod are near to each and therefore likely perceive similar things). The ocean is divided into grid cells using *spatial hashing*, a technique often used to improve computational efficiency in flock and crowd simulations [14].

1. If there is no current target school, identify a target school and obtains its position, width, and height
2. Perform message passing between whales in order to:
   (a) If circling, arrange location in circle with other whales (by role)
   (b) Ensure that all bubble blowers are ready before beginning to blow bubbles
   (c) Determine whose turn it is to lunge (if they are taking turns lunging)
3. If not close to target school, calculate hunting position
   (a) Increase depth while traveling
   (b) Bubble blowers may start forming an approach curtain before reaching the school
4. If close to school, determine hunting location
   (a) If bubble blower, calculate depth using Equation 3
   (b) If prey herder, calculate depth using Equation 4
   (c) Calculate the circling radius and position further along circle
5. If it is time to lunge
   (a) If taking turns, only the selected whale lunges
   (b) If not taking turns, all whales lunge together towards the target school center
6. Update velocity based on separate, momentum, and hunting location preferences

Fig. 1. BNH-Whale algorithm used to model whale behaviour.

The fish in each cell are stored in a list. The target school is found by first identifying the grid cell containing the most fish. Then from nearby grid cells are added to the school, as long as the density of the fish in the next neighbouring cells is sufficiently high. The center of the target school is calculated by averaging the positions of all fish in it. The width (δ) of the target school is calculated as the maximum horizontal distance between any two fish in the school along either the x or z axis, and the height (λ) of the target school is calculated as the maximum vertical distance between any two fish in the school.

The whales are guided by flocking behaviour with three factors: separation, momentum, and hunting location. Separation ($P_{1,i}$) is a vector representing the preference of whale h_i to avoid other whales up to a defined distance c_{hh}.

$$\text{Separation: } P_{1,i} = \sum_{h_n \in H} \left(\left(c_{hh} - d(h_i, h_n) \right) / c_{hh} \right)^2 u \left(p(h_i) - p(h_n) \right) \quad (1)$$

Momentum ($P_{2,i}$) reflects the preference for continuing with the same velocity.

$$\text{Momentum: } P_{2,i} = m(h_i) * v(h_i) \quad (2)$$

The *hunting location* (γ) for a whale is the location it wants to be with respect to its target school of fish. This location depends on the role of the whale and its distance from the school. If it is a bubble blower and far from the school, the hunting location is the nearest location at the starting depth for blowing bubbles and at its circling radius from the center of the school. If it is a prey herder and far from the school, its hunting location is below the school at a smaller circling radius. If a whale is at its circling radius from the school, its hunting location is further along the circle.

While circling, the depth of a bubble blower is based on the depth of the school and is calculated using Eq. 3 [1]; this choice of depth causes the bubbles to arrive at the surface at the same time as or before the fish arrive. Thus, the fish are unable to escape by swimming over the net.

$$\gamma_y = 6.380 * \log\left(p_y(F_T)\right) - 2.686 \tag{3}$$

The depth of a prey herder is calculated using Eq. 4 to be just below the bottom of the school and within the distance required to startle the fish.

$$\gamma_y = p_y(F_T) - 0.5 * \lambda - c_{\text{height fraction}} * c_{fh} \tag{4}$$

The circling radius of a bubble blower depends on both the *horizontal span* (greatest horizontal width in any direction) of the school [4] and how close a whale must be to startle a fish. Equation 5 is used to find the preliminary radius.

$$r_{\text{pre}} = 0.5 * \delta + c_{\text{width fraction}} * c_{fh} \tag{5}$$

The circling radius of a bubble blower also depends on its depth because the bubbles spread out more at the surface when they are emitted at greater depths [1]. Additionally, in the cases where the target school is located below the bubble net, the radius is increased to simplify the task of the prey herders. Thus, the circling radius r is the preliminary radius multiplied by a factor that depends on the depths of the bubble blower and school and the maximum allowed depth of a bubble blower using Eq. 6.

$$r = \beta\left(\gamma_y, p_y(F_T), c_{\text{maxDepth}}\right) * r_{\text{pre}} \tag{6}$$

The circling radius of a prey herder is found similarly with a different β function.

Circling behaviour is produced by first finding the *outward vector*, which is a unit vector pointing from the target school to the whale (Eq. 7).

$$\alpha_{xz} = u_{xz}\left(p_{xz}(h_i) - p_{xz}(F_T)\right) \tag{7}$$

By multiplying the outward vector by the circling radius, a location on the circle is obtained, and then by adding an offset perpendicular to the outward vector and the y axis using Eq. 8, a location further along on the circle is found.

$$\gamma_{xz} = p_{xz}(F_T) + r * \alpha_{xz} + c * \alpha_{xz} \times \hat{j} \tag{8}$$

Hunting location $P_{3,i} = \gamma$ is found by combining depth γ_y and horizontal γ_{xz}.

4   Conclusions

The BNH-Whale algorithm is designed to simulate the bubble net hunting behaviour of humpback whales. It simulates the herding maneuvers executed by whales in 3D and the escape behaviour of the fish. Future research could measure the effectiveness of the method by measuring the number of fish caught. It could also include further aspects of bubble net hunting such as the fatiguing of fish, varied sizes of fish and whales, modelling of the sound used to provoke the startle reflex in the fish [6], and improving the model of the bubbles to incorporate fluid dynamics. The simulation could also be designed to be GPU based.

Acknowledgments. The authors acknowledge financial assistance from the Natural Sciences and Engineering Research Council of Canada via a Discovery Grant to Hamilton and an Undergraduate Student Research Award to Hala.

References

1. Sharpe, F.A.: Social foraging of the southeast Alaskan humpback whale, *Megaptera novaeangliae*. Ph.D. thesis, Simon Fraser University (2001)
2. Domenici, P.: The scaling of locomotor performance in predator-prey encounters: from fish to killer whales. Comp. Biochem. Physiol. A **131**, 169–182 (2001). doi:10.1016/S1095-6433(01)00465-2
3. Kieckhefer, T.R.: Feeding ecology of humpback whales in continental shelf waters near Cordell Bank, California. Master's thesis, San Jose State University (1992)
4. Hazen, E.L., Friedlaender, A.S., Thompson, M.A., Ware, C.R., Weinrich, M.T., Halpin, P.N., Wiley, D.N.: Fine-scale prey aggregations and foraging ecology of humpback whales *Megaptera novaeangliae*. Mar. Ecol. Prog. Ser. **395**, 75–89 (2009). doi:10.3354/meps08108
5. Nøttestad, L., Axelsen, B.E.: Herring schooling manoeuvres in response to killer whale attacks. Can. J. Zool. **77**, 1540–1546 (1999). doi:10.1139/cjz-77-10-1540
6. Leighton, T., Finfer, D., Grover, E., White, P.: An acoustical hypothesis for the spiral bubble nets of humpback whales, and the implications for whale feeding. Acoust. Bull. **32**(1), 17–21 (2007)
7. Parent, R.: Computer Animation: Algorithms and Techniques, 3rd edn. Morgan Kaufmann Publishers, San Francisco (2012)
8. Potvin, J., Goldbogen, J.A., Shadwick, R.E.: Passive versus active engulfment: verdict from trajectory simulations of lunge-feeding fin whales *Balaenoptera physalus*. J. R. Soc. Interface **6**, 1005–1025 (2009). doi:10.1098/rsif.2008.0492
9. Goldbogen, J.A., Calambokidis, J., Oleson, E., Potvin, J., Pyenson, N.D., Schorr, G., Shadwick, R.E.: Mechanics, hydrodynamics and energetics of blue whale lunge feeding: efficiency dependence on krill density. J. Exp. Biol. **214**, 131–146 (2011). doi:10.1242/jeb.048157
10. Ware, C., Arsenault, R., Plumlee, M., Wiley, D.: Visualizing the underwater behavior of humpback whales. IEEE Comput. Graphics Appl. **26**(4), 14–18 (2006)
11. Reynolds, C.W.: Flocks, herds and schools: a distributed behavioral model. Comput. Graph. **21**, 25–34 (1987)
12. Huse, G., Railsback, S., Fern, A.: Modelling changes in migration patterns of herring by numerical domination. J. Fish Biol. **60**, 571–582 (2002)
13. Terzopoulos, D., Tu, X., Grzeszczuk, R.: Artificial fishes: autonomous locomotion, perception, behavior, and learning in a simulated physical world. Artif. Life **1**(4), 327–351 (1994)
14. Reynolds, C.: Big fast crowds on PS3. In: Sandbox 2006: Proceedings of the 2006 ACM SIGGRAPH Symposium on Videogames, pp. 113–121 (2006). doi:10.1145/1183316.1183333

Gaussian Neuron in Deep Belief Network
for Sentiment Prediction

Yong Jin[1(✉)], Donglei Du[2], and Harry Zhang[1]

[1] Faculty of Computer Science, University of New Brunswick, Fredericton E3B 5A3, Canada
{yjin1,hzhang}@unb.ca
[2] Faculty of Business Administration, University of New Brunswick,
Fredericton E3B 5A3, Canada
ddu@unb.ca

Abstract. Deep learning has been widely applied in natural language processing. The neuron model in a deep belief network is important for its performance, and so more attention should be paid to investigate how much influence the neuron will play on its results. In this paper we investigate the neuron's effect for sentiment prediction, and then apply both total accuracy and F-measure to evaluate the performance. Finally, our experimental results show the idea of Gaussian neuron performs relatively better on the Stanford Twitter Sentiment corpus, which further proves the neuron model should be considered for a specific problem.

Keywords: Sentiment prediction · Gaussian neuron · Deep belief network

1 Introduction

Sentiment analysis, or opinion mining, popular both in academic and engineering fields, is to identify the attitude of a speaker or a writer on a topic. Due to the increasingly large amount of online opinion resources, such as discussion forums, product review sites and social media websites, people are inclined to develop a system that can automatically determine the polarities of the opinions. The recently developed Deep Learning method described in [1] can be applied in opinion mining.

In Deep Belief Network (DBN), Restricted Boltzmann Machine (RBM) [2] makes it different from the traditional feed-forward neural networks because the units between different layers (except the output layer) are bidirectional and there are no connections within the same layer [3]. Specifically, the two layers' units in an RBM are connected through the energy function by Eq. (1):

$$Energy(x, h) = -b^T x - c^T h - h^T W x \qquad (1)$$

where x represents the vector of visible units, h is hidden unit vector, h^T is the transposition of h, W represents the weights connecting hidden and visible units, and b, c are the bias vectors of the visible and hidden layers respectively. To train RBM, the Gibbs sampling plays a significant role in which the conditional probabilities $p(h|x)$

© Springer International Publishing Switzerland 2016
R. Khoury and C. Drummond (Eds.): Canadian AI 2016, LNAI 9673, pp. 46–51, 2016.
DOI: 10.1007/978-3-319-34111-8_6

and $p(x|h)$ need to be calculated [3]. Hinton also proposes a detailed practical guide for training RBMs [4].

Some neuron models have been introduced into deep learning. Nair and Hinton use noisy rectified linear units to approximate the stepped sigmoid units [5], but it only proves better on face verification data. Glorot *et al.* propose rectifying neuron: $max(0, x)$ to create sparse representations with true zeros [6], but it is unfortunately non-differentiable at zero. Besides, some deep learning models have been proposed in sentiment analysis area. Socher *et al.* integrate d-dimensional word vectors into multiple models, such as recursive auto-encoders [7], and deep recursive network [8], which perform effective but require a very large labelled tree bank for training.

This work is to investigate a more effective neuron model in DBN for Twitter sentiment prediction through a variety of empirical studies. From the point view of the traditional "bag of words" representation for each sentence, not from the grammar parsing angle, this paper explores the validity of DBN on the neuron effect for sentiment prediction. The rest of the paper is organized as follows: In Sect. 2, we propose our idea of the Gaussian neuron model in DBN; In Sect. 3, we will describe the experiments and results analysis; In Sect. 4, the conclusion and future research direction are discussed.

2   Gaussian Neuron in Deep Belief Network

2.1   Gaussian Neuron

In DBN, the conditional probabilities of each hidden unit and visible unit, $p(h|x)$ and $p(x|h)$, are actually computed via the neuron's activation function φ. Generally, the sigmoid neuron function Eq. (2) with the independent variable t is widely used in neural networks.

$$\varphi(t) = sigmoid(t) = \frac{1}{1 + e^{-t}}, \tag{2}$$

The Eq. (2) represents the conditional probability of hidden unit h by computing the activation function φ with respect to the linear combination t of input x. On the other hand, considering the activation function is actually to compute a probability, it naturally inspires us to come up with some other feasible probability functions. Especially in sentiment analysis, the neuron activation function needs further investigation, because the t value in an activation function comes from a linear combination of the previous layer's units, which indeed represents a sequence of words (visible layer), or a sequence of phrases (hidden layers). How this combination contributes to the sentiment of the next layer, probably has different effect curves (not only sigmoid or tangent hyperbolic functions.).

Hence, different cumulative distribution functions could be used here, such as t-distribution, *beta*-distribution, and Gaussian distribution, etc. These neuron functions offer the flexibility for sentiment analysis using the deep learning approach. In this paper, we propose different neuron activation functions for Twitter sentiment prediction. The Gaussian hidden units [9] have been discussed in neural networks, while the Gaussian neuron needs further investigation for sentiment prediction using DBN. Specifically, the Gaussian neuron is given by Eq. (3).

$$\varphi(t) = Gaussian(t;\mu,\sigma^2) = \frac{1}{\sigma\sqrt{2\pi}} \int_{-\infty}^{t} e^{-\frac{(z-\mu)^2}{2\sigma^2}} dz, \tag{3}$$

where t is the independent variable, and φ is actually the cumulative distribution function of Gaussian distribution with mean and variance (μ, σ^2). This neuron gives diversified activation function curves just by setting different values of (μ, σ^2).

Learning the DBN is divided into two phases [3]: unsupervised RBM training and supervised neural network training. The first part follows the practical guide written by Hinton [4], and the second one is actually traditional back propagation, a common method to train artificial neural network, introduced by Rumelhart *et al.* [10]. Specifically, the derivative of the Gaussian cumulative distribution function can be easily calculated via the Gaussian density function. Hence, the Gaussian neuron in DBN has similar time complexity as the sigmoid neuron, and therefore we can only pay attention to their classification accuracy performance.

2.2 Evaluation

In this paper, two main evaluation metrics are employed to compare the performances: total accuracy *(TA)* and F-measure (F_1 score) [11]. Firstly, the predictive accuracy is typically used for evaluating the performance of almost all machine learning algorithms. Besides, for the purpose to combine precision and recall into evaluation together, F_1 score is also used. Suppose in the binary classification task, the four variables, *tp*, *tn*, *fp*, and *fn*, represent the true positives, true negatives, false positives, and false negatives respectively. Then the two measures are calculated by Eqs. (4) and (5).

$$TA = \frac{tp + tn}{tp + tn + fp + fn}, \tag{4}$$

$$F_1 = \frac{2tp}{2tp + fp + fn}. \tag{5}$$

3 Experiments and Results

3.1 Data Collection and Pre-processing

This work will focus on Twitter sentiment prediction. The labelled twitter corpus we use is the Stanford Twitter Sentiment (STS) [12]. In order to construct some independent data sets for more comparisons and also to speed up our training time, we randomly sample ten non-overlapped small data sets from this training set of STS.

The raw data is pre-processed for the model inference. We ignore case sensitive for all texts; delete the numbers and some special symbols, such as @, /, |, $; and replace some acronyms and abbreviations to their expanded forms. For example, "I've" is for

"I have", "can't" is "can not", "won't" is "will not", etc. This is done so that some meaningful words especially the negation word "not" are extracted because they will play an important role for sentiment. Finally, the vocabulary of the data set is extracted as attributes, each word token is denoted by its presence or not, and then each sentence (example) can be represented by a vector where the element is one if the word exists in it, otherwise zero. And the sentence's label is denoted by a vector with value one at corresponding position and zeros elsewhere.

The randomly sampled ten non-overlapped data sets, including four balanced data sets and six imbalanced data sets, are described in Table 1. Here 'attributes' means the number of word tokens for each data set, IR is the imbalanced ratio which equals the ratio of majority over minority.

Table 1. Description of data sets used in the experiments

Data set	Size	Attributes	Classes	Positive	Negative	IR
pos500neg500	1000	1015	2	500	500	1.00
pos500neg5000	5500	1060	2	500	5000	10.00
pos1000neg1000	2000	1107	2	1000	1000	1.00
pos3000neg3000	6000	1016	2	3000	3000	1.00
pos3000neg10000	13000	995	2	3000	10000	3.33
pos3000neg20000	23000	988	2	3000	20000	6.67
pos5000neg500	5500	1033	2	5000	500	10.00
pos5000neg5000	10000	990	2	5000	5000	1.00
pos10000neg3000	13000	988	2	10000	3000	3.33
pos20000neg3000	23000	978	2	20000	3000	6.67

3.2 Experimental Results

In our experiments, the widely used sigmoid neuron is implemented for baseline. Then we perform two DBNs, DBN-Gaussian and DBN-sigmoid. Different DBN architectures have been tested, including the number of hidden layers (1, 2, 3) and hidden units (100, 200, 300, 400, 500), the functions from the penultimate layer to output layer (Gaussian, sigmoid, softmax), and the settings of (μ, σ) with combinations of $(\mu = 0, 0.1, 0.2, 0.5, 0.8, 1, 2; \sigma = 0.1, 0.5, 1, 1.5, 2, 3, 4, 5)$, we then manually set the system as: input (visible units)→200 hidden units→100 hidden units→output layer (class labels). And $(\mu = 0.8, \sigma = 3)$ is selected for DBN-Gaussian. Our platform is MATLAB 2014a in the PC of 64-bit OS, Intel Core i5-5200U, CPU 2.20 GHz, and RAM 8.0 GB.

Each data set is performed using ten-fold cross validation (nine for training and one for testing) to obtain an average accuracy. Firstly, in RBM training, the momentum is 0.1, the learning rate is 0.9, and the number of epochs is 50. Besides, for supervised training, the maximum number of epochs is 1000 with 10^{-6} as the convergence control based on early stopping rule.

Table 2 displays the total accuracies TA and F_1 score of the two neuron DBN models on each data set, and each cell of TA is represented by the mean and standard deviation of the ten-fold cross validation results, and F_1 score is the average value. Here we assume the accuracy has won (lost) if it passes a two-tailed t-test within 95 % confidence level with the sign (*) marked. Based on the comparisons of sigmoid and Gaussian neuron in DBN, the ordinary accuracy TA and F_1 score do make some differences among these data sets.

Table 2. Experimentsal results on classification accuracy (%)

Data set	DBN-Gaussian		DBN-sigmoid	
	TA	F_1	TA	F_1
pos500neg500	69.10 ± 4.10 *	0.670	67.20 ± 4.00	0.668
pos500neg5000	89.60 ± 1.42	0.253 *	90.96 ± 1.15 *	0.101
pos1000neg1000	66.90 ± 2.97	0.657	68.00 ± 3.06	0.682 *
pos3000neg3000	71.32 ± 1.84 *	0.719 *	70.05 ± 2.04	0.705
pos3000neg10000	80.72 ± 1.26 *	0.533 *	78.34 ± 0.94	0.511
pos3000neg20000	88.02 ± 0.76 *	0.419 *	86.49 ± 1.24	0.395
pos5000neg500	88.20 ± 2.08	0.936	90.87 ± 0.98 *	0.951 *
pos5000neg5000	72.29 ± 1.77	0.726	71.21 ± 1.87	0.717
pos10000neg3000	80.68 ± 0.81 *	0.879 *	78.46 ± 1.34	0.863
pos20000neg3000	87.70 ± 0.72 *	0.932 *	86.10 ± 0.93	0.921
average	79.45 ± 1.77	0.672	78.77 ± 1.76	0.651

In Table 3, an explicit comparison between Gaussian DBN and sigmoid DBN is summarized, in which $w/l/t$ means the method wins in w data sets, losses in l data sets, and ties in t data sets. The results show that the Gaussian DBN has relatively better TA and F_1 score than sigmoid DBN in both balanced and imbalanced data sets. Hence, compared to sigmoid DBN, it indicates that: (1) the mean μ of Gaussian DBN should be greater than zero, but not too big; and (2) the standard deviation σ should be relatively large. These two indications provide us some useful guidance towards the activation function settings for sentiment prediction.

Table 3. Summary of evaluation comparisons

Model	All data sets (10)		Balanced data sets (4)		Imbalanced data sets (6)	
	TA	F_1 score	TA	F_1 score	TA	F_1 score
DBN-Gaussian	6/2/2	6/2/2	2/0/2	1/1/2	4/2/0	5/1/0
DBN-sigmoid	2/6/2	2/6/2	0/2/2	1/1/2	2/4/0	1/5/0

4 Conclusions

In this paper, we propose an idea of Gaussian neuron by changing the neuron model that can flexibly adapt the DBN model into sentiment prediction, and then we accommodate both the total accuracy and F_1 score to evaluate the performance. Our experimental results reveal that there exists some significant improvements on the total accuracy and F_1 score compared to the widely used sigmoid neuron in DBN.

The Gaussian neuron is inspired by the contribution of word combinations to the sentiment. Then we give some suggestions regarding how to choose the activation function for Twitter sentiment prediction, preferring the activation function with the positive mean value and relative large standard deviation value (compared to sigmoid function). In this system, the form of Gaussian neuron (mean and standard deviation) is manually set in the experiments, while in the future, it can be set to learn during the training which will best fit the problem. However, some more sophisticated neuron functions or mixture of cumulative functions could have better effect.

References

1. Bengio, Y.: Learning deep architectures for AI. Found. Trends Mach. Learn. **2**(1), 1–127 (2009)
2. Ackley, D.H., Hinton, G.E., Sejnowski, T.J.: A learning algorithm for boltzmann machines. Cogn. Sci. **9**(1), 147–169 (1985)
3. Hinton, G., Osindero, S., Teh, Y.: A fast learning algorithm for deep belief nets. Neural Comput. **18**(7), 1527–1554 (2006)
4. Hinton, G.: A practical guide to training restricted boltzmann machines. Momentum **9**(1), 926 (2010)
5. Nair, V., Hinton, G.: Rectified linear units improve restricted boltzmann machines. In: Proceedings of the 27th International Conference on Machine Learning (ICML 2010) (2010)
6. Glorot, X., Bordes, A., Bengio, Y.: Deep sparse rectifier networks. In: Proceedings of the 14th International Conference on Artificial Intelligence and Statistics. JMLR W&CP Volume (2011)
7. Socher, R., Pennington, J., Huang, E.H., Ng, A.Y., Manning, C.D.: Semi-supervised recursive autoencoders for predicting sentiment distributions. In: Proceedings of the Conference on Empirical Methods in Natural Language Processing (EMNLP) (2011)
8. Socher, R., Perelygin, A., Wu, J.Y., Chuang, J., Manning, C.D., Ng, A.Y., Potts, C.: Recursive deep models for semantic compositionality over a sentiment treebank. In: Proceedings of the Conference on Empirical Methods in Natural Language Processing (EMNLP) (2013)
9. Hartman, E.J., Keeler, J.D., Kowalski, J.M.: Layered neural networks with Gaussian hidden units as universal approximations. Neural Comput. **2**(2), 210–215 (1990)
10. Rumelhart, D.E., Hinton, G., Williams, R.J.: Learning representations by back-propagating errors. Cogn. Model. **5**(3), 1 (1988)
11. Olson, D.L., Delen, D.: Advanced Data Mining Techniques. Springer Science & Business Media, Heidelberg (2008)
12. Go, A., Bhayani, R., Huang, L.: Twitter sentiment classification using distant supervision. Technical report, Stanford University (2009)

Grounding Social Interaction
with Affective Intelligence

Joshua D.A. Jung[1]([✉]), Jesse Hoey[1], Jonathan H. Morgan[2], Tobias Schröder[3],
and Ingo Wolf[3]

[1] Department of Computer Science,
University of Waterloo, Waterloo, ON, Canada
{j35jung,jhoey}@cs.uwaterloo.ca
[2] Department of Sociology, Duke University, Durham, NC, USA
jonathan.h.morgan@duke.edu
[3] Institute for Urban Futures, University of Applied Sciences, Potsdam, Germany
{schroeder,wolf}@fh-potsdam.de

Abstract. Symbolic interactionist principles of sociology are based on
the idea that human action is guided by culturally shared symbolic representations of identities, behaviours, situations and emotions. Shared
linguistic, paralinguistic, or kinesic elements allow humans to coordinate
action by enacting *identities* in social situations. Structures of identity-
based interactions can lead to the enactment of social orders that solve
social dilemmas (e.g., by promoting cooperation). Our goal is to build
an artificial agent that mimics the identity-based interactions of humans.
This paper describes a study in which humans played a repeated prisoner's dilemma game against other humans or one of three artificial
agents (bots). One of the bots has an explicit representation of identity and demonstrates more human-like behaviour than the other bots.

1 Introduction

The prisoner's dilemma has long been studied, starting with the work of Axelrod
and Hamilton [3]. Recent work has looked at modelling both rational choice
and social imitation to simulate more human-like behaviour in networked PD
games [16]. Others have looked at using emotional signals to influence play in
PD games, for example by changing expectations of future games [4]. Emotions
have also been linked with intrinsic reward and exploration bonuses [14]. It has
become increasingly clear that human handling of an infinite action space (not
limited to the realm of the prisoner's dilemma) may be governed largely by affective processes [1,10]. Shared affective structures allow agents to focus on the subset
of possibilities that provide interactions aligning with the shared structure. This
subset of possibilities forms the set of "cultural expectations" for behaviours that
are "rational relative to the social conventions and ethics" ([1], p. 200).

The original version of this chapter was revised. An erratum to this chapter can be
found at DOI 10.1007/978-3-319-34111-8_40

R. Khoury and C. Drummond (Eds.): Canadian AI 2016, LNAI 9673, pp. 52–57, 2016.
DOI: 10.1007/978-3-319-34111-8_7

A recent product of these ideas is *BayesAct* [8], which models the emotional control of social interaction by humans and can explain the emergence of stable role relations and patterns of interaction [13]. Here, we empirically study the class of interactions in the iterated prisoner's dilemma, a fundamental paradigm in the social sciences aimed at understanding the dynamics of human cooperation vs. competition. Our results are encouraging in terms of supporting the validity of the *BayesAct* agent as a mechanistic model of human social interactions.

BayesAct [2,7,8,13] is a partially observable Markov decision process model of affective interactions between a human and an artificial agent. *BayesAct* arises from the sociological (symbolic interactionist) "Affect Control Theory" (ACT) [6]. *BayesAct* generalises this theory by modeling affective states as probability distributions, and allowing decision-theoretic reasoning about affect. *BayesAct* proposes that humans learn and maintain a set of *shared* cultural affective *sentiments* about people, objects, behaviours, and about the dynamics of interpersonal events. Humans use a simple affective mapping to appraise individuals, situations, and events as sentiments in a three dimensional vector space of evaluation (E: good vs. bad), potency (P: strong vs. weak) and activity (A: active vs. inactive). These "EPA" mappings can be measured, and the culturally shared consistency has repeatedly been demonstrated to be extremely robust in large cross-cultural studies [12]. Many believe this consistency "gestalt" is a keystone of human intelligence. Humans use it to make predictions about what others will do, and to guide their own behaviour. Further, it defines an affective heuristic (a *prescription*) for making decisions quickly in interactions. Humans strive to achieve consistency by choosing actions that maximally increase *alignment* (decrease *deflection* in ACT terms) in shared affective cultural sentiments. The shared sentiments and dynamics, the affective prescriptions, and the resulting *affective ecosystem* of vector mappings, result in an equilibrium or *social order* [5], which is optimal for the group as a whole, rather than for individual members. Humans living at the equilibrium "feel" good and want to stay there, with positive evolutionary consequences. However, agents with sufficient resources can plan beyond the prescription, allowing them to manipulate other agents to achieve individual profit in collaborative games [2].

For example, in the repeated prisoner's dilemma, cooperation has a different emotional signature than defection: it is usually viewed as nicer (higher evaluation). Rationality predicts an agent will try to optimize over his expected total payout, perhaps modifying this payout by some additional intrinsic reward for altruism. The *BayesAct* view is quite different: it says that an agent will take the most aligned action given her estimates of her own and her partner's affective *identity*. Thus, *friends* will do nice things to *friends* and cooperate, but will be more likely to defect against a *scrooge* or a *traitor*. *Scrooges* will defect, as this is consistent with a more negative identity, but may cooperate to manipulate.

As elucidated by Squazzoni et al. [15], models of social networks must take into account the heterogeneity of individuals, behaviours, and dynamics in order to better account for the available evidence. In this paper we argue that the principles encoded in *BayesAct* can capture this heterogeneity. As evidence, here we present results from an experiment in which participants played a repeated prisoner's dilemma (PD) game against each other and against a set of computer programs, one of them *BayesAct*.

2 Experiments and Results

The prisoner's dilemma is a classic two-person game in which each person can either *defect* by taking \$1 from a (common) pile, or *cooperate* by giving \$2 from the same pile to the other person. There is one Nash equilibrium in which both players defect, but when humans play the game they often are able to achieve cooperation. A rational agent will optimise over his expected long-term payoffs, possibly by averaging over his expectations of his opponent's type (or strategy).

A *BayesAct* agent computes what *affective* action (an EPA vector) is prescribed in the situation, given his estimates of his and the other's (called the *client*) identities, and of the affective dynamics, and then seeks the propositional action ($\in \{$*cooperate, defect*$\}$) that, according to a stored cultural definition, is most consistent with the prescribed affective action. As the game is repeated, the *BayesAct* agent updates his estimates of identity (for self and other/*client*), and adjusts his play accordingly.

For example, if *agent* thought of himself as a *friend* (EPA:$\{2.75, 1.88, 1.38\}$) and knew the other agent to be a *friend*, the deflection minimizing action would likely be something good (high E). Indeed, a simulation shows that one would expect a behaviour with EPA $= \{1.98, 1.09, 0.96\}$, with closest labels such as *treat* or *toast*. Intuitively, cooperate seems like a more aligned propositional action than defect. This intuition is confirmed by the distances from the predicted (affectively aligned) behaviour to *collaborate with* (EPA:$\{1.44, 1.11, 0.61\}$) and *abandon* (EPA:$\{-2.28, -0.48, -0.84\}$)[1] of 0.4 and 23.9, respectively, clearly showing the closer proximity of collaboration to this affectively aligned action.

The *agent* will predict the *client*'s behavior using the same principle: compute the deflection minimising affective action, then deduce the propositional action based on that. Thus, a *friend* would predict that a *scrooge* would defect, but would still want to cooperate in order to *reform* or *befriend* the other agent. If a *BayesAct* agent has sufficient resources, he could search for an affective action near to his optimal one, but that would still allow him to defect. Importantly, he is *not trading off costs in the game with costs of disobeying the social prescriptions*: his resource bounds and action search strategy are preventing him from finding the more optimal (individual) strategy, implicitly favouring those actions that benefit the group and solve the social dilemma.

In order to compare the predictions of *BayesAct* to human play, we recruited 70 students (55 male and 15 female) from a senior undergraduate class on artificial intelligence at the University of Waterloo[2]. The participants played a total of 360 games in a computer lab environment. The length of each game was randomly chosen between 12–18 rounds (plays of cooperation or defection). Each

---

[1] These are representative of the affective meaning of the actions in the game [2].

[2] The study was reviewed and approved by the UW Office of Research Ethics. For further discussion of experiment procedures, see [9].

game a participant played was against either (1) another randomly chosen participant; (2) an automated *tit-for-tat* player; (3) a *BayesAct* agent as described above; or (4) a fixed strategy of cooperate three times followed by always defect, hereafter referred to as *jerkbot*. The *BayesAct* agent reward is only over the game (e.g. 2, 1, or 0), and we use a two time-step game in which both *agent* and *client* choose their actions at the first time step, and then communicate this to each other on the second step.

Participants were assigned some order in which to play each opponent, but that order was randomized for each participant. Further, participants were told that all of their opponents were human. Upon sign-up, and after each game (of between 12–18 rounds), participants were asked the following by providing them with a slider for each E,P,A dimension, known as a semantic differential [6]:

- how they felt about the plays in the game (take 1 or give 2), out of context. *BayesAct* agents then interpret the affective signature of actions in the game by comparing the EPA vectors to these two vectors.
- how they felt about themselves (their self identity). This gives *BayesAct* its self-identity, as we want it to replicate a participant. We use the raw data from all student responses across all questions as this self-identity *BayesAct*.
- how they felt about their opponent in the game they just played. Before the first game we asked they how they felt about a generalised opponent in this game, giving the *BayesAct* client identity.

A total of 89 samples were used for identities (resampled to get $N = 2000$ samples used in the *BayesAct* particle filter) and an average of 89 samples used for the SCB. From this sample, we measured for *Give 2* an EPA of $\{1.4, 0.10, 0.18\}$, and for *Take 1*, $\{-0.65, 0.85, 0.70\}$. *Take 1* is seen as more negative and more powerful and active. Additionally, the self is seen as more positive than the opponent or "other" (with average E value 1/0.25 for self/other), but about the same power (0.56/0.64) and activity (0.41/0.33).

Table 1. Summary statistics. Coops: number of cooperations after 10th game.

Opponent	Num. avg. game		Agent (human)		Client (human or bot)	
	Games	Length	Payoff	Coops	Payoff	Coops
Jerkbot	83	15.01	15.86 ± 3.00	0.09 ± 0.24	22.33 ± 6.00	0.00 ± 0.00
Bayesact	73	14.85	27.05 ± 5.92	0.54 ± 0.40	22.19 ± 7.87	0.69 ± 0.32
Human	35	15.43	24.11 ± 7.55	0.56 ± 0.45	26.00 ± 5.92	0.51 ± 0.47
Titfortat	82	14.82	27.66 ± 5.39	0.81 ± 0.35	26.96 ± 6.00	0.83 ± 0.34

Table 1 (cols 2, 3) shows the statistics of game numbers and lengths against the different opponents. Figure 1 shows the mean, standard deviation, and median reward gathered at each step of the game, for each of the opponents. The blue lines show the human play, while the red lines show the opponent (one of human, *BayesAct*, tit-for-tat, or *jerkbot*). We see that humans mostly manage to

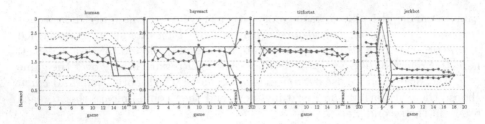

Fig. 1. Blue = human; Red = agent (human, Bayesact, titfortat and *jerkbot*); dashed = std.dev.; solid (thin, with markers): mean; solid (thick): median. (Color figure online)

cooperate together until about 4–5 games before the end. The *tit-for-tat* strategy ensures more even cooperation, but is significantly different from humans. *Jerkbot* is obvious, as a few defections after three games convinces the human to defect thereafter. The *BayesAct* agent play is very similar to the human play, but the human participants take advantage of the *BayesAct* agents late in the game. This may be because the *BayesAct* agent is using a short (5 s) planning timeout, and we would need to compare to a zero timeout (so only using the ACT prescriptions) and to longer timeouts to see how this behaviour changes.

Table 2. Means of pre-game and post-game impressions for each opponent type.

Opponent	Give 2			Take 1			Self (human)			Other (human/bot)		
	E	P	A	E	P	A	E	P	A	E	P	A
(Initial)	1.4	0.1	0.2	−0.6	0.9	0.6	1.1	0.6	0.3	0.2	0.6	0.3
Jerkbot	1.3	−0.3	−0.1	−1.3	0.8	0.7	1.3	−0.1	0.9	−1.9	0.4	0.5
Bayesact	1.3	0.1	0.0	−0.9	1.1	1.0	0.7	1.4	1.2	0.4	−0.1	−0.3
Human	1.7	0.7	0.3	−1.2	0.4	0.3	1.5	1.2	1.0	0.5	0.0	0.1
Titfortat	2.3	1.2	1.1	−1.2	0.5	0.3	1.9	1.7	1.7	2.2	1.1	1.1

To further investigate the differences between the different opponents, we measure the mean fraction of cooperative actions on the part of the human after (and including) the 10th game (see Table 1). We find that, when playing against another human, humans cooperate in 0.56 ± 0.45 of these last games. This number was almost the same when playing *BayesAct* agent at 0.54 ± 0.40. Against *tit-for-tat*, there was much more cooperation (0.81 ± 0.35). Finally, against *jerkbot*, it was very low 0.09 ± 0.24. We also computed the mean EPA ratings of the self and other after each game, as shown in Table 2. We found that *jerkbot* (EPA:$\{-1.9, 0.4, 0.5\}$) is seen as much more negative, and *tit-for-tat* (EPA:$\{2.2, 1.1, 1.1\}$) much more positive, than human (EPA:$\{0.5, 0.0, 0.1\}$) or *BayesAct* (EPA:$\{0.4, -0.1, -0.3\}$), and that the human participants felt less powerful when playing *jerkbot* (EPA of self:$\{1.3, -0.1, 0.9\}$) than when playing *BayesAct* (EPA of self:$\{0.7, 1.4, .2\}$), or another human (EPA of self:$\{1.5, 1.2, 1.0\}$). Human participants felt more powerful, positive and active when playing *tit-for-tat* (EPA of self:$\{1.9, 1.7, 1.7\}$).

3   Conclusion

We have presented a model for affectively guided play in the prisoner's dilemma. Our aim is to design agents that are human-like in their behaviours using symbolic interactionist principles, which prescribe socially expected actions given the identities of the actor and her opponent. In this paper, we have shown how these principles result in more human-like play in the iterated prisoner's dilemma. We are currently running simulations of *BayesAct* agents (learned from human data) in a networked prisoner's dilemma setting. Other research avenues include assistive technologies [11], intelligent tutoring [8] and other games [2].

References

1. Antonio, D.: Descartes' Error: Emotion, Reason, and the Human Brain. GP Putnam's Sons, New York (1994)
2. Asghar, N., Hoey, J.: Monte-Carlo planning for socially aligned agents using Bayesian affect control theory. In: Proceedings of the Uncertainty in Artificial Intelligence (UAI), pp. 72–81 (2015)
3. Axelrod, R., Hamilton, W.D.: The evolution of cooperation. Science **211**(4489), 1390–1396 (1981)
4. De Melo, C.M., Carnevale, P., Read, S., Antos, D., Gratch, J.: Bayesian model of the social effects of emotion in decision-making in multiagent systems. In: Proceedings of AAMAS, vol. 1, pp. 55–62 (2012)
5. Goffman, E.: Behavior in Public Places. The Free Press, New York (1963)
6. Heise, D.R.: Expressive Order: Confirming Sentiments in Social Actions. Springer Science & Business Media, Heidelberg (2007)
7. Hoey, J., Schröder, T.: Bayesian affect control theory of self. In: AAAI, pp. 529–536. Citeseer (2015)
8. Hoey, J., Schröder, T., Alhothali, A.: Affect control processes: intelligent affective interaction using a partially observable Markov decision process. Artif. Intell. **230**, 134–172 (2016)
9. Jung, J.D.A., Hoey, J., Morgan, J.H., Schröder, T., Wolf, I.: Comparison of affect-control theoretic agents and humans in the prisoner's dilemma. Technical report CS-2015-18, University of Waterloo School of Computer Science (2015)
10. LeDoux, J.: The Emotional Brain: The Mysterious Underpinnings of Emotional Life. Simon and Schuster, New York (1998)
11. Malhotra, A., Hoey, J., König, A., van Vuuren, S.: A study of elderly people's emotional understanding of prompts given by virtual humans. In: Proceedings of the 10th EAI Conference on Pervasive Computing Technologies for Healthcare (2015)
12. Osgood, C.E., May, W.H., Miron, M.S.: Cross-Cultural Universals of Affective Meaning. University of Illinois Press, Champaign (1975)
13. Schröder, T., Hoey, J., Rogers, K.B.: Modeling dynamic identities and uncertainty in social interactions: Bayesian affect control theory. Am. Soc. Rev. (2016, in press)
14. Sequeira, P., Melo, F.S., Paiva, A.: Learning by appraising: an emotion-based approach to intrinsic reward design. Adapt. Behav. **22**(5), 330–349 (2014)
15. Squazzoni, F., Jager, W., Edmonds, B.: Social simulation in the social sciences a brief overview. Soc. Sci. Comput. Rev. **32**(3), 279–294 (2014)
16. Vilone, D., Ramasco, J.J., Sánchez, A., San Miguel, M.: Social imitation versus strategic choice, or consensus versus cooperation, in the networked prisoner's dilemma. Phys. Rev. E **90**(2), 022810 (2014)

The Mismeasure of Machines

Eric Neufeld[(✉)] and Sonje Finnestad

Department of Computer Science, University of Saskatchewan,
176 Thorvaldson Bldg. 110 Science Place, Saskatoon, SK S7N 5C9, Canada
`{eric.neufeld,sonje.f}usask.ca`

Abstract. We reply to Hector Levesque's critique of the Turing Test and his proposal for "a new type of Turing Test", a Winograd Schema Test. We question whether the role of deception in the Turing Test is, as Levesque asserts, "a serious problem". We argue that the Levesque Test specifies the nature of intelligence in a way that Turing wished to avoid. We conclude that the Turing Test appeals to the collective judgment of humankind, a judgment that, in the case of AI, has yet to be rendered.

Keywords: Turing test · Winograd schema · Artificial intelligence testing

1 Introduction: The Turing Test

In 1950 [1], Turing asked, "Can machines think?" If we are going to ask this question, he said, we need to define our terms. What is the definition of 'machine'? What is the definition of 'think'? His next move is well known. He proposed to replace, "Can machines think?" with "a new form of the problem", the *imitation game*.

Turing described several versions of this *game* or *test*; see [2] for details. The small differences do not matter, for our purposes, except for the fact, to which we shall return later, that in the final version of the test [3] the single interrogator of the original has become a jury. Once the original question has been replaced by the test, the issue becomes whether, in the context of a text-based conversation or question and answer session, a computer can pass for human.

Turing made a couple of surmises. In [1], he famously remarks, "I believe that in about fifty years' time it will be possible to programme computers … to make them play the imitation game so well that an average interrogator will not have more than 70 percent chance of making the right identification after five minutes of questioning". Less famously, in a 1952 radio broadcast [3], he says, in response to a question from Max Newman, that it will be "at least 100 years" before a machine will, in Newman's words, "stand any chance with no questions barred".

Turing's account of this game or test is as close as he comes to a definition of thinking or intelligence and these surmises are as close as he comes to answering the question, "Can machines think?" We might paraphrase Turing's position, albeit rather colourfully, as "they will walk among us".

R. Khoury and C. Drummond (Eds.): Canadian AI 2016, LNAI 9673, pp. 58–63, 2016.
DOI: 10.1007/978-3-319-34111-8_8

2 Leveque's Critique

2.1 Too Much Lying

According to Levesque [4, 5], the Turing Test "has a serious problem: it relies too much on deception". Levesque uses loaded language: "a computer program passes the test iff it can *fool* an interrogator into thinking she is dealing with a person not a computer." A program "will either have to be evasive ... or manufacture some sort of false identity (and be prepared to lie convincingly)." "All other things being equal," says Levesque, "we should much prefer a test that did not depend on chicanery of this sort". "Is intelligence just a bag of tricks?" he asks.

The Turing test undoubtedly involves deception, for the same reason it involves communication by text. Turing [1] explains: "The new problem has the advantage of drawing a fairly sharp line between the physical and the intellectual capacities of a man." It's the intellectual that interests us. "We do not wish to penalise the machine for its inability to shine in beauty competitions ... The conditions of our game make these disabilities irrelevant".

For the same reason, questions that amount to asking whether a computer possesses physical characteristics of humanness cannot be answered truthfully. If the interrogator asks, "are you a machine?" a computer cannot say, "yes". If asked, "which tastes better: dark chocolate or milk chocolate?" a computer must give an answer consistent with an ability to taste. Anything else amounts to replying, "yes" to "are you a machine?" or communicating in an electronic voice or being seen to have a body composed of metal and plastic.

Such lying as there is, is there to protect the integrity of the test as a test of intelligence rather than the possession of human physical characteristics. It is not clear to us that this constitutes a problem, let alone a problem serious enough to justify a rejection of the Turing Test.

2.2 The Loebner Competition Chatterbots

Levesque's discussion of this "serious problem" [4] focuses on the kind of tactics seen in the Loebner competition. The Loebner chatterbots, he says, "rely heavily on wordplay, jokes, quotations, asides, emotional outbursts, points of order, and so on. Everything, it would appear, except clear and direct answers to questions!"

Strategems of this kind surely reflect the fact that, in the current state of the technology, the chatterbots are unable to answer the questions. Does Levesque believe that any winner of the Loebner competition has passed the Turing test?

These remarks are not intended to disparage the efforts of those who host or compete in the Loebner competition. At this point in history, it may be that the best way to win that competition - which is, in any case and in our view, not the test Turing envisaged - involves tactics of the kind Levesque decries. None of this precludes future development of truly remarkable AI that answers questions directly, intelligently, and engagingly, winning Loebner Competitions right and left and decisively passing the Turing Test.

For these and other reasons [2], we are not convinced that the behaviour of the Loebner chatterbots warrants a rejection of the Turing test.

2.3   Imitating a Person

A related problem, according to Levesque, is that the Turing Test requires a computer to pass, not just as intelligent, but as an intelligent *person* [4, 5]. Sometimes this objection seems to be an elaboration of the deception objection; at other times, however, Levesque appears to suggest that the Turing Test requires more than is necessary. "A machine," he says [5], should be able to show us it is thinking without having to pretend to be somebody or to have some property (like being tall) that it does not have". A test that, like the one he proposes (described below), does not require "the ability to generate 'credible' English" is preferable.

We can imagine that Turing might reply to this objection as he did to another [1]: "May not machines carry out something which ought to be described as thinking but which is very different from what a man does?" Turing acknowledged that "This objection is a very strong one, but at least we can say that if, nevertheless, a machine can be constructed to play the imitation game satisfactorily, we need not be troubled by this objection". Turing never maintained that a machine must pass his test to be considered intelligent, just that a machine that passed his test must be considered intelligent; in other words, to repeat an oft-repeated point, the Turing Test is a sufficient, not a necessary, condition of machine intelligence.

Nevertheless, Levesque might say, why not just cut to the chase? Why not avoid all the deception and evasions and the unnecessary requirements and just focus on demonstrating intelligence? He proposes [4], as "a new type of Turing Test", a *Winograd Schema Test*.

3   The Levesque Test

3.1   Winograd Schemas and Thinking

A *Winograd Schema* (*WS*) is an anaphoric disambiguation test [5]. An essential feature is the inclusion of a *special* word that, when replaced by an *alternate*, flips the answer. In the examples below, the *special* word is italicized and its alternate appears, likewise italicized, in parentheses:

The trophy would not fit in the brown suitcase because it was too *big* (*small*). What was too big?

Answer 0: the trophy
Answer 1: the suitcase.

The town councillors refused to give the angry demonstrators a permit because they *feared* (*advocated*) violence. Who feared violence?

Answer 0: the town councillors
Answer 1: the angry demonstrators

Levesque [5] claims that "doing better than guessing requires subjects to figure out what is going on." It requires background knowledge (not expressed in the question) and "it is precisely bringing this background knowledge to bear" – "using what we know" – "that we informally call *thinking*". So, "with a very high probability, anything that answers correctly is engaging in behaviour that we would say shows thinking in people".

This, Levesque assures us, is not an attempt to enter into "the philosophical question that Turing sidesteps". Turing, according to Levesque, does not wish to sidestep an account of "what we call thinking in people", an account Levesque is more than happy to provide; the issue he wants to sidestep is the philosophical question of whether "a subject that passes the test is really and truly thinking".

We respectfully disagree. Turing [1], when he expresses his disinclination to define thinking and proposes, instead, the imitation game, remarks that it would be dangerous to take our definitions from "the normal use of words". While we strongly suspect that Turing wished to sidestep philosophical accounts as well, this sounds much closer to Levesque's, "what we call thinking" than philosophy. We maintain that Levesque has gone where Turing (with good reason) did not wish to go.

Hence, Levesque is concerned not just with how a machine answers but that the answers are obtained by *thinking* [4]: "Getting the answer right but for dubious reasons" does not count. One reason he favours a *WS* test [5] is that "clever tricks involving word order or other features of words or groups of words will not work. Having access to a large corpus of English text would likely not help much".

This is in marked contrast to Turing. In 1948 [6] describing a chess-playing precursor to the Turing Test, he specifies that the judges "are to be rather poor chess players". Several years later [3], in 1952, describing the jury version of his test, Turing says that a jury "should not be expert about machines." It appears that, for Turing, considerations of what sorts of operations might be performed to generate this or that response should not be a factor in the assessment of a conversation (or chess) partner as intelligent. This is consistent with Turing's refusal to define thinking; likewise Levesque's concern with how the answers are derived is consistent with his apparent willingness to define it, at least in part.

3.2 Some Historical Perspective

A test that presupposes something like a definition of intelligence is liable to meet the same fate as previous test beds, e.g., tic-tac-toe, checkers, chess, Jeopardy, Go, not to mention other tests of intelligence. We can illustrate aspects of this danger by looking a little more closely at the *WS* examples given in Sect. 4.1 above.

The suitcase question could be answered either way, depending on how "The trophy would not fit" is understood. If it is understood to mean that the trophy could not be be contained in the suitcase, then the answer seems obvious enough. But if it can also mean that the trophy could not be contained without extra space, so that it bumped and rattled around, the whole thing begins to seem less obvious.

In the second example, the answers are likewise supposed to be obvious. "Informally," Levesque [5] says, "a good question for a *WS* is one that an untrained subject (your Aunt Edna, say) can answer immediately". We wonder how Martin Luther King, or all those 'Aunt Edna's who marched with him, would have answered this question. It is entirely possible that protestors would fear violence and that town councillors would – for that reason – be unwilling to grant them a permit, though the protestors wish to protest nonetheless. Likewise, it is not inconceivable that town councillors would advocate violence – there are certainly historical precedents – and that they would refuse to grant a permit, to further their own ends or because they believe it would entail a diversion of resources that could be more usefully employed.

Levesque [5] claims that such interpretations "are farfetched and will not trouble your Aunt Edna". However, this is a very specific Aunt Edna, a white, middle-class American Aunt Edna with conventional American views. The point we wish to make here is that these are precisely the sort of issues and debates that have dogged human intelligence testing, as chronicled by Gould [7]. The reader may find our discussion of the two *WS* examples a stretch, but some old IQ test questions are truly startling now, though they must have seemed obvious to someone's Aunt Edna at the time, e.g. [8]:

> Crisco is a: patent medicine, disinfectant, toothpaste, food product
> Washington is to Adams as first is to …
> Christy Mathewson is famous as a: writer, artist, baseball player, comedian.

In sum, Levesque ventures down a path strewn with the wreckage of accounts of intelligence rejected or abandoned. We do not say that it is pointless to go down this path - far from it; we only wish to point out that this is an area in flux and anyone who ventures so specific a test of intelligence, and thereby some sort of account, explicit or implicit, of what intelligence is, can expect it to take its place, sooner or later, amongst the detritus.

4   The Wisdom of the Turing Test

We might informally sum up the differences between the Levesque Test and the Turing Test by saying that the Aunt Edna of the Turing Test is not some particular standard but, rather, an evolving, and collective, judge.

Turing, as we noted earlier, insisted that his test be judged by non-experts. This, together with his refusal to put forward his own account of intelligence, suggests that he envisages something like a collective consensus; this is perhaps implicit in the 'jury version' of the test and it seems even clearer when he says [1]: "… at the end of the century the use of words and general educated opinion will have altered so much that one will be able to speak of machines thinking without expecting to be contradicted."

"Passing as a person" as criterion of intelligence appeals to the ordinary criteria whereby we judge, "I am dealing with another intelligent human". This is an indirect criterion: it appeals to criteria without specifying them. As such – and crucially – it accommodates different – and changing – views as well as technological developments.

Views of what intelligence is, how it is assessed, and who is intelligent had undergone significant changes by Turing's time and there have been changes since. We can discern an expanding circle of perceived intelligence: women, the poor, and those of other races and cultures are now understood to be as intelligent as property-owning, educated, European males. As mentioned earlier, there has been an ongoing, often ferocious, debate over intelligence testing. In a related development, many have come to believe in a theory of multiple intelligences [7]. Scientists and other experts have certainly contributed to these changes, yet what we are describing is a collective – and ongoing – discernment by people who are not 'experts' in the relevant sense.

Many would maintain that this is as it should be: that such consequential judgments of value as well as of fact can and must be made by humankind. We believe that Turing's test is rooted in such a mindset. Turing's brilliant insight – like so much of what he said, so far ahead of its time – is that humankind can and will decide what counts as intelligence in a given time and place. What he called 'my test' is an application of this insight in the domain of machine intelligence.

References

1. Turing, A.M.: Computing machinery and intelligence. Mind **59**, 433–460 (1950)
2. Neufeld, E., Finnestad, S.: Proceedings of FLAIRS-29, Key Largo (to appear)
3. Turing, A.M., Braithwaite, R., Jefferson, G., Newman, M.: Can automatic calculating machines be said to think? (1952). In: Copeland, B.J. (ed.) The Essential Turing, p. 487. Clarendon Press, Oxford (2004)
4. Levesque, H.J.: On our best behaviour. Artif. Intell. **212**, 27–35 (2014)
5. Levesque, H.J., Davis, E., Morgenstern, L.: The Winograd schema challenge. In: KR, May 2012
6. Turing, A.M.: Intelligent machinery (1948). In: Copeland, B.J. (ed.) The Essential Turing, p. 410. Clarendon Press, Oxford (2004)
7. Gould, S.J.: The Mismeasure of Man. WW Norton and Company, New York (1996)
8. Gould, S.J.: A nation of morons. New Sci. **6**, 349–352 (1982)

Uncovering Hidden Sentiment in Meetings

Gabriel Murray[✉]

University of the Fraser Valley, Abbotsford, BC, Canada
gabriel.murray@ufv.ca
http://www.ufv.ca/cis/gabriel-murray/

Abstract. The sentiment expressed by a meeting participant in their face-to-face comments may differ from the sentiment contained in their private summary of the meeting. In this work, we investigate whether we can predict the sentiment score of a participant's private post-meeting summary, based on multi-modal features derived from the group interaction during the meeting. We describe several effective prediction models, all of which outperform a baseline that assumes the sentiment score of the summary will be the same as the sentiment score of the participant's comments during the meeting.

Keywords: Sentiment detection · Subjectivity · Multi-modal interaction

1 Introduction

Being able to predict group members' positive or negative sentiment based on their interaction in a meeting could be valuable for improving group efficiency, productivity, and social cohesion. However, there are obstacles to being able to accurately predict the sentiment held by meeting participants. For example, a group member might refrain from making highly negative comments during the meeting even though they have negative opinions about items under discussion. Or a group member might make little vocal contribution during the meeting, despite having strong positive or negative opinions.

In this research, we study meeting data in which participants have been asked to write a short, private summary after each meeting. The summaries can also include any problems or issues that occurred during the meeting. The private summaries are not seen by the other participants. We show that we can predict the sentiment scores of these private summaries, based on multi-modal features from the meeting itself. These prediction models outperform a baseline in which it is assumed that the sentiment score for a participant's private summary will be the same as the sentiment score for their comments during the meeting.

The structure of the paper is as follows. In Sect. 2, we describe related work on sentiment detection in meetings and on meeting analysis in general. In Sect. 3 we describe our sentiment prediction system, including the dataset and features, sentiment scoring method, and the prediction models. The results are presented in Sect. 4 and we conclude in Sect. 5.

© Springer International Publishing Switzerland 2016
R. Khoury and C. Drummond (Eds.): Canadian AI 2016, LNAI 9673, pp. 64–72, 2016.
DOI: 10.1007/978-3-319-34111-8_9

2   Related Work

Closely related work has aimed to detect meeting sentences containing positive or negative sentiment. Raaijmakers et al. [1] and Murray and Carenini [2] both use multi-modal features to classify whether dialogue acts segments (sentence-like units in meetings) contain positive or negative subjectivity.[1] Our work differs from theirs in two ways. First, we are predicting the sentiment score of post-meeting participant summaries rather than the sentiment of meeting sentences. Second, our score prediction is a regression, rather than classification, task.

Several recent books survey the more general field of sentiment analysis, including detection of opinions and emotions [3–5].

Much work has been done on studying multi-modal interaction in meetings more generally [6], including the use of machine learning models to learn about and improve group efficiency and productivity in meetings [7,8]. There has also been a rich vein of research on modelling group interaction and small group dynamics, including phenomena such as dominance and influence [9–14]. Much of that work has focused on non-verbal cues, while we incorporate both verbal and non-verbal features in these experiments.

To our knowledge, this is the first work to use participant summaries to analyze sentiment amongst group members. The only other work we have seen that uses participant summaries is by Kim and Shah [15], who use self-reported summaries to assess whether a group has achieved "consensus of understanding."

3   Hidden Sentiment Prediction

The goal of our system is to predict the sentiment that will be contained in the private post-meeting summary written by a participant, based on the meeting and the participant's interaction in the meeting. The participant summaries must therefore be scored according to their sentiment. We rely on the sentiment lexicon supplied by Taboada et al. [16] as part of their SO-Cal sentiment detection system. The lexicon contains lists of sentiment-bearing adjectives, adverbs, nouns and verbs, each of which is associated with a positive or negative score. Positive scores range from 1 to 5, and negative scores range from -1 to -5.

Taboada et al., citing Boucher and Osgood [17], note that many texts seem to have a positive bias, with positive words being much more frequent than negative words. That is certainly the case with meeting transcripts, where negative sentences are relatively rare [18] and difficult to detect [2]. This may be due to participants refraining from stating negative opinions in face-to-face interactions, particularly in meetings where the participants do not know each other, as is the case in the corpus we describe below. This could also be due to the use of euphemisms, where mildly negative words are indicative of strong negative sentiment. Whatever the underlying cause for the imbalance, Taboada et al. assume that negative words carry more cognitive weight and they found that increasing the sentiment weights of negative words by 50 % improved their sentiment

---

[1] The terms *subjectivity* and *sentiment* are very closely related, and we use the latter.

prediction performance in comparison with gold-standard sentiment labels. We carry out the same 50 % adjustment of negative word scores in this work.

Having carried out the negative score adjustment just described, the sentiment score for a document is then the average sentiment score for all sentiment-bearing words in the document.

3.1   The AMI Meeting Corpus

The meeting data and associated participant summaries are from the AMI meeting corpus [19]. We use the scenario portion of the corpus, where participants are role-playing as members of a company designing a remote control. Each group consists of four members, assigned the roles of project manager, user interface designer, industrial design expert, and marketing expert. Each group goes through a series of four meetings, wherein they discuss different phases of design, finance, and production. After each meeting, the participants were asked to write individual summaries of what happened during the meeting, including any problems that occurred.

Below we show a sample of the types of comments participants make in these post-meeting summaries:

- "We have no feel for the strengths and weaknesses of the team and what our particular roles are for this project."
- "Lack of familarity with each other personally and socially as a team."
- "A lack of direction in the meetings."
- "I was not convinced myself that some of the trends were desirable to incorporate, and the group confirmed this."
- "Industrial Designer, Alima, who was originally frustrated because she could not find enough information, presented a very coherent explanation of how the remote works."
- "We decided to focus on fashion, usability, and simplicity in our design."

Figure 1 shows the distribution of sentiment scores for participant comments in meetings and for participant summaries. Each meeting is treated as four separate documents, each document consisting of a single participant's comments. One surprising finding is that when the meetings and summaries are scored in the manner described above, meetings tend to be more negative while the summaries tend to be more positive.

Figure 2 shows a scatterplot, where each point corresponds to the sentiment score of an individual participant's meeting comments and the sentiment score of their subsequent private summary. We can see that a participant's sentiment in the meeting is not always a good predictor of their sentiment in the corresponding summary. In many cases, they are relatively neutral in the meeting but positive in the summary, and in a few cases they are positive in the meeting but relatively negative in the summary.

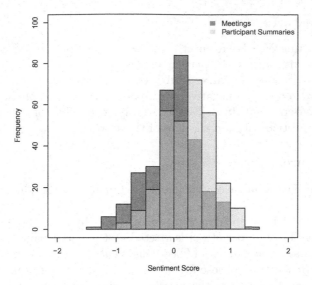

Fig. 1. Sentiment distribution

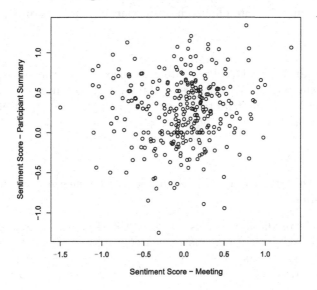

Fig. 2. Sentiment in meetings vs. summaries

3.2   Sentiment Features and Models

All of our prediction models use the same set of verbal and non-verbal features derived from the meetings. We group them into the following broad classes:

Sentiment Features

- **posWords, negWords:** Respectively, the number of positive and negative sentiment words used by the participant in the meeting.
- **totalPosSubjScore, totalNegSubjScore:** Respectively, the sum of the positive word scores and negative word scores used by the participant.
- **totalSubjWords, totalSubjScore:** The total number of subjective words used by the participant, and the sum of those word scores.

Activity Features

- **totalDacts:** The number of dialogue act segments by the participant in the meeting.
- **totalTime:** The total speaking time of the participant in the meeting.
- **totalWords:** The number of words uttered by the participant in the meeting.
- **totalFillPause:** The number of filled pauses (*uh, um*, etc.) by the participant.
- **first, last:** Respectively, these features indicate whether the participant was the first person to speak in the meeting or the last person to speak in the meeting.
- **rateOfSpeech:** The rate-of-speech of the participant, in words per second.

Meeting Features

- **meetA, meetB, meetC:** There are four meetings in the series, A–D. The position in the series is encoded using three binary features.
- **allmeetwords, allmeetdacts:** Respectively, the total number of words and dialogue acts in the meeting, across all participants.

Speaker Features

- **PM, UI, ME** There are four assigned roles in the meeting, encoded by three binary features.

We use three prediction models for this task. The first is a multiple linear regression. The second is a multi-layer neural network, with two hidden layers each containing two units, as shown in Fig. 3. The third system is a random forest with 500 trees and seven variables tried at each split.

3.3 Experimental Setup

Each meeting yields four datapoints, one for each participant. However, not all AMI meetings contain participant summaries. We ultimately ended up with 302 datapoints. For the multiple regression and neural network predictions, we report results using 10-fold cross-validation. For the random forest regression, we report out-of-bag prediction results.

The evaluation metric used is mean-squared error (MSE).

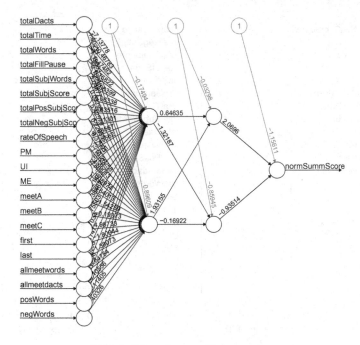

Fig. 3. Neural network structure

4   Results

The MSE scores are shown in Table 1. The best-performing predictions were using random forests and multiple regression, which were comparable to each other. The multi-layer neural network did not perform as well, and adding hidden layers and units only resulted in further degraded results. All systems performed better than a baseline prediction that assumes the summary score will be the same as the meeting score.

Table 1. MSE Scores

System	MSE
Baseline (Score Same as Meeting)	0.416
Multi-Layer Neural Network	0.243
Multiple Regression	0.177
Random Forest	0.175

For analyzing the most useful features, we consider just the best-performing system, random forests. Figure 4 shows two measures of variable importance in the random forest regression. The "%IncMSE" plot shows the percentage that

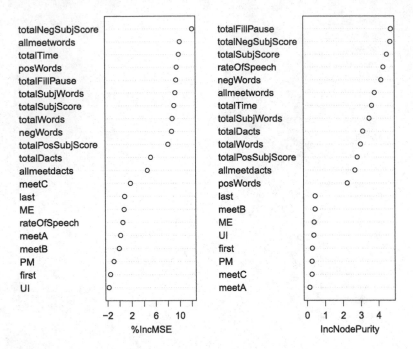

Fig. 4. Variable importance

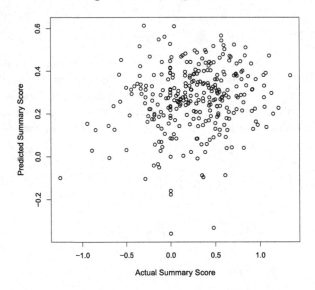

Fig. 5. Actual summary scores vs. predictions

the MSE increases when the variable is removed. "IncNodPurity" shows the increase in node purity when splitting on that variable. Many of the sentiment features from the meeting are useful predictors of the sentiment in the resultant summary. However, non-sentiment features that relate to the length of the meeting are also very good predictors. To highlight one feature, the number of filled pauses is a very useful indicator according to both metrics.

Despite the in-meeting sentiment features being amongst the most useful predictors, we can achieve very good performance using only the non-sentiment features for prediction. A random forest regression using the non-sentiment predictors yields only a slightly higher MSE of 0.18, compared with 0.175 for the full feature set.

5 Conclusion

In this work, we investigated whether we can successfully predict the sentiment contained in a meeting participant's private summary, based on characteristics of the meeting and the participant's interaction in the meeting. Using a variety of verbal and non-verbal cues, we showed that three prediction models can outperform a baseline where the sentiment of the summary is predicted to be the same as in the meeting. Of the three prediction models, multiple regression and random forests performed the best.

There is still room for improvement in our sentiment predictions, as evidenced by Fig. 5 showing the actual summary sentiment scores plotted against the predicted sentiment scores. In particular, there is a positive bias in the predictions, with the predicted scores generally being more positive than the actual scores.

An unexpected finding is that the participant summaries are not more negative than the participants' comments in the meeting. In fact, the summaries tend to be slightly more positive than the corresponding meeting comments. Our assumption that participant's true opinions would tend to be more negative than they indicated in the meeting was not supported by this data.

In future work, we plan to incorporate intensification, diminishment, and negation, which may be improve our sentiment modelling. We also plan to incorporate additional non-verbal features such as prosody and head gestures, in order to improve prediction performance.

References

1. Raaijmakers, S., Truong, K., Wilson, T.: Multimodal subjectivity analysis of multiparty conversation. In: Proceedings of EMNLP 2008, Honolulu, HI, USA (2008)
2. Murray, G., Carenini, G.: Subjectivity detection in spoken and written conversations. Nat. Lang. Eng. **17**(03), 397–418 (2011)
3. Pang, B., Lee, L.: Opinion Mining and Sentiment Analysis. Now Publishers, Vancouver (2008)
4. Liu, B.: Sentiment Analysis and Opinion Mining. Morgan & Claypool Publishers, San Rafael (2012)
5. Liu, B.: Sentiment Analysis: Mining Opinions, Sentiments, and Emotions. Cambridge University Press, Cambridge (2015)

6. Renals, S., Bourlard, H., Carletta, J., Popescu-Belis, A.: Multimodal Signal Processing: Human Interactions in Meetings, 1st edn. Cambridge University Press, New York (2012)
7. Murray, G.: Analyzing productivity shifts in meetings. In: Barbosa, D., Milios, E. (eds.) Canadian AI 2015. LNCS, vol. 9091, pp. 141–154. Springer, Heidelberg (2015)
8. Kim, B., Rudin, C.: Learning about meetings. Data Min. Knowl. Discov. **28**(5–6), 1134–1157 (2014)
9. Rienks, R., Zhang, D., Gatica-Perez, D., Post, W.: Detection and application of influence rankings in small group meetings. In: Proceedings of ICMI 2006, Banff, Canada (2006)
10. Pentland, A., Heibeck, T.: Honest Signals. MIT press, Cambridge (2008)
11. Jayagopi, D., Hung, H., Yeo, C., Gatica-Perez, D.: Modeling dominance in group conversations from non-verbal activity cues. IEEE Trans. Audio Speech Lang. Process. **17**(3), 501–513 (2009)
12. op den Akker, R.: Multi-modal analysis of small-group conversational dynamics. In: Renals, S., Bourlard, H., Carletta, J., Popescu-Belis, A. (eds.) Multimodal Signal Processing, pp. 155–169. Cambridge University Press, New York (2012)
13. Dong, W., Lepri, B., Pianesi, F., Pentland, A.: Modeling functional roles dynamics in small group interactions. IEEE Trans. Multimed. **15**(1), 83–95 (2013)
14. Frauendorfer, D., Mast, M.S., Sanchez-Cortes, D., Gatica-Perez, D.: Emergent power hierarchies and group performance. Int. J. Psychol. **50**(5), 392–396 (2014)
15. Kim, J.H., Shah, J.: Automatic prediction of consistency among team members' understanding of group decisions in meetings. In: Proceedings of IEEE SMC, pp. 3702–3708 (2014)
16. Taboada, M., Brooke, J., Tofiloski, M., Voll, K., Stede, M.: Lexicon-based methods for sentiment analysis. Comput. Linguist. **37**(2), 267–307 (2011)
17. Boucher, J., Osgood, C.: The pollyanna hypothesis. J. Verbal Learn. Verbal Behav. **8**(1), 1–8 (1969)
18. Wilson, T.: Annotating subjective content in meetings. In: Proceedings of LREC, pp. 2738–2745 (2008)
19. Carletta, J.: Unleashing the killer corpus: experiences in creating the multi-everything ami meeting corpus. In: Proceedings of LREC 2006, Genoa, Italy, pp. 181–190 (2006)

Fuzzy Computational Model for Emotions Originated in Workplace Events

Ahmad Soleimani[(✉)] and Ziad Kobti

School of Computer Science, University of Windsor, Windsor, ON, Canada
{soleima,kobti}@uwindsor.ca

Abstract. This article investigates the relationship between affect-relevant events that are associated with the workplace of a human agent and the set of emotions that are elicited in the agent in response to those occurred events. The proposed model uses a hybrid appraisal and dimensional approach that strives to deeply analyze emotion triggering events in workplaces and to project their affective impacts onto the three dimensions of evaluation, potency, and activity introduced in the Affect Control Theory. A detailed test-case with a blend of realistic and simulated data was used to verify the performance of the proposed model.

1 Introduction

Despite the relatively young age of *Affective Computing (AC)* [12], it has managed to make huge achievements in emotion detection (e.g., [3]), emotion modeling (e.g., [5]), and emotion exhibition in artificial agents (e.g., [2]).

A key element in an AC-based system is the mechanism taken to model various processes involved in an emotional experience. Different approaches such as dimensional (e.g., [5]), core-affect (e.g., [13]), fuzzy (e.g., [4]), etc., were considered by researchers for this purpose. Undoubtedly, appraisal models (e.g., [10]) are among the most widely used approaches for emotion modeling [8].

The core component in appraisal theories of emotion is the fact that emotions are physiological-mental reactions to the lived situations. A major factor that creates a situation or alters a current situation is the occurrence of events. Hence, direct impacts of a relevant event or its consequences as well as the potentials for adaptation and available coping strategies play a key role in eliciting different emotions in the individual under study.

In this research work, we propose a mechanism that analyzes and maps workplace events to their corresponding emotional states. The analysis and appraisal processes is being performed based on the principles of *Affect Control Theory (ACT)* [11].

Considering the important non-deterministic factor in emotional dynamics, a fuzzy state machine framework was used to best reflect the transitioning processes between different emotional states. Where applicable, results generated by the proposed model were partially validated using corresponding data obtained from two related case studies (i.e., [1,9]).

© Springer International Publishing Switzerland 2016
R. Khoury and C. Drummond (Eds.): Canadian AI 2016, LNAI 9673, pp. 73–79, 2016.
DOI: 10.1007/978-3-319-34111-8_10

2 Proposed Computational Model

The proposed computational model is aimed at achieving two major goals: first, to establish event-emotion correlations that links affect relevant events to the corresponding elicited emotions; and second, to come up with an emotion dynamics mechanism that is capable of tracking the changes in the emotional states.

In order to achieve the first goal, a special approach based on a hybrid of dimensional and appraisal methods was utilized. Accordingly, occurred events were assessed and measured using a set of appraisal variables and then were projected onto the three dimensions of the ACT, i.e., *Evaluation, Potency,* and *Activity.*

The first dimension of this ternary space, evaluation, determines the overall positiveness or negativeness of the occurred event along with its valence intensity; Potency is linked to the degree of adaptation and coping strategies associated with the occurred event; Activity, on the other hand measures the degree of activeness or passiveness that the agent experience as a result of the occurrence of the event.

In a numerical scale, all of these three dimensions are measured using an interval of $[-1,1]$. For instance, an event with an EPA vector of $\overrightarrow{EPA} = (-0.78, -0.68, -0.45)$, represents a highly undesirable event that has the potential of creating hyper tension in the agent with a considerably low degree of control over the outcomes of the event.

Considering the fact that this paper is mostly intended to validate the proposed model for event-emotion matrix, for brevity, we refrain from dissecting the entire model here and only a short summary is provided. Interested users are referred to [15] for the complete general model.

In order to quantify the dimension of evaluation, a fuzzy system that describes events in terms of their impact on the set of goals of the agent was created. Accordingly, the desirability of an event is being determined based on the impact of the event on the set of goals of the agent as well as the importance of each goal.

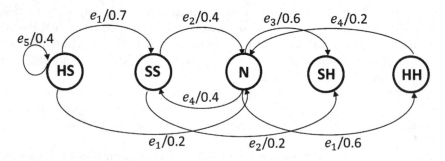

Fig. 1. Fuzzy automaton for the five fuzzy states of the sad-happy emotional channel. Here, the fuzzy states are HighlySad, SlightlySad, Neutral, SlightlyHappy and Highly-Happy. The stimuli are the set of five events, $e_1 - e_5$. The number beside each event indicates the fuzzy membership value for the associated inter-state transition

Potency as the second dimension of ACT is the sentiment of being either dominant or submissive towards the event or its outcomes. Potency in fact represents a post appraisal process of the occurred event and it acts as a regulation mechanism based on the available coping strategies and the notion of emotion regulation [6].

With respect to finding an acceptable approximation for the third dimension of ACT, activity, it was concluded that activity is mostly associated with the three appraisal variables of *Likelihood*, *Unexpectedness* and *Urgency* of the occurred event.

In order to achieve the second goal of this research work, i.e., modeling emotion dynamics, a fuzzy state-machine framework was considered at which the fuzzy states are the emotional states within the related emotional channel and the fuzzy transitioning functions are the EPA vectors for each event that were obtained through the first stage of event's assessments. For example, Fig. 1 depicts a partial automaton for sad-happy bipolar emotional channel. For instance, the occurrence of event e_2 at the state of SS (*SlightlySad*) can either take the agent to state N (*Neutral*) with a fuzzy membership value of 0.4 or directly to the state of SH (*SlightlyHappy*) with a fuzzy membership value of 0.2.

3   Experiments and Discussion

In order to evaluate the performance of the proposed computational model and to verify its functionality, several simulation experiments were conducted. Here, one of these experiments is discussed. The environment for this experiment is a healthcare unit (e.g., hospital). The purpose of this experiment was to study the affective behavior of the nurse agent in response to some workplace events that are associated with this environment. It is worth noting that the conditions of experiment, set of goals, and other relevant parameters were set in a way that considers the environment as merely a workplace for the nurse agent, whereas patients were looked at as customers to nurse agents. Such settings were essential in order to distract the scenario from sensitive health related aspects. A set of four goals, $G_n = \{G_1, G_2, G_3, G_4\}$ was considered for the nurse agent. (see Table 1). Table 2 lists 10 events that have affective impact on nurse agents, $E_n = \{e_1, e_2, ..., e_9, e_{10}\}$ along with the impact of each event on the set of goals. The entries in this table were represented using a fuzzy scale. SN stands for *SlightlyNegative*, HN stands for *HighlyNegative*, SP stands for *SlightlyPositive*, HP stands for *HighlyPositive*, and finally NI stands for *NoImpact or neutral*.

Table 1. List of relevant goals along with their importance

G_1 : Being professional	G_2 : Being humanistic	G_3 : Minimum stress	G_4 : Job affairs
HighlyImportant	HighlyImportant	SlightlyImportant	HighlyImportant

Table 2. List of events and their impact on each goal

Events/Impact on goals	G_1	G_2	G_3	G_4
e_1 : Misbehavior/abuse of a patient agent	SN	NI	HN	NI
e_2 : Wrong prescription/service	HN	HN	HN	HN
e_3 : Appreciation from a patient	SP	HP	SP	SP
e_4 : Appreciation from supervisor	HP	NI	NI	HP
e_5 : Unpleasant coworker/supervisor left work unit	SP	NI	HP	SP
e_6 : Problems getting along with a coworker/supervisor	HN	NI	NH	SN
e_7 : Benefits were reduced	NI	NI	SN	HN
e_8 : Received negative performance evaluation	HN	SN	HN	HN
e_9 : Complain from a patient	SN	NI	HN	SN
e_{10} : Assigned to a trouble-some patient	SN	SN	HN	NI

The first step in modeling the emotional behavior of patient and nurse agents would be to determine the affective impact of each applicable event. Such a step entails determining the $\overrightarrow{EPA}$ for all events of E by calculating the three components of evaluation, potency and activity for each event. As discussed in Sect. 3, with respect to the dimension of evaluation, a system of fuzzy rules that links the fuzzy variable of *Desirability* to other fuzzy variables of *Importance* and *Impact* will be created. Once defuzzified, the solution of the aforementioned fuzzy system would be the value for *Evaluation* dimension. Accordingly, similar approaches would be taken to calculate the values of *Potency* and *Activity*. The chart in Fig. 2 represents the EPA vectors calculated for all events of E_n.

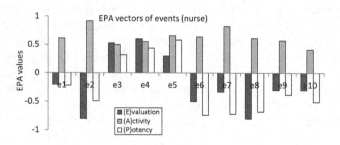

Fig. 2. EPA values of all events for the nurse agent

Discussion. The aim of this experiment is to study the affective impact and to track the changes in the emotional response level of the agent as a result of the occurrence of some stochastic events from E_n. The bipolar emotional channel under study is *distress*-joy. In order to study the impact of applying the above sequence of events on the Emotional Response Level (ERL) of the agent within

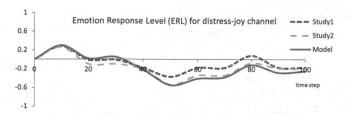

Fig. 3. ERL for nurse agent in distress-joy channel

the emotional channel of distress-joy and the way that the emotional level of the agent transitions between different sates, it would be necessary to generate the EA and P relational matrices for all occurred events. Next, it would be required to apply the related EA and P relational matrices of each participating event to the current emotional fuzzy state of the agent. The results generated from the model were compared against those obtained from two realistic case studies. The first study, referred as study1 in this article investigates the relationship between events and the affective states of a worker agent in a workspace. Study1 included 101 participants at which they briefly described organizational events or situations that caused them to recently experience the specified emotions at work. Participants were hotel employees from functional and administrative departments of ten international hotels in Australia and the Asia/Pacific region. The full description of the study can be found in [1].

On the other hand, the purpose of the second study referred as study2 in this article, was to investigate the causes and consequences of emotions at work by identifying several job-related events likely to produce affective states and then to study the impact of the latter on work attitudes. The hypothesis argued in this study was: "experiencing certain work events leads to affective reactions, which in turn influence work attitudes". The hypothesis was supported by results obtained from an empirical study which included 203 questionnaires performed on a sample of French managers. More details about study2 can be found in [9].

The performance of the ERL for the agent according to the proposed model as well as the two case studies are all reflected in the chart of Fig. 3. A subset of E_p that includes the sequence of events $< e_3, e_{10}, e_6, e_8, e_5, e_4, e_7 >$ which took place at time steps $< 0, 10, 30, 40, 50, 70, 80 >$ was considered. Initially, the occurrence of the positive event of e_3 (i.e., appreciation from a patient agent) managed to significantly improve the ERL of the nurse agent and to take it to above SlightlyJoyful from the initial state of Neutral. Being assigned to a trouble-some patient at step $= 10$, caused the ERL of the nurse to drop quickly to its initial value of Neutral. During the time steps of $20 - 30$, and at the absence of any affective relevant event, the agent started to cop with the new situation and that can be seen through the smooth gradual increase in the ERL. A highly undesirable event of e_6 that occurred at step $= 30$ caused the ERL of the agent to drop to the Distressed zone. The changes in the ERL of the nurse agent continued as more events took place in the system. The ultimate emotional state of the nurse at the end of the simulation was SD or SlightlyDistressed. The sequence of all transitions for the entire simulation time was: N, SJ, N, SD, HD, HD, SD

Figure 3 shows that the affective behavior of the agent according to the proposed model was either between the ERL curves of the two studies or was moving slightly above or below them. Therefore, it can be argued that the performance of the model-generated ERL was inline with that obtained from the case studies yielding in a limited validation of the proposed model.

4 Conclusion and Future Directions

In this article, a computational model for predicting elicited emotions in a human agent as a result of some common affect relevant events that take place at workplaces was proposed. The proposed model benefited from a hybrid architecture of appraisal and dimensional processes based on affect control theory. Furthermore, it utilized a fuzzy automata framework for the purpose of modeling emotion dynamics and in particular the transitions between different states within the same emotional channel.

The performance of the proposed model was evaluated using a test-case with simulated data. The affective behavior of the agents in response to occurred events found to be compliant with the results obtained from two relative case studies.

An important possible extension to the proposed model has to do with the concept of affective interventions and reverse engineering of events. The idea here is to generate and purposefully enforce a set of pro-regulation events in order to neutralize or reverse a hyper negative affective state that was reached at as a result of the occurrence of some adverse events. It would appear that such an extension could be promising within the field of psychopathology and in particular the treatment of event-related traumas.

References

1. Basch, J., Fisher, C.D.: Affective events - emotions matrix: a classification of work events and associated emotions. School of Business Discussion Papers. Paper 65 (1998)
2. Baylor, A.L.: A virtual change agent: Motivating pre-service teachers to integrate technology in their future classrooms. Edu. Technol. Soc. 11(2), 309–321 (2008)
3. Calvo, R.A., I.B., Scheding, S.: Effect of experimental factors on the recognition of affective mental states through physiological measures. In: Proceedings of 22nd Australasian Joint Conference Artificial Intelligence (2009)
4. El Nasr, M.S., Yen, J.: Flame: Fuzzy logic adaptive model of emotions. Auton. Agents Multi-Agent Syst. 3(3), 219–257 (2000)
5. Gebhard, P.: Alma - a layered model of affect. In: Fourth International Joint Conference on Autonomous Agents and Multiagent Systems (2005)
6. Gross, J.J.: Antecedent and response-focused emotion regulation: Divergent consequences for experience, expression, an physiology. J. Person. Soc. Psychol. 74, 224–237 (1998)

7. Kimberly, R., Schröder, T., Scheve, C.: Dissecting the sociality of emotion: A multilevel approach. Emot. Rev. **6**(2), 124–133 (2014)
8. Marsella, S., G.J., Petta, P.: Computational Models of Emotion, for Affective Computing. In: Scherer, K.R., Bänziger, T., Roesch, E. (eds.) A blueprint for a affective computing: A sourcebook and manual (2010). http://ict.usc.edu
9. Mignonac, K., Herrbach, O.: Linking work events, affective states, and attitudes: An empirical study of managers' emotions. J. Bus. Psychol. **19**, 221–240 (2004)
10. Ortony, A., Clore, G., Collins, A.: The Cognitive Structure of Emotions. Cambridge University Press, Cambridge (1988)
11. Osgood, C.E.: Studies of the generality of affective meaning systems. Am. Psychol. **17**, 10–28 (1962)
12. Picard, R.W.: Affective Computing. The MIT Press, New York (1997)
13. Russell, J.A.: Core affect and the psychological construction of emotion. Psychol. Rev. **110**(1), 140–172 (2003)
14. Skowron, M., Rank, S., Swiderska, A., Kuster, D., Kappas, A.: Applying a text-based affective dialogue system in psychological research: Case studies on the effects of system behaviour, interaction context and social exclusion. Cogn. Comput. **6**, 872–891 (2014)
15. Soleimani, A., Kobti, Z.: Event-driven fuzzy automata for tracking changes in the emotional behavior of affective agents. In: Conference on Affective Computing and Intelligent Interaction (ACII 2013), Humaine Association (2013)

Audio and Visual Recognition

Tolerance-Based Approach
to Audio Signal Classification

Sheela Ramanna$^{(\boxtimes)}$ and Ashmeet Singh

Department of Applied Computer Science,
University of Winnipeg, Winnipeg, MB R3B 2E9, Canada
s.ramanna@uwinnipeg.ca, singh-a3@webmail.uwinnipeg.ca

Abstract. In this paper, we propose a supervised learning algorithm to classify audio signals based on the tolerance near sets model (TNS). In the TNS method, tolerance classes are directly induced from the data set using the tolerance level ε and a distance function. Preliminary experiments with an audio signal data set show promising results in terms of the accuracy of classifier. Overall, TCL is able to demonstrate similar performance in terms of accuracy with Fuzzy IDT algorithm [1] and comparable performance with a rough set based classifier as well as classical machine learning algorithms based on decision trees, rules, bayesian learning and support vector machines.

Keywords: Audio signal classification · Granular computing · Machine learning · Rough sets · Tolerance near sets

1 Introduction

Granular methods typically include fuzzy sets [16], rough sets [7] and near sets [8, 9]. A granule is a clump of objects (points), in the universe of discourse, drawn together by indistinguishability, similarity, proximity, or functionality [16]. The tolerance class learner (TCL) algorithm based on the tolerance near sets model is a modified form of the algorithm that was introduced in [12] to detect solar flare images.

In the TNS method, tolerance classes are directly induced from the data set using the tolerance level ε and a distance function. The TNS method lends itself to applications where features are real-valued such as image data, audio and video signals. In this paper, we compare the proposed TCL algorithm to two granular models: one based on the classical rough sets and a second one based on fuzzy decision trees as well as other general machine learning algorithms.

The contribution of the paper is a Tolerant Class Learner (TCL) algorithm based on the TNS model and is a supervised learning algorithm. Overall, TCL is able to demonstrate similar performance in terms of accuracy with Fuzzy

This research has been supported by the Natural Sciences and Engineering Research Council of Canada (NSERC) Discovery grant 194376. Special thanks to Dr. Rajen Bhatt, Robert Bosch Technology Research Center, US for sharing this data set.

R. Khoury and C. Drummond (Eds.): Canadian AI 2016, LNAI 9673, pp. 83–88, 2016.
DOI: 10.1007/978-3-319-34111-8_11

IDT algorithm [1] and comparable performance with standard machine learning algorithms. It should be noted that unlike the highly optimized machine learning algorithms available in WEKA[1] and RSES[2], TCL has not yet been optimized.

2   Related Work

Automatic music genre recognition (or classification) from acoustic features is a very popular music information retrieval (MIR) task where machine learning algorithms have found widespread use [14,15]. The broader field of MIR research is a multidisciplinary study that encompasses audio engineering, cognitive science, computer science and perception (psychology). A survey of MIR systems can be found in [13]. In addition, improving genre classification by annotation is also a very important research topic since it is critical to any supervised learning task. A more recent paper on the million song data set and the challenges of improving genre annotation can be found in [3]. Since the focus of this work is strictly on a specific learning method, we restrict our discussion to only granular based methods (rough, fuzzy, and near) rather than the general machine learning methods applied to MIR. In [2] fuzzy-c means has been used for classification of audio signals. More recently, music genre recognition based on rough sets can be found in [4].

3   Formal Models: Rough Sets and Near Sets

We briefly introduce tolerance forms of granular computing that are closely related to each other: rough sets and near sets. In rough set theory a universe U of objects is partitioned into indiscernible object granules i.e. *equivalence classes* by using an equivalence relation $R \subseteq U \times U$. Then it employs those classes to define three regions for a given concept $X \subseteq U$: *positive*, *negative* and *boundary*. A *tolerance model* uses a tolerance relation instead of equivalence where the transitivity property is relaxed which enables overlapping classes (in other words, objects can belong to more than one class). Tolerance relations can be considered generalizations of equivalence relations.

Near set theory was influenced by rough set theory. The basic structure which underlies near set theory is a perceptual system which consists of perceptual objects (i.e., objects that have their origin in the physical world). However, the key difference between these two approaches one based on tolerance rough sets (TRS) and tolerance near sets model (TNS) is that in the TRS model, the data set is approximated into lower and upper approximations using a tolerance relation. Some preliminaries of near sets will be discussed in the following sections.

Definition 1 Perceptual System [8,9]. A perceptual system is a pair $\langle O, F \rangle$, where O is a nonempty set of perceptual objects and F is a countable set of real-valued probe functions $\phi_i : O \to \mathbb{R}$.

---

[1] http://www.cs.waikato.ac.nz/ml/weka/.
[2] http://www.mimuw.edu.pl/~szczuka/rses/start.html.

An object description is defined by means of a tuple of probe function values $\Phi(x)$ associated with an object $x \in X$, where $X \subseteq O$ as defined by Eq. 1.

$$\Phi(x) = (\phi_1(x), \phi_2(x) \ldots, \phi_n(x)) \tag{1}$$

where $\phi_i : O \rightarrow \mathbb{R}$ is a probe function of a single feature. The following is a typical example of probe function used in audio classification.

Example 1 (Spectral Centroid Probe Function).

$$\phi_1 = C_t = \frac{\sum_{n=1}^{N} M_t[n] \times n}{\sum_{n=1}^{N} M_t[n]}$$

where $M_t[n]$ is the magnitude of the Fourier transform at frame t and frequency bin n. The centroid is a measure of spectral shape and higher centroid values correspond to brighter textures with more high frequencies [14].

Definition 2 Tolerance (Nearness) Relation [10,11]. Let $\langle O, F \rangle$ be a perceptual system and let $B \subseteq F$,

$$\cong_{B,\epsilon} = \{(x, y) \in O \times O : \| \phi(x) - \phi(y) \|_2 \leq \varepsilon\} \tag{2}$$

where $\| \cdot \|_2$ denotes the L^2 norm of a vector. For an audio classification system, Definition 2 implies that the idea is to find objects (ex: data set elements) that resemble each other with a tolerable level of error based on their feature values.

Definition 3 Tolerance Near Sets [10,11]. Let $\langle O, F \rangle$ be a perceptual system and let $X \subseteq O$. A set X is a tolerance near set iff there is $Y \subseteq O$ such that $X \bowtie_F Y$.

In effect, tolerance near sets are those sets that are defined by the nearness relation $\bowtie_F$.

4   Tolerance Class Learner - TCL

In this section, we present our proposed Tolerant Class Learner (TCL) in terms of a supervised learning algorithm. Classification is performed in two phases in a supervised form. In the *learning phase*, given a tolerance level ε, tolerance classes are induced from the training set, and the representative of each tolerance class is computed as well as a category determination is performed based on majority voting. In the *testing phase*, the L^2 distance is computed for each element in test set with all of tolerance class representatives obtained in the first phase and assigned the category of the tolerance class representative based on the lowest distance value. The algorithms use the following notation:

$procedure\_name(input_1, ..., input_n; output_1, ..., output_m)$. The complexity of the algorithms is fairly straight forward to compute. In phase 1, the complexity of $computeMinMaxNorm$ and $computeNormL2$ functions is $\mathcal{O}(n^2)$. The complexity of $defineToleranceRelation$ and $defineClass$ functions is $\mathcal{O}(n^3)$. In phase 2, the complexity of $DetermineCat$ function is $\mathcal{O}(n^2)$.

Algorithm 1. Phase I: Learning Categories

Input   : $\varepsilon > 0$, // Tolerance level
          Φ, // Probe functions or Features
          $TR = \{TR_1, \ldots, TR_M\}$, // Training Data Set
Output: $(NC, \{R_1, \ldots, R_{NC}\}, \{Cat_1, \ldots, Cat_{NC}\})$

1 **for** $i \leftarrow 1$ **to** M **do**
2 $\quad$ computeMinMaxNorm(TR_i);

3 **for** $i \leftarrow 1$ **to** M **do**
4 $\quad$ **for** $j \leftarrow i+1$ **to** M **do**
5 $\quad\quad$ computeNormL$_2$($TR_i, TR_j, ProbeValue_{ij}$; NormL2$_{ij}$);

6 **for** $i \leftarrow 1$ **to** M **do**
7 $\quad$ **for** $j \leftarrow i+1$ **to** M **do**
8 $\quad\quad$ defineToleranceRelation(NormL2$_{ij}, \varepsilon$; SetOfPairs);

9 // Begin of defineClass(SetOfPairs; NC, H, Cat_i);
10 $H \leftarrow \emptyset$; // Set of tolerance classes
11 **for** $i \leftarrow 1$ **to** M **do**
12 $\quad$ computeNeighbour(SetOfPairs, i, TR; N_i); // Compute the neighbourhood
     $\quad N_i$ of ith training data TR_i
13 $\quad C_i \leftarrow N_i$; // Start the class C_i as the set containing all N_i
14 $\quad$ **for** *all* $x, y \in N_i$ **do**
15 $\quad\quad$ **if** $x, y \notin SetOfPairs$ **then**
16 $\quad\quad\quad C_i \leftarrow C_i - \{y\}$; // Exclude y from class C_i
17 $\quad H \leftarrow H \cup \{C_i\}$;
18 $\quad$ // C_i is one tolerance class induced by the tolerance relation
19 $\quad$ computeMajorityCat(C_i; Cat_i); // Determine Category by majority voting
     $\quad$ for each C_i
20 $NC \leftarrow |H|$; // Number of classes
21 // End of defineClass
22 defineClassRepresentative(NC, H; $\{R_1, \ldots, R_{NC}\}, \{Cat_1, \ldots, Cat_{NC}\}$);

---

5   Experiments and Discussion

The TCL algorithms were implemented in C++. Our dataset for training and testing has been restricted to a two-class problem with speech or non-speech categories. It consists of 3.5 hours of audio data with a total of 12,928 segments (elements) with each segment marked as either speech or non-speech manually. The following nine (9) features were used in our experiments which were also used by Fuzzy Decision Tree Classifier [1]: Normalized zero-crossing rate, Variance of differential entropy computed from Mel-filter banks, Mean of Spectral flux, Variance of differential spectral centroid, Variance of first 5 Mel-Frequency cepstral coefficients (MFCC). MFCCs are perceptually motivated features commonly used in speech recognition research [5,6,14]. We have used 10-Fold Cross Validation (CV) for all our experiments except for FDT (whose results were reported in [1]). We have chosen a representative set of classical machine learning

---

Algorithm 2. Phase II: Assigning categories

Input   : $\varepsilon > 0$, // Tolerance level
       Φ, // Probe functions or Features
       $TS = \{TS_1, \ldots, TS_M\}$, // Testing Data Set
       $\{R_1, \ldots, R_{NC}\}, \{Cat_1, \ldots, Cat_{NC}\}$ // Representative Class and their
associated categories
Output: $(TS' = \{TS'_1, \ldots, TS'_M\})$ // Testing Data Set with assigned categories

1 **for** $i \leftarrow 1$ **to** M **do**
2   computeMinMaxNorm(TS_i);

3 **for** $i \leftarrow 1$ **to** M **do**
4   **for** $j \leftarrow i+1$ **to** NC **do**
5   computeNormL$_2$($TS_i, RC_j, ProbeValue_{ij}$; NormL2$_{ij}$);

6 DetermineCat($NormL2_{ij}$; TS') // Computes min. distance and assigns category

---

algorithms such as Decision Trees, Rule-based Systems, Support Vector Machines and Bayesian Learning implemented in WEKA as well the classical rough sets model implemented in RSES. For consistency, we created ten sets of training and testing pairs which were then used for experimentation across the three systems: RSES, WEKA and TCL.

The results reported for TCL are for the best value of ε. Due to lack of space, we are reporting only the total average accuracy for each method: RSES(90 %), J48(91 %), NaiveBayes(88 %), JRIP(91 %), SVM(90 %), FDT(87 %) and TCL(87 %).

The average number of rules used by some of the appropriate classifiers are as follows: RSES(30,809), J48(97), JRIP(16) and FDT(2). The rough set classifier uses a form of genetic algorithm to generate the rules. The average execution time (in minutes) used by the classifiers are: RSES(210), WEKA*(2), FDT(2 ms) and TCL(48).

Overall, TCL is able to demonstrate similar performance in terms of accuracy with Fuzzy IDT algorithm and comparable performance with algorithms implemented in RSES and WEKA. It should be noted that all except the Bayesian, SVM, TCL classifiers *discretize* the data set. In our experiments, it can be observed that the best accuracy that can be obtained with this data set is *91 %*. One of the key issues in TCL is determining the most optimal value of ε since this value is key to determining representative classes. This is similar to neighbourhood-based classifiers such as k-means. However, determining this parameter can be optimized, a process similar to the ones used for determining k-values.

6   Conclusion and Future Work

We have proposed a supervised learning algorithm based on a tolerance form of near sets for classification learning. Preliminary experiments with an audio signal data set show promising results in terms of the accuracy of classifier. Based on the results, an obvious optimization is to remove outliers from tolerance classes since this has a bearing on the average feature vector value. The second optimization task that is also common to all learning problems is the determination of best features. As future work, we plan to experiment with a larger data set that includes m-class audio signals data such as music and different music genres.

References

1. Bhatt, R., Krishnamoorthy, P., Kumar, S.: Efficient general genre video abstraction scheme for embedded devices using pure audio cues. In: 7th International Conference on ICT and Knowledge Engineering, pp. 63–67, Dec 2009
2. Haque, M.A., Kim, J.M.: An analysis of content-based classification of audio signals using a fuzzy c-means algorithm. Multimed. Tools Appl. **63**(1), 77–92 (2013)
3. Hendrik, S.: Improving genre annotations for the million song dataset. In: 16th International Society for Music Information Retrieval Conference, pp. 63–70 (2015)
4. Hoffmann, P., Kostek, B.: Music genre recognition in the rough set-based environment. In: Kryszkiewicz, M., Bandyopadhyay, S., Rybinski, H., Pal, S.K. (eds.) PReMI 2015. LNCS, vol. 9124, pp. 377–386. Springer, Heidelberg (2015)
5. Hunt, M., L.M., Mermelstein, P.: Experiments in syllable-based recognition of continuous speech. In: Proceedings of International Conference on Acoustics, Speech and Signal Processing, pp. 880–883 (1996)
6. Logan, B.: Mel frequency cepstral coefficients for music modeling. In: Proceedings of 1st International Conference on Music Information Retrieval, Plymouth, MA (2000)
7. Pawlak, Z.: Rough sets. Int. J. Comput. Inf. Sci. **11**(5), 341–356 (1982)
8. Peters, J.: Near sets, general theory about nearness of objects. Appl. Math. Sci. **1**(53), 2029–2609 (2007)
9. Peters, J.: Near sets, special theory about nearness of objects. Fundamenta Informaticae **75**(1–4), 407–433 (2007)
10. Peters, J.: Tolerance near sets and image correspondence. Int. J. Bio-inspired Comput. **1**(4), 239–245 (2009)
11. Peters, J.: Corrigenda and addenda: tolerance near sets and image correspondence. Int. J. Bio-Inspired Comput. **2**(5), 310–318 (2010)
12. Poli, G., Llapa, E., Cecatto, J., Saito, J., Peters, J., Ramanna, S., Nicoletti, M.: Solar flare detection system based on tolerance near sets in a GPU-CUDA framework. Knowl.-Based Syst. J. **70**, 345–360 (2014). Elsevier
13. Typke, R., Wiering, F., Veltkamp, R.C.: A survey of music information retrieval systems. In: ISMIR 2005, 6th International Conference on Music Information Retrieval, London, UK, Proceedings, pp. 153–160, 11–15 September 2005
14. Tzanetakis, G., Cook, P.: Musical genre classification of audio signals. IEEE Trans. Speech Audio Process. **10**(5), 293–302 (2002)
15. Wold, E., Blum, T., Keislar, D., Wheaton, J.: Content-based classification, search, and retrieval of audio. IEEE Multimed. **3**(2), 27–36 (1996)
16. Zadeh, L.: Towards a theory of fuzzy information granulation and its centrality in human reasoning and fuzzy logic. Fuzzy Sets Syst. **177**(19), 111–127 (1997)

A Novel Dataset for Real-Life Evaluation of Facial Expression Recognition Methodologies

Muhammad Hameed Siddiqi[1], Maqbool Ali[2], Muhammad Idris[2],
Oresti Banos[2], Sungyoung Lee[2], and Hyunseung Choo[1(✉)]

[1] Department of Computer Science and Engineering,
Sungkyunkwan University, Suwon, Korea
{siddiqi,choo}@skku.edu
[2] Department of Computer Engineering, Kyung Hee University, Suwon, Korea
{maqbool.ali,idris,oresti,sylee}@oslab.khu.ac.kr

Abstract. One limitation seen among most of the previous methods is that they were evaluated under settings that are far from real-life scenarios. The reason is that the existing facial expression recognition (FER) datasets are mostly pose-based and assume a predefined setup. The expressions in these datasets are recorded using a fixed camera deployment with a constant background and static ambient settings. In a real-life scenario, FER systems are expected to deal with changing ambient conditions, dynamic background, varying camera angles, different face size, and other human-related variations. Accordingly, in this work, three FER datasets are collected over a period of six months, keeping in view the limitations of existing datasets. These datasets are collected from YouTube, real world talk shows, and real world interviews. The most widely used FER methodologies are implemented, and evaluated using these datasets to analyze their performance in real-life situations.

Keywords: Facial expression recognition · Feature extraction · Feature selection · Recognition · YouTube · Real-world

1 Introduction

Existing FER methodologies utilized previous datasets and did not consider the real world challenges in their respective systems. For instance, two most commonly used expression datasets used for evaluating FER systems are Cohn-Kanade (CK) [5] dataset, and JAFFE dataset [7]. JAFFE dataset is collected from 10 different subjects (Japanese female), where CK dataset is collected from 97 subjects (university students). Both datasets are collected under controlled laboratory settings with constant lighting effects, camera stetting, and

H. Choo—Supported by the MSIP, Korea, under the G-ITRC support program (IITP-2015-R6812-15-0001) supervised by the IITP, and by the Priority Research Centers Program through the National Research Foundation of Korea (NRF) funded by the Ministry of Education, Science and Technology (NRF-2010-0020210).

R. Khoury and C. Drummond (Eds.): Canadian AI 2016, LNAI 9673, pp. 89–95, 2016.
DOI: 10.1007/978-3-319-34111-8_12

background. All of the images are taken from the frontal view of the camera, with tied hair in the case of JAFFE dataset, in order to expose all the sensitive regions. Furthermore, these datasets are pose-based, i.e., subjects performed the expressions exactly when they are asked to. Limited efforts have been put into designing a new dataset that is closer to real-life situations, probably because creating such a dataset is a very difficult and time consuming task. And this is where the contribution of this work lies.

Accordingly, in this work, we have defined a realistic and innovative dataset collected from YouTube, some real world talk shows, and some interviews that considered the above-mentioned limitations. From lab settings to a real-life environment, we defined three cases with increasing complexity. In all three cases, a large number of different subjects of different gender, race, and age were included. Also, the defined datasets have various sizes of the face that is related to proximity. From existing works, more recent methodologies were implemented.

2   Existing Standard FER Methodologies and Datasets

Standard Methods: For feature extraction, LBP [4], LDP [8], curvelet transform [14], and wavelet transform [10]; for feature selection, (LDA) [9], kernel discriminant analysis (KDA) [17], and generalized discriminant analysis (GDA) [15]; and for recognition, SVM(s) [8], HMM(s) [9] and HCRF(s) [11] were used.

Existing Datasets: The extended version of CMU-PIE dataset [12] was collected named Multi-PIE [3] that covered the limitations of CMU-PIE. However, multi-PIE is a pose-based dataset collected under a static illumination conditions. Similarly, the extended Cohn-Kanade (CK+) dataset [6] is the extension of CK dataset [5] which covered the limitations of CK dataset. This dataset consists of both pose-based and spontaneous expressions. However, this dataset has been collected under a controlled environment and though some subjects were at a 30-degree angle with the camera, the remaining subjects were with frontal view to the camera. Georgia Tech Face dataset [1] contains images of 50 people, which is a pose-based dataset and the images show frontal and/or tilted faces with different expressions. USTC-NVIE [16] consists of both pose and spontaneous expressions collected by more than 100 people. However, a visible and infrared thermal camera was used for dataset collection with a predefined lighting setup. Another important real world dataset named VADANA: Vims Appearance dataset [13] was collected to consider the research problem of gender and age in the area of FER. However, this dataset is a pose-based dataset, too. Likewise, another real world dataset named YMU (YouTube Makeup) dataset [2] was collected from the YouTube makeup tutorials consisting of 151 subjects. However, in this dataset, only females (having makeup) are involved that may cause the gender problem.

3   Novel Datasets to Benchmark Real-World FER Systems

Emulated Dataset: In this dataset, the ordinary subjects performed expressions in a pose-based manner in a controlled lab environment. The subjects belonged to different colors, age, and ethnicities. The subject age ranges from 4 years to 60 years. In some of the cases, the images in some expressions are rotated using the camera for better accuracy of the system. The subjects include both males and females. Each expression has at least 165 images. The images used in the dataset are of size 240 × 320 and 320 × 240 pixels with facial frame.

Semi-naturalistic Dataset: In this dataset, the expressions are collected from the actors and actresses of Hollywood and Bollywood in their respective movies, where we had no control on expression timings, camera, lighting and background settings. The expressions have different views from different angles with glasses, hair open and close, and other obvious actions are collected in this dataset with dynamic settings. Each expression consists of at least 165 images. The dataset has the images of size 240 × 320 and 320 × 240 pixels with facial frame.

Naturalistic Dataset: In this dataset, subjects from various parts of the world, races, and ethnicities have been selected. The expressions are spontaneous that have been captured in natural and dynamic settings from real world talk-shows, interviews, and YouTube natural videos such as news and real world incidents. The total of 165 images have been considered for each expression. The age range of the subjects are from 18 to 50 years. Images used in the dataset are of size 240 × 320 and 320 × 240 pixels with facial frame. All the datasets include six basic expressions such as happy, sad, angry, normal, disgust, and fear. These datasets will be made available for future research to the research community.

4   Experimental Results

A comprehensive set of experiments were performed, in which the performance of each method was tested and validated using 10-fold cross-validation rule for each dataset. All the experiments were performed in Matlab using an Intel® Pentium® Dual-Core^{TM} (2.5 GHz) with a RAM capacity of 3 GB.

Experimental Analysis Using Emulated Dataset: The average recognition rates for each method when benchmarked on the emulated dataset are shown in Fig. 1 (first image). As it can be seen, the vast majority of the evaluated methods yield an average accuracy within the range 65 to 80 %, thus far from perfect recognition capabilities. Some combinations of feature extraction and selection methods provide better results than others, especially, LBP + KDA, LDP + GDA, Curvelet + KDA and Wavelets + KDA and GDA. The sort of feature extraction and selection technique used in the FER model turns to have a less clear impact than classification paradigm. In fact, highest accuracies are generally obtained by using HCRF, while poorest results are obtained for systems based on SVM.

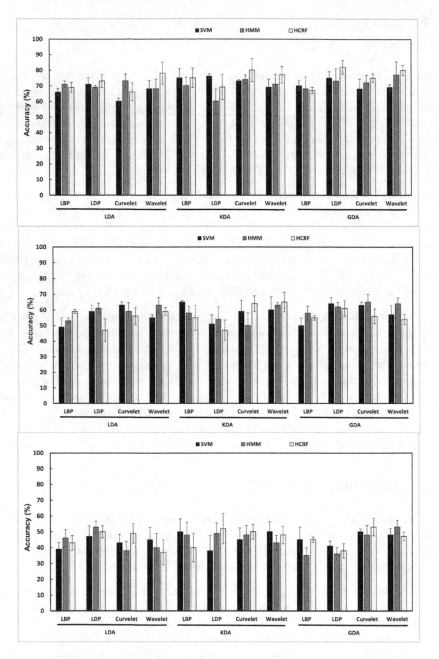

Fig. 1. The average (bar and standard deviation whiskers) classification rates from the evaluation of the standard FER methods using the defined emulated (first image), semi-naturalistic (middle image), and naturalistic (last image) datasets. The top legend presents the recognition paradigm. The horizontal axis labels present the standard feature extraction methods used for each experiment, while, the underlined shows respectively the standard feature selection methods.

Experimental Analysis Using Semi-naturalistic Dataset: Figure 1 (middle image) depicts the performance values of each standard method for the semi-naturalistic case. At first sight, a significant drop in the performance is observed with respect to the ideal scenario. Here, the performance of the evaluated models span from 45 % to 65 %, which is unacceptable for realistic FER applications. The combinations of feature extraction and selection methods that yield best results are different to the ones highlighted for the emulated case. Concretely, Wavelet + KDA and LDP + GDA seem to provide the best performance for all classification paradigms. Conversely to the emulated scenario, no classification paradigm is observed to prevail over the others for the semi-naturalistic case.

Experimental Analysis Using Naturalistic Dataset: The accuracy results corresponding to the third evaluation scenario, i.e., the naturalistic case, are shown in Fig. 1 (last image). As it could be expected, the performance of all models is dramatically reduced with respect to the emulated or ideal scenario, with accuracies that range between less than 40 % to 55 %. Although marginalizing across feature extraction, selection, and classification techniques is of arguable value given the low accuracy values, it may be said that best combinations are LDP + LDA, and Curvelet/Wavelet + KDA/GDA. Similarly, no clear conclusions can be derived from the analysis of the prevalence of the classification models, although highest results tend to be obtained by using the HCRF. Despite, the combinations of feature extraction and selection methods that yield best results are different to the ones highlighted for the emulated case.

5   Conclusion

Human FER has emerged as a fascinating research area during the last two decades. However, accurate FER in real world scenarios is still a challenging work. Most of the previous FER methodologies achieved high recognition rate using all the previous datasets. However, most of these datasets were collected under predefined setups. And, these methodologies showed poor performance when applied on real world datasets. Several factors that effects the accuracy of the FER methodologies include varying light conditions and dynamic variation of the background.

In this work, we have defined three kinds of datasets named emulated, semi-naturalistic, and naturalistic datasets. The defined datasets considered most of the limitations of the existing datasets in real world scenarios. These datasets are collected from real world talk shows, interviews, and YouTube. We have evaluated some well-known existing standard FER methodologies using the defined datasets. All the standard methodologies were tested and validated using 10-fold cross-validation rule. It can be seen that all the methodologies showed least performance on semi-naturalistic and naturalistic datasets.

Therefore, it is desirable that in future we will propose new methods to improve the accuracy of FER systems in real-life scenarios.

References

1. Chen, L., Man, H., Nefian, A.V.: Face recognition based on multi-class mapping of fisher scores. Pattern Recogn. **38**(6), 799–811 (2005)
2. Dantcheva, A., Chen, C., Ross, A.: Can facial cosmetics affect the matching accuracy of face recognition systems? In: 2012 IEEE Fifth International Conference on Biometrics: Theory, Applications and Systems (BTAS), pp. 391–398. IEEE (2012)
3. Gross, R., Matthews, I., Cohn, J., Kanade, T., Baker, S.: Multi-pie. Image Vis. Comput. **28**(5), 807–813 (2010)
4. Hablani, R., Chaudhari, N., Tanwani, S.: Recognition of facial expressions using local binary patterns of important facial parts. Int. J. Image Process. (IJIP) **7**(2), 163 (2013)
5. Kanade, T., Cohn, J.F., Tian, Y.: Comprehensive database for facial expression analysis. In: Fourth IEEE International Conference on Automatic Face and Gesture Recognition, Proceedings, pp. 46–53. IEEE (2000)
6. Lucey, P., Cohn, J.F., Kanade, T., Saragih, J., Ambadar, Z., Matthews, I.: The extended cohn-kanade dataset (ck+): a complete dataset for action unit and emotion-specified expression. In: 2010 IEEE Computer Society Conference on Computer Vision and Pattern Recognition Workshops (CVPRW), pp. 94–101. IEEE (2010)
7. Lyons, M., Akamatsu, S., Kamachi, M., Gyoba, J.: Coding facial expressions with gabor wavelets. In: Third IEEE International Conference on Automatic Face and Gesture Recognition, Proceedings, pp. 200–205. IEEE (1998)
8. Rivera, A.R., Castillo, R., Chae, O.: Local directional number pattern for face analysis: face and expression recognition. IEEE Trans. Image Process. **22**(5), 1740–1752 (2013)
9. Siddiqi, M.H., Ali, R., Idris, M., Khan, A.M., Kim, E.S., Whang, M.C., Lee, S.: Human facial expression recognition using curvelet feature extraction and normalized mutual information feature selection. Multimedia Tools Appl. **75**(2), 935–959 (2016)
10. Siddiqi, M.H., Ali, R., Khan, A.M., Kim, E.S., Kim, G.J., Lee, S.: Facial expression recognition using active contour-based face detection, facial movement-based feature extraction, and non-linear feature selection. Multimedia Syst. **21**(6), 541–555 (2015)
11. Siddiqi, M.H., Ali, R., Khan, A.M., Park, Y.-T., Lee, S.: Human facial expression recognition using stepwise linear discriminant analysis and hidden conditional random fields. IEEE Trans. Image Process. **24**(4), 1386–1398 (2015)
12. Sim, T., Baker, S., Bsat, M.: The cmu pose, illumination, and expression database. IEEE Trans. Pattern Anal. Mach. Intell. **25**(12), 1615–1618 (2003)
13. Somanath, G., Rohith, M.V., Kambhamettu, C.: Vadana: a dense dataset for facial image analysis. In: 2011 IEEE International Conference on Computer Vision Workshops (ICCV Workshops), pp. 2175–2182. IEEE (2011)
14. Tang, M., Chen, F.: Facial expression recognition and its application based on curvelet transform and pso-svm. Optik-Int. J. Light Electron Opt. **124**(22), 5401–5406 (2013)
15. Uddin, M.Z., Kim, T.-S., Song, B.C.: An optical flow feature-based robust facial expression recognition with hmm from video. Int. J. Innovative Comput. Inf. Control **9**(4), 1409–1421 (2013)

16. Wang, S., Liu, Z., Lv, S., Lv, Y., Wu, G., Peng, P., Chen, F., Wang, X.: A natural visible and infrared facial expression database for expression recognition and emotion inference. IEEE Trans. Multimedia **12**(7), 682–691 (2010)
17. Wu, Q., Zhou, X., Zheng, W.: Facial expression recognition using fuzzy kernel discriminant analysis. In: Wang, L., Jiao, L., Shi, G., Li, X., Liu, J. (eds.) FSKD 2006. LNCS (LNAI), vol. 4223, pp. 780–783. Springer, Heidelberg (2006)

Visual Perception Similarities to Improve the Quality of User Cold Start Recommendations

Crícia Z. Felício[1,3]([✉]), Claudianne M.M. de Almeida[2,3], Guilherme Alves[3], Fabíola S.F. Pereira[3], Klérisson V.R. Paixão[3], and Sandra de Amo[3]

[1] Federal Institute of Triângulo Mineiro, Uberlândia, Brazil
cricia@iftm.edu.br
[2] Federal Institute of Norte de Minas, Arinos, Brazil
claudianne.almeida@ifnmg.edu.br
[3] Federal University of Uberlândia, Uberlândia, Brazil
guilhermealves@mestrado.ufu.br, fabfernandes@comp.ufu.br,
klerisson@doutorado.ufu.br, deamo@ufu.br

Abstract. Recommender systems are well-know for taking advantage of available personal data to provide us information that best fit our interests. However, even after the explosion of social media on the web, hence personal information, we are still facing new users without any information. This problem is known as user cold start and is one of the most challenging problems in this field. We propose a novel approach, VP-Similarity, based on human visual attention for addressing this problem. Our algorithm computes visual perception's similarities among users to build a visual perception network. Then, this networked information is provided to recommender system to generate recommendations. Experimental results validated that VP-Similarity achieves high-quality ranking results for user cold start recommendation.

Keywords: Visual perception · Cold start · Recommender system

1 Introduction

Traditional recommenders usually take advantage of users' previous choices and ratings to predict products (items) that would fulfill their wishes. When it comes to new users without rating records, however, the performance of these approaches fall a great deal. This is known as user cold start and it is one of the most challenging problems in this field [1]. Remarkably, with the emerging of social networks, social information has been largely explored to mitigate cold start problem in item recommendation [2–4]. However, it is important to notice that social information is not always available or it is costly to be accessed.

An emerging interesting source of information is *human visual perception.* Tracking users eyes movements to capture users' behaviors has been becoming increasingly tangible. Herein, we aim to explore the visual perception as an important source of knowledge to address the user cold start problem. As a

© Springer International Publishing Switzerland 2016
R. Khoury and C. Drummond (Eds.): Canadian AI 2016, LNAI 9673, pp. 96–101, 2016.
DOI: 10.1007/978-3-319-34111-8_13

motivation example, let us imagine the domain of painting's recommendation. Consider a painting containing two main scenes, one describing a cat and other describing a child. Some people, looking at the painting, will focus their attention to the cat (U_{cat}). Others, to the child (U_{child}). Note that, in each group, there are people who like the painting and also the ones who do not like it. The reasons for which some people in U_{cat} or U_{child} do not like the painting are different, but we can affirm that people in U_{cat} (or U_{child}) are similar given their visual perception focuses. Now, imagine a new user u in the system, that has never expressed his preferences or tastes about paintings – u is a cold start user. Let us suppose that u focused on the cat. Then, we could assume that u has a lot in common with people from U_{cat} as regarding the way they perceive paintings. Thus, we hypothesize that it is reasonable to use people's preferences from U_{cat} to provide a recommendation to our new user.

In this work, we propose a method to identify similarities among users based on visual perception. We take advantage of the volume of existent approaches that address cold start problem through social network information [3–5] and innovate, by proposing to use a *Visual Perception Network*, inferred from the similarities on users visual perception. We evaluate our approach against the innovative Paintings dataset[1].

2   VP-Similarity Method

To adopt visual perception as contextual information for recommendation systems, first, we define the *VP-Similarity Method*. It is responsible to handle visual perception and infer visual perception similarities among users. Then, we incorporate visual perception network into social recommender systems and evaluate the performance in cold start scenario.

Definition 1 (Visual Fixation). *Let $\mathcal{I} = \{\mathcal{I}_1, ..., \mathcal{I}_m\}$ be a set of images. Let $\mathcal{U} = \{u_1, ..., u_n\}$ be a set of users. A visual fixation of an user u_j over an image $\mathcal{I}_k$ is a pair (p, f) where p is the position, represented by the pixels cluster centroid of that fixation, and f is the duration. We denominate $\mathcal{F}_{jk} = \{(p_1, f_1), ..., (p_z, f_z)\}$ the set of visual fixations of u_j over $\mathcal{I}_k$ (Fig. 1).*

Definition 2 (Visual Perception). *Let the images in $\mathcal{I}$ be divided in r equal parts $Q = \{q_1, ..., q_r\}$ as illustrated in Fig. 2. From the positions and durations described in the set of visual fixations $\mathcal{F}_{jk}$, we call v_s the percentage of time that u_j fixed to $\mathcal{I}_k$ in each quadrant q_s, for $1 \leq s \leq r$. The visual perception of an user u_j over an image $\mathcal{I}_k$ is defined as the vector $\mathcal{P}_{jk} = (v_1, ..., v_r)$. Finally, the visual perception of u_j over all images $\mathcal{I}$ is represented by the concatenation of all visual perceptions vectors from u_j: $\mathcal{P}_j = \mathcal{P}_{j1} \parallel ... \parallel \mathcal{P}_{jm}$.*

Table 1 shows an example of visual perception. There are visual perceptions from 7 users over 2 images. Images are divided in 4 parts. For each image, we have the percentage of time each user fixed his visual attention in a corresponding image part.

---

[1] Publicly available at: www.lsi.facom.ufu.br/datasets.

We denominate *VP-similarity score* between two users u_1 and u_2 as the distance between their respective visual perceptions vectors $\mathcal{P}_1$ and $\mathcal{P}_2$. This distance is defined by the function $l(u_1, u_2)$, where $l : \mathcal{P}_1 \times \mathcal{P}_2 \to \mathbb{R}$ and $l(u_1, u_2)$ can assume any classic similarity function like Euclidean distance, cosine similarity or Pearson distance correlation. By abuse of notation, we will write $l(u_1, u_2)$ as $l_{1,2}$. Table 1 highlights the VP-similarity score between u_4 and u_5, considering l as cosine similarity is 0.76 (* has been assumed as 0).

Fig. 1. Gaze positions and fixation length captured of an user.

Fig. 2. Example of a painting split in four equal parts.

Table 1. Users visual perception over two images of paintings dataset.

	$\mathcal{I}_1$				$\mathcal{I}_2$			
	q_1	q_2	q_3	q_4	q_1	q_2	q_3	q_4
u_1	0.50	0.10	0.40	0.00	*	*	*	*
u_2	0.60	0.20	0.10	0.10	0.10	0.70	0.10	0.10
u_3	0.40	0.40	0.20	0.00	0.00	0.90	0.10	0.00
C_1	**0.50**	**0.23**	**0.23**	**0.03**	**0.05**	**0.80**	**0.10**	**0.05**
u_4	*	*	*	*	0.75	0.08	0.05	0.12
u_5	0.05	0.25	0.20	0.50	0.70	0.05	0.10	0.15
u_6	0.15	0.20	0.25	0.40	0.82	0.02	0.10	0.06
u_7	*	*	*	*	*	*	*	*
C_2	**0.10**	**0.23**	**0.23**	**0.45**	**0.76**	**0.05**	**0.08**	**0.11**

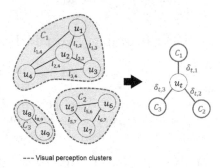

--- Visual perception clusters

Fig. 3. Inferred visual perception networks (left) and selection of visual perception cluster for target user u_t (right).

Inferring Visual Perception Network. Our hypothesis is that users with similar visual perceptions are a good source for new user recommendation. Thus, we propose to cluster users according to their VP-similarity scores. In this work, we use *K-means* as classical clustering algorithm. Inside each resultant cluster we have a *Visual Perception Network*, defined as a complete graph, where nodes are users and the edges are labeled with respective VP-similarity scores. This process is shown in the left side of Fig. 3.

We define as *cluster consensual vector* the vector containing the averages of all visual perceptions from users inside the cluster. Table 1 illustrates two clusters C_1 and C_2 and their respective consensual vectors. This notion is specially important to perform a recommendation: when a target new user u_t is added to the system, our VP-Similarity method generates his visual perception $\mathcal{P}_t$ of u_t.

Also, the VP-similarity score between u_t and each cluster C_k is computed. We denote $\delta_{t,k}$ as the VP-similarity score between an user u_t and a cluster C_k. This notation is similar to l, previously defined. The goal is to determine the most similar cluster concerning the target user and use the respective Visual Perception Network to perform recommendations. Thus, VP-Similarity finds the nearest visual perception cluster for a target user.

3   Experimental Setup

Dataset. In order to obtain visual perception information and ratings over items, we recruited 194 volunteers for rating 200 paintings, which were randomly chosen between 605 paintings public available at *Ciudad de la Pintura* (http://pintura.aut.org). For each volunteer, an eye tracker device captures eye movements on each painting displayed on the 22' monitor with image resolution of 500×700 pixels. The volunteer should rate each painting in a 1–5 scale according to its preference. The dataset is composed by 194 users, 605 items, 38,753 ratings, 67 % of sparsity and 28,992 links among users.

Recommendation Baseline. We provide visual perception to social recommender systems based on matrix factorization. We compare the effectiveness of this approach among the following state-of-the-art recommenders:

- *Global Average:* A standard "popular" baseline, which recommends using the global average rate for an item.
- *SoRec:* A matrix factorization approach that combining social information and ratings to build a recommendation model [3].
- *SocialMF:* It is a model-based matrix factorization approach for recommendation in social networks with trust propagation mechanism [4].
- *TrustMF:* This approach combines a truster and trustee model mixing information of both, the users who trust the target user and those who are trusted by the user to build a recommendation model [5].

Parameter Settings. We use LibRec [6] library implementation of SoRec, SocialMF, TrustMF and GlobalAverage methods with default parameters. For social matrix factorization approaches the experiments were executed with 10 latent factors and number of interactions equal to 100.

Evaluation Protocol. We simulate a cold start scenario, using an evaluation protocol called **1-rating protocol**. To do that we employ the *"All but One"* protocol. Instead give as much as possible user's ratings for training, we provide only 1 rate from the target user per iteration. Note that all linked users' preferences in VP-Network are provided to build the recommendation model. We validate the model over the target user's remain rates.

4   Results and Discussions

We assess the effectiveness of our proposed Visual Perception Network model for item recommendation. In particular, we aim to answer the follow research question: *How effective is visual perception networks for item recommendation?* We also evaluate the prediction quality of visual perception approaches

Table 2. nDCG results for cold start scenario using 1-rating protocol.

Rank size	SocialMF	SoRec	TrustMF	GlobalAvg
5	0.6655 ± 0.1731	**0.8330** $\pm$ 0.1273	0.6281 ± 0.1468	0.7167 ± 0.1532
10	0.6726 ± 0.1550	**0.8297** $\pm$ 0.1103	0.6330 ± 0.1302	0.6954 ± 0.1305
15	0.6724 ± 0.1457	**0.8259** $\pm$ 0.1006	0.6346 ± 0.1249	0.6812 ± 0.1178
20	0.6749 ± 0.1379	**0.8218** $\pm$ 0.0988	0.6409 ± 0.1183	0.6656 ± 0.1142

among the state-of-art recommenders presented in Sect. 3. Table 2 shows the result of this comparison in terms of $nDCG$ rank size of 5, 10, 15, and 20 for items recommended in our Paintings dataset.

We note that the experimental results show the superiority of SoRec over the others social approaches and the baseline. In particular, its performance might be explained because it work better in cold start scenario, since all social recommenders use the same VP-Network and are tested over the same conditions. The SoRec result attests the effectiveness of apply visual perception to leading with cold start users recommendation in contrast to others social approaches that have similar or inferior results than the baseline method.

5   Related Work

Solutions for User Cold Start Problem. Generally, state-of-the-art works explore the same directions of our proposal: using contextual information to mitigate user cold start problem. The *similarity-networks* and *kinds of contextual information* to handle that are varied. Remarking on the use of a similarity-based networks, the work [7] uses a network among the users based on trust. It is not a social or a visual perception network. It is a method that generates a trust network based on ratings and users' similarities. A substantial body of literature explores *hybrid recommender systems* as a way to provide accurately recommendations on user cold start scenario. Those systems take advantage of different kinds of contextual information: [2] (preference and social data), [8] (item and preference data), [4] (demographic and social data).

Visual Perception as Implicit User Feedback. Few works investigated the use of eye tracker for recommending task [9,10]. In these works, the user preference is not measured from ratings but from the way he/she looks at different images, such as eye fixation time and location. In [9] human visual perception data are adopted to build a gaze-based classifier for the image preference mining. User visual perception and preference data have been taken as a knowledge source to recommend images. However, while [9] proposed a classifier, our method is cluster-based. In [10] a content-based filtering enhanced by human visual attention was applied to clothing recommendation. This approach is specific for clothes domain and relays on visual attention similarity combined with the measures conventionally used in content-based image recommendation systems.

6 Final Remarks

In this paper, we have devised and evaluated VP-SIMILARITY, a method whose ultimate goal is to incorporate visual perception in social recommender systems to deal with the user cold start problem. Our main contributions were: (1) A method to compute visual perception similarities among users; and (2) evaluating the use of visual perception similarities in social recommender systems to leading with user cold start problem. We have several ideas to extend this work in the future. First, it is worth exploring other datasets. Second, work to extend matrix factorization methods, and hybrid models approaches using visual perception similarities. Finally, we intend to evaluate if visual perception could be complementary to explicit preferences for general users recommendation.

References

1. Sahebi, S., Cohen, W.: Community-based recommendations: a solution to the cold start problem. In: Workshop on Recommender Systems and the Social Web (RSWEB), held in conjunction with ACM RecSys 2011 (2011)
2. Felício, C.Z., Paixão, K.V.R., Alves, G., de Amo, S.: Social prefrec framework: leveraging recommender systems based on social information. In: Proceedings of the 3rd Symposium on Knowledge Discovery, Mining and Learning, pp. 66–73 (2015)
3. Ma, H., Yang, H., Lyu, M.R., King, I.: Sorec: social recommendation using probabilistic matrix factorization. In: Proceedings of the 17th ACM Conference on Information and Knowledge Management, CIKM 2008, pp. 931–940. ACM (2008)
4. Jamali, M., Ester, M.: A matrix factorization technique with trust propagation for recommendation in social networks. In: Proceedings of the Fourth ACM Conference on Recommender Systems, RecSys 2010, pp. 135–142. ACM (2010)
5. Yang, B., Lei, Y., Liu, D., Liu, J.: Social collaborative filtering by trust. In: Proceedings of the 23rd International Joint Conference on Artificial Intelligence, IJCAI, pp. 2747–2753 (2013)
6. Guo, G., Zhang, J., Sun, Z., Yorke-Smith, N.: Librec: a java library for recommender systems. In: 23rd User Modeling, Adaptation, and Personalization (2015)
7. Victor, P., Cornelis, C., Teredesai, A.M., De Cock, M.: Whom should i trust?: the impact of key figures on cold start recommendations. In: Proceedings of the 2008 ACM Symposium on Applied Computing, SAC 2008, pp. 2014–2018. ACM (2008)
8. de Amo, S., Oliveira, C.G.: Towards a tunable framework for recommendation systems based on pairwise preference mining algorithms. In: Sokolova, M., van Beek, P. (eds.) Canadian AI. LNCS, vol. 8436, pp. 282–288. Springer, Heidelberg (2014)
9. Sugano, Y., Ozaki, Y., Kasai, H., Ogaki, K., Sato, Y.: Image preference estimation with a data-driven approach: a comparative study between gaze and image features. J. Eye Mov. Res. **7**(3), 1–9 (2014)
10. de Melo, V.E., Nogueira, E.A., Guliato, D.: Content-based filtering enhanced by human visual attention applied to clothing recommendation. In: IEEE 27th International Conference on Tools with Artificial Intelligence (ICTAI), pp. 644–651 (2015)

Object-Based Representation for Scene Classification

Xuhui Luo and Jinhua Xu[✉]

Shanghai Key Laboratory of Multidimensional Information Processing,
Department of Computer Science and Technology, East China Normal University,
500 Dongchuan Rd, Shanghai, China
51131201034@ecnu.cn, jhxu@cs.ecnu.edu.cn

Abstract. How to encode and represent a scene remains a critical problem in both human and computer vision. Traditional local and global features are useful and have some successes; however, many observations on human scene perception seem to point to an object-based representation. In this paper, we propose a high-level representation for scene categorization. First, we utilize semantic segmentation to get semantic regions. Then we obtain an object histogram representation of a scene by summation pooling over all regions. Second, we build spatial and geometrical priors for each object and each pair of co-occurrent objects from training scenes, and integrate the spatial and geometrical information of objects into the scene representation. Experimental results on two datasets demonstrate that the proposed representation is effective and competitive.

1 Introduction

Scene classification is one of the most fundamental problems in computer vision, and has a wide range of applications, for example, video surveillance, image and video retrieval, web content analysis, human-computer interaction.

Bag of words (BOW) is the most popular model for Scene classification. In BOW, there are three major steps, local feature extraction, local feature coding and global representation by pooling. There are two major drawbacks for BOW model. One is lack of spatial information, since the scene is represented by a histogram of local features and the spatial information of local features is lost. The other drawback of BOW is lack of semantic information, which is useful in scene understanding and categorization.

To solve the two problems of BOW, we propose a high level representation for scene categorization. We employ a semantic segmentation to partition a scene into semantic regions. Different from the traditional segmentation, where each region is assigned a single label, we assign a probability distribution over all possible objects to each region, and then pool over all regions to obtain an object histogram for the scene, which is refereed to as bag of objects representation. To encode the spatial and geometrical information including location, scale and perspective, we build the spatial and geometrical priors for each object and each

© Springer International Publishing Switzerland 2016
R. Khoury and C. Drummond (Eds.): Canadian AI 2016, LNAI 9673, pp. 102–108, 2016.
DOI: 10.1007/978-3-319-34111-8_14

pair of co-occurrent objects using gaussian mixture model (GMM) from training images and integrate the spatial information into the scene representation.

The rest of the paper is organized as follows. In the next section, the related work is briefly reviewed. Section 3 describes the details of our proposed model. Section 4 represents our experimental results on two datasets. The last section concludes the paper.

2   Related Work

Some intermediate or high level representations have been proposed to provide semantic information of a scene. In [9], a semantic modeling was proposed, in which local image regions were extracted on a regular 10*10 grid and annotated manually as nine semantic concepts such as sky, water and grass. In [7], the scenes were also partitioned into non-overlapping 16*16 patches and a latent semantic variable was learned for each patch, replacing the arduous manual annotation in [9]. Semantic attribute was used to represent images in [5,8]. In recent work [6], object bank was proposed, which represented an image based on its response to a large number of object filters. In [3,4,11], deep convolutional neural networks (CNNs) were successfully applied to scene categorization. It was found in [12] that the CNN for scene classification automatically discovers meaningful object detectors, which are representative of the learned scene categories. In the latest work [2,10], with the help of a convolutional neural network trained to recognize objects, a scene is represented as a bag of semantics.

3   The Proposed Model

3.1   Semantic Segmentation

We choose second-order-pooling (O2P) [1] to perform segmentation, which was amongst the winners of PASCAL VOC Segmentation challenge. In a general semantic segmentation, where a scene is segmented into semantic regions and a unique label is assigned to each region. In our object histogram model, we need the statistics of semantic information in a scene. Therefore, we calculate the probability of each region belonging to each object. If the number of objects considered is N_o, then the region i is assigned to a vector:

$$P^i = [p_1^i, p_2^i, \cdots, p_{N_o}^i], \tag{1}$$

where p_j^i is the probability that the region i is the object j, and $\sum_{j=1}^{N_o} p_j^i = 1$.

3.2   High-Level Representation

We will describe three different image representations, which are object histogram, spatial and geometrical histogram (SG hist) and co-occurrence histogram (Cooc hist).

Object Histogram. A couple of prominent objects are diagnostic of a scenes category, for example, refrigerators and stoves are diagnostic for kitchens, while toilets and bathtubs are diagnostic for bathrooms. Since the objects in a scene are unknown, we build the object histogram based on the segmented regions.

We represent a scene as an object histogram by summation pooling over all segments in a scene. If there are N_s semantic regions in a scene, its object histogram can be obtained as

$$H_{obj} = \sum_{i=1}^{N_s} P^i, \tag{2}$$

where P^i is defined in (1).

Spatial and Geometrical Histogram. To represent the spatial and geometrical information of objects, we use the features including location, scale and perspective. We use a regions's center coordinate to represent its location. Also we use the number of pixels within a region to represent its scale. Perspective can be represented by the ratio of height and width. Since the scenes may have various resolutions, center coordinates and scales need to be normalized, $x^c, y^c \in [0, 1]$, $s = NP_o/NP_s$, where NP_o and NP_s are the number of pixels of a region and of a scene respectively. For perspective, we include both the height/width ratio and width/height ratio. Then the spatial and geometrical features of objects can be summarized as:

$$o = [x^c, y^c, s, r_{hw}, r_{wh}]. \tag{3}$$

We build prior models for objects using Gaussian Mixture Model (GMM). Since different objects may have different modes, variational Bayesian Inference instead of the common EM algorithm is used to learn the GMM parameters, and the number of components of GMM can be determined automatically. For the ith object, the GMM model is defined as follows:

$$P(o_i) = \sum_{k=1}^{K_i} \pi_{ik} N(o_i; \mu_{ik}, \sigma_{ik}), \tag{4}$$

where K_i is the number of components of the ith object, μ_{ik} and σ_{ik} are the mean and standard deviation respectively, π_{ik} is the prior probability of the kth component ($0 \le \pi_{ik} \le 1$ and $\sum_{k=1}^{K_i} \pi_{ik} = 1$). Each object instance is represented by $[\gamma_{i1}, \gamma_{i2}, \ldots, \gamma_{iK_i}]$. Here γ_{ik} represents the probability that the ith object has the kth mode, $0 \le \gamma_{ik} \le 1$ and $\sum_{k=1}^{K_i} \gamma_{ik} = 1$. If there are more than one instance of an object in a scene, we sum up all instances, $[\sum \gamma_{i1}, \sum \gamma_{i2}, \ldots, \sum \gamma_{iK_i}]$. The spatial and geometrical histogram of a scene can be represented by concatenating all the N_o most frequent objects:

$$H_{SG} = [\sum \gamma_{11}, \ldots, \sum \gamma_{1K_1}, \ldots, \sum \gamma_{N_o1}, \ldots, \sum \gamma_{N_oK_{N_o}}]. \tag{5}$$

If there does not exist any instance of an object in a scene, the corresponding components of the object are all zeros.

Co-occurrence Histogram. Some objects may co-occur in a scene, and the co-occurrent objects may have some kinds of relative spatial relationship. Therefore, we build a spatial model for each pair of co-occurrent objects. The features we use include normalized center coordinate difference, scale ratio, horizontal overlap and vertical overlap.

$$o_{ij} = [x_{ij}^c, y_{ij}^c, s_{ij}, x_{ij}^{ov}, y_{ij}^{ov}].$$

Let (x_i^c, y_i^c) and s_i are the normalised center coordinate and scale of the ith object defined in the last subsection, then

$$x_{ij}^c = x_i^c - x_j^c, \quad y_{ij}^c = y_i^c - y_j^c, \quad s_{ij} = \frac{\min(s_i, s_j)}{\max(s_i, s_j)}.$$

The bounding box of an object is given by the left-top point (x^{BB_1}, y^{BB_1}) and right-bottom point (x^{BB_2}, y^{BB_2}), then the horizontal overlap is defined as

$$x_{ij}^{ov} = \frac{\max(0, \min(x_i^{BB_2}, x_j^{BB_2}) - \max(x_i^{BB_1}, x_j^{BB_1}))}{\min(x_i^{BB_2} - x_i^{BB_1}, x_j^{BB_2} - x_j^{BB_1})}. \tag{6}$$

If the ith object and the jth object do not overlap horizontally, $x_{ij}^{ov} = 0$. If one object is completely overlapped by the other object horizontally, $x_{ij}^{ov} = 1$. For partial overlap, $0 < x_{ij}^{ov} < 1$. Vertical overlap is defined similarly.

We build a spatial relationship prior for each pair of co-occurrent objects using Gaussian Mixture Model (GMM) in the same way as in the spatial and geometrical histogram. For the ith and the jth object, the GMM model is defined as follows:

$$P(o_{ij}) = \sum_{k=1}^{K_{ij}} \pi_{ij,k} N(o_{ij}; \mu_{ij,k}, \sigma_{ij,k}). \tag{7}$$

We build the co-occurrence histogram of a scene by concatenating all the co-occurrent objects, as in (5).

4   Experiments

We tested our model on two datasets, LabelMe Outdoor (LMO) Dataset and a subset of MIT Indoor-67 Dataset. The total number of objects in natural scenes may be hundreds of thousands. In experiments, we only used those objects occurring frequently in a dataset. Support vector Machine (SVM) with χ^2 kernel was used for classification.

4.1   LabelMe Outdoor (LMO) Dataset

This dataset contains 8 outdoor scene categories. We used 17 most frequent object to perform the O2P segmentation.

For each category, we randomly selected 100 images for training, and the rest for testing. We build the object histogram model, spatial and geometrical histogram and co-occurrence histogram for the training and test images using the segmented objects. The categorization results were shown in Table 1. On this dataset, the object histogram achieved 76.14 %. The co-occurrence histogram improved to 80.54 %. We can see that the co-occurrence of pairs of object can make some improvement. The spatial and geometrical histogram obtained the best result 90.83 % among the three single components. If we combined the object histogram with the spatial and geometrical histogram, the accuracy was improved further to 91.57 %. When all three components were combined, the full model achieved 94.86 %.

4.2   Sub-dataset of MIT Indoor-67 Dataset

MIT Indoor-67 database contains 67 Indoor categories. We chose 5 indoor categories to test our model. The five categories correspond to the 5 indoor categories of the scene 15 dataset, including bedroom, kitchen, living room, office and store. The store category covered 6 categories of MIT indoor-67 dataset, that is grocery store, bakery, bookstore, clothing store, shoe shop, and toy store. There are 1858 annotated object categories for the above indoor scene categories, however we used only 39 most frequent objects among these categories.

For each category, we randomly selected 80 images for training, and the rest for testing. Since the objects in indoor scenes are relatively small, the average segmentation accuracy on this dataset was only 5.46 %.

On this dataset, the object accuracies for the object histogram, spatial and geometrical histogram and co-occurrence histogram were 56.53 %, 54.62 %, 53.82 % respectively. It can be seen that the object histogram can provide useful information for scene categorization, although the segment accuracy was low. However the results of the co-occurrence histogram and spatial histogram were even worse than that of the object histogram. This is because in indoor scenes, there are usually many various small objects, it is hard for the automatic segmentation algorithms like O2P to segment the objects accurately. Therefore the spatial and geometrical information may not be correct. If we combined the object histogram with the spatial and geometrical histogram, the accuracy was improved to 66.08 %.

Table 1. Accuracy of model components with segmented objects

	Object list	SG hist	Cooc hist	Object + SG	Full
LMO	76.14	90.83	80.54	91.57	**94.86**
Indoor5	56.53	54.62	53.82	**66.08**	65.92

4.3   Comparison with Other Methods

We compared our model with some low-level representation models, including Gist, Bag of words (BOW), spatial pyramid matching (SPM), sparse coding SPM (ScSPM) in Table 2. In outdoor scenes, our high level representation is much better than all low level representation models, while in indoor scenes, our model is comparable to SPM and ScSPM.

We also compared our model with the latent semantic representation (LSR) in [7] on LMO dataset in Table 2. It can be seen the LSR slightly improved the performance over the low-level representations, but significantly worse than our model.

Table 2. Comparison with other models on all datasets

	Gist	BOW	SPM	ScSPM	LSR [7]	Ours
LMO	82.83	66.78	86.42	88.86	89.8	94.86
Indoor5	61.20	54.57	67.80	69.97	-	66.08

5   Conclusion

In this paper, we proposed a new high-level object-based model which utilized the semantic segmentation. In experiments, it was demonstrated that object histogram can provide useful information for scene categorization. Integration of spatial and geometrical information of single object and pairs of co-occurrent objects can improve the categorization accuracy. Compared with low-level representation models and some semantic representation, our model had better performance, especially on outdoor scenes.

Acknowledgements. This work is supported by the National Natural Science Foundation of China under Project 61175116, the Science and Technology Commission of Shanghai Municipality under research grant no. 14DZ2260800 and Shanghai Knowledge Service Platform for Trustworthy Internet of Things (No. ZF1213).

References

1. Carreira, J., Caseiro, R., Batista, J., Sminchisescu, C.: Semantic segmentation with second-order pooling. In: Fitzgibbon, A., Lazebnik, S., Perona, P., Sato, Y., Schmid, C. (eds.) ECCV 2012, Part VII. LNCS, vol. 7578, pp. 430–443. Springer, Heidelberg (2012)
2. Dixit, M., Chen, S., Gao, D., Rasiwasia, N., Vasconcelos, N.: Scene classification with semantic fisher vectors. In: Proceedings of IEEE Conference on Computer Vision and Pattern Recognition (CVPR) (2015)
3. Gong, Y., Wang, L., Guo, R., Lazebnik, S.: Multi-scale orderless pooling of deep convolutional activation features. In: Fleet, D., Pajdla, T., Schiele, B., Tuytelaars, T. (eds.) ECCV 2014, Part VII. LNCS, vol. 8695, pp. 392–407. Springer, Heidelberg (2014)

4. Krizhevsky, A., Sutskever, I., Hinton, G.E.: Imagenet classification with deep convolutional neural networks. In: Proceedings of Advances in Neural Information Processing Systems (NIPS), pp. 1097–1105 (2012)
5. Kwitt, R., Vasconcelos, N., Rasiwasia, N.: Scene recognition on the semantic manifold. In: Fitzgibbon, A., Lazebnik, S., Perona, P., Sato, Y., Schmid, C. (eds.) ECCV 2012, Part IV. LNCS, vol. 7575, pp. 359–372. Springer, Heidelberg (2012)
6. Li, L.J., Su, H., Lim, Y., Li, F.F.: Object bank: an object-level image representation for high-level visual recognition. Int. J. Comput. Vision 107, 20–39 (2014)
7. Li, X., Guo, Y.: Latent semantic representation learning for scene classification. In: Proceedings of the 31st International Conference on Machine Learning (2014)
8. Su, Y., Jurie, F.: Improving image classification using semantic attributes. Int. J. Comput. Vision 100, 59–77 (2012)
9. Vogel, J., Schiele, B.: Semantic modeling of natural scenes for content-based image retrieval. Int. J. Comput. Vision 72, 133–157 (2007)
10. Wu, R., Wang, B., Wang, W., Yu, Y.: Harvesting discriminative meta objects with deep CNN features for scene classification. In: Proceedings of International Conference on Computer Vision (ICCV) (2015)
11. Zhou, B., Lapedriza, A., Xiao, J., Torralba, A., Oliva, A.: Learning deep features for scene recognition using places database. In: Proceedings of Advances in Neural Information Processing Systems (NIPS) (2014)
12. Zhou, B., Khosla, A., Lapedriza, A., Oliva, A., Torralba, A.: Object detectors emerge in deep scene CNNS. In: Proceedings of International Conference on Learning Representations (ICLR) (2015)

Salient Object Detection in Noisy Images

Nitin Kumar[1]([✉]), Maheep Singh[1], M.C. Govil[2], E.S. Pilli[2], and Ajay Jaiswal[3]

[1] Department of Computer Science and Engineering,
National Institute of Technology, Uttarakhand, India
nitin2689@gmail.com, maheep.singh@rediffmail.com
[2] Department of Computer Science and Engineering,
Malviya National Institute of Technology, Jaipur, India
govilmc@yahoo.com, espilli.cse@mnit.ac.in
[3] Shaheed Sukhdev College of Business Studies, University of Delhi, Delhi, India
a_ajayjaiswal@yahoo.com

Abstract. Salient Object Detection (SOD) has several applications including image and video compression, video summarization, image segmentation and object discovery *etc.* Several Methods have been suggested in literature for detecting salient object in digital images. Most of these methods aim at detecting salient objects in images which does not contain any artifact such as noise. In this paper, we have evaluated several salient object detection methods in noisy environment on publicly available ASD Dataset. The performance of the methods is evaluated in terms of Precision, Recall and F-measure and Area under the curve (AUC). It has been observed that there is no clear winner but the methods proposed by Liu et al. and Harel et al. are better in comparison to other methods.

Keywords: Salient · Detection · Noise · Multi-scale

1 Introduction

Humans can detect visually dissimilar (salient) areas in an image quickly and without much effort. These areas are perceived by human brain which can determine the areas more important than others. This increases the amount of visual attention paid by humans to some part of the image than others and is called Salient Object Detection (SOD). SOD can be carried out with the help of digital images and computational power. This renders possibility of several applications of SOD such as object detection and recognition [1,2], image and video compression [3,4], video summarization [5], image segmentation [6], object discovery [7,8], content based image retrieval [9] and image collection browsing [10] *etc.* Several methods have been published in literature for salient object detection. One of the earlier methods for salient object detection methods is proposed by Itti et al. [11]. Their research work was inspired by the neural architecture existing in the visual cortex of the humans. Salient Object Detection methods are broadly classified into two categories [11]: (a) bottom-up methods (b) Top-down methods. In bottom-up models, low level features from an image such

© Springer International Publishing Switzerland 2016
R. Khoury and C. Drummond (Eds.): Canadian AI 2016, LNAI 9673, pp. 109–114, 2016.
DOI: 10.1007/978-3-319-34111-8_15

as color, intensity and orientation *etc.* are extracted while in top-down models, existing knowledge of the visual system of the human beings is used for salient object detection. The bottom-up models are rapid, saliency-driven and task-independent while top-down models are slower, volition-controlled and task-dependent [11].

All the methods proposed in literature for salient object detection assume the input image to be of good quality *i.e.* free from any discrepancies. But it is highly probable that the input image is corrupted with some noise. To the best of our knowledge, no method has been proposed in literature for salient object detection in noisy images. Thus, there is need to investigate which salient object detection method performs best when images are corrupted with noise. In this paper, we have compared eight state-of-the-art salient object detection methods in terms of various criteria. The rest of the paper is organized as follows: Sect. 2 provides a brief overview of the state-of-the-art methods for salient object detection in literature. In Sect. 3, Experimental Setup and Results are presented while some concluding remarks and future work are given at the end in Sect. 4.

2   Salient Object Detection Methods

Saliency based on Visual Attention (IT): Itti et al. [11] have presented the first popular research work in the field of salient object detection. Their model was based on the implementation of psychological theories in the visual cortex of the human beings. In this research work [11], three maps were obtained viz. intensity map ($\bar{I}$), color map ($\bar{C}$) and orientation map ($\bar{O}$). These feature maps are generated across different scales and a total of 42 feature maps are computed. These maps are normalized to [0,1] and combined to form final saliency map **S**.

Frequency-Tuned Salient Region Detection (AC): The drawback of the Itti et al. [11] method was that all the maps were combined at scale 4 which produced very small saliency maps. This drawback was overcome by Achanta et al. [12]. They proposed salient object detection method based on the frequency domain analysis of an image. For a given image **I** of size $w \times h$, the saliency map S can be found with the following equation [12]:

$$\mathbf{S}(i,j) = ||\mathbf{I}_\mu - \mathbf{I}_{whc}(i,j)|| \tag{1}$$

where $\mathbf{I}_\mu$ is the mean image and $\mathbf{I}_{whc}$ is the Gaussian blurred image.

Saliency Detection using Maximum Symmetric Surround (AM): In images where the percentage of salient region is high and the percentage of background region is less, the saliency detection methods highlight the background instead of the salient object. To overcome this limitation, Achanta and Susstrunk [13] proposed saliency detection using maximum symmetric surround. For an image **I** of size $w \times h$, the saliency map is computed with the help of following equation:

$$\mathbf{S}_{ss}(i,j) = ||\mathbf{I}_\mu(i,j) - \mathbf{I}_f(i,j)|| \tag{2}$$

where $\mathbf{I}_f(i,j)$ is the mean CIELAB vector of the subimage whose center pixel is at position (i,j).

Saliency based on Information Maximization (BT): Bruce and Tsotsos [14] proposed a bottom-up saliency detection method based on maximizing the information contained in a captured scene in a local manner. The experimental results show that this method is plausible via implementation to the primary visual cortex in human beings. In this method [14], first a representation based on independent component analysis is employed on a large number of natural images to determine a suitable basis. The images are mapped onto this basis and then the Shannon's self information measure is computed to find the saliency maps.

Graph based Visual Saliency (GB): Graph based visual saliency (GBVS) is another bottom-up saliency detection method proposed by Harel et al. [15]. This model is based on graph computations which help in computing the activation maps and finally normalizing/combining these activation maps. This model is based on the idea that the salient pixels in the image possess high values of activation function while background pixels possess low values. The activation maps are generated using a Markov model and the combination/normalization of activation maps is again performed by employing Markov model.

Context-Aware Saliency Detection (CA): Goferman et al. [16] have proposed context-aware saliency which aims at detecting the regions in an image that represent the scene. This model is based on the following four basic principles of human visual attention [16]: (i) Local low-level considerations including factors such as contrast and color. (ii) Global considerations which suppress frequently occuring features, while maintaining features that deviate the norm. (iii) Visual organization rules, which state that visual forms may possess one or several centers of gravity about which the form is organized. (iv) High level factors such as human faces.

Saliency Detection: A Spectral Residual Approach (SR): Saliency detection using spectral residual is proposed by Hou and Zhang [17]. This method is free from assumptions regarding the features, categories and other forms of prior knowledge of salient objects in an image. The spectral residual of an image is extracted by analyzing the log-spectrum of the image. This is a fast method based on the spatial domain in contarst to the frequency domain methods.

Learning to Detect a Salient Object (Liu): Liu et al. [18] have proposed another bottom-up popular method in literature for salient object detection. In this method, a novel set of features i.e. multi-scale contrast, center-surround histogram and color-spatial distribution corresponding to local, regional and global salient objects is proposed. Three feature maps ($\mathbf{F}_i$; $i = 1, 2, 3$) obtained corresponding to the three features are then combined to form final saliency map $\mathbf{w}$ as given below:

$$S = \sum_{i=1}^{3} w_i \mathbf{F}_i \tag{3}$$

The feature maps are combined by a weight vector leaned using Conditional Random Field (CRF) on a training set. In our implementation, we have used the same weights as given by [18] i.e. $\{0.24, 0.54, 0.22\}$.

3   Experimental Setup and Results

The performance of eight state-of-the-art methods (*i.e.* IT, FT, AM, BT, GB, CA, SR, Liu) is compared on publicly available ASD [12] Dataset. ASD dataset contains 1000 natural images. Each of these images are of size 300×400 or 400×300. These images are down sampled by a factor of 4. Afterwards, Gaussian noise is added to images with mean 0 and variance 0.01. A sample image with its corresponding noisy images is shown in Fig. 1. The ground truth for all these images is downloaded from http://ivrg.epfl.ch/supplementary_material/RK_CVPR09/index.html. All the experiments have been performed on a PC with Intel(R) Xeon(R) 3.4 GHz running Windows 7 with 16 GB RAM.

If $\mathbf{S}_k$ is the resultant saliency map of k-th image, the binary saliency map $\mathbf{S}_k^b$ is obtained from $\mathbf{S}_k$ using adaptive threshold ($threshold_k = 255 - mean(\mathbf{S}_k)$):

$$\mathbf{S}_k^b(i,j) = \begin{cases} 1 \text{ if } \mathbf{S}_k(i,j) > threshold_k \\ 0 \text{ if } \mathbf{S}_k(i,j) < threshold_k \end{cases} \tag{4}$$

(a)   (b)

Fig. 1. (a) Sample image and (b) Corresponding noisy image

The performance of the methods is compared in terms of Precision (P), Recall (R), F-measure (F) and area under the curve (AUC). Let us suppose that $\mathbf{D}$ represents the binary saliency map produced by a method and $\mathbf{D}^{GT}$ represents the ground truth, then performance measures are given by the equations: $P = \frac{TP}{TP+FP}$, $R = \frac{TP}{TP+FN}$ and $F = \frac{(1+\beta^2)*P*R}{\beta^2*P+R}$. Here, $TP = \sum_{\mathbf{D}_{i,j}^{GT}=1} \mathbf{D}_{i,j}$, $FP = \sum_{\mathbf{D}_{i,j}^{GT}=0} \mathbf{D}_{i,j}$ and $FN = \sum_{\mathbf{D}_{i,j}=0} \mathbf{D}_{i,j}^{GT}$ are the number of true positive, false positive and false negative pixels respectively.

The variable β is set to 0.3 to give more weight to Precision than Recall. The experimental results of the various methods is given in Table 1. The Receiver Operating Characteristic (ROC) Curve shown in Fig. 2. is obtained by varying the threshold from 0 to 255.

It can be easily observed from Table 1 that:

1. There is no clear winner among the compared methods.
2. The method proposed by Liu et al. [18] (Liu) outperforms all other methods in terms of Precision (P) and Area under the curve (AUC).
3. The performance of method by Harel et al. [15] (GB) is better in comparison to all other methods in terms of Area under the Curve (AUC).
4. Overall performance of AM method in terms of F-Measure is worst.

Table 1. Performance comparison of Salient Object Detection Methods

	Precision	Recall	F-measure	AUC
IT	0.4624	**0.2642**	0.4355	0.6658
FT	0.6902	0.0463	0.3213	0.7448
AM	0.8088	0.0089	0.0960	0.7596
BT	0.6539	0.1989	0.5500	0.3524
GB	0.8224	0.2517	**0.6927**	0.8694
CA	0.8147	0.1171	0.5460	0.8622
SR	0.6802	0.0765	0.4118	0.7845
Liu	**0.8533**	0.1684	0.6388	**0.8761**

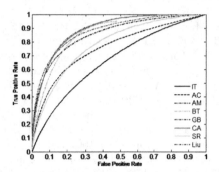

Fig. 2. ROC Curve for the compared methods

4   Conclusion and Future Work

In this paper, we have evaluated the performance of eight state-of-the-art methods for salient object detection in noisy images. Gaussian Noise is artificially added to images. The performance of the methods is evaluated in terms of Precision, Recall, F-measure and Area under the curve (AUC). The performance is evaluated on ASD Dataset. Experimental results demonstrate that there is no clear winner. The performance of the method proposed by Liu et al. (Liu) and Harel et al. (GB) are better than other methods. In future work, we shall propose some feature extraction technique for salient object detection which gives better performance than state-of-the-art methods even in noisy images.

References

1. Kanan, C., Cottrell, G.: Robust classification of objects, faces, and flowers using natural image statistics. In: Proceedings of the IEEE Computer Scoiecty Conference on Computer Vision and Pattern Recognition, pp. 2472–2479 (2010)
2. Ren, Z., Gao, S., Chia, L.-T., Tsang, I.: Region-based saliency detection and its application in object recognition. IEEE TCSVT **24**, 769–779 (2013)

3. Guo, C., Zhang, L.: A novel multiresolution spatiotemporal saliency detection model and its applications in image and video compression. IEEE TIP **19**, 185–198 (2010)

4. Itti, L.: Automatic foveation for video compression using a neurobiological model of visual attention. IEEE TIP **13**, 1304–1318 (2004)

5. Ma, Y.-F., Hua, X.-S., Lu, L., Zhang, H.-J.: A generic framework of user attention model and its application in video summarization. IEEE TMM **7**, 907–919 (2005)

6. Donoser, M., Urschler, M., Hirzer, M., Bischof, H.: Saliency driven total variation segmentation. In: ICCV, pp. 817–824 (2009)

7. Karpathy, A., Miller, S., Fei-Fei, L.: Object discovery in 3d scenes via shape analysis. In: Proceedings of the International Conference on Robotics and Automation, pp. 2088–2095 (2013)

8. Frintrop, S., Garca, G.M., Cremers, A.B.: A cognitive approach for object discovery. In: Proceedings of the International Conference on Pattern Recognition, pp. 2329–2334 (2014)

9. Chen, T., Cheng, M.-M., Tan, P., Shamir, A., Hu, S.-M.: Sketch2photo: internet image montage. ACM TOG **28**, 124:1–124:10 (2009)

10. Sun, J., Xie, J., Liu, J., Sikora, T.: Image adaptation and dynamic browsing based on two-layer saliency combination. IEEE Trans. Broadcast. **59**, 602–613 (2013)

11. Itti, L., Koch, C., Niebur, E.: A model of saliency-based visual attention for rapid scene analysis. IEEE Trnas. PAMI **20**, 1254–1259 (1998)

12. Achanta, R., Hemami, S., Estrada, F., Susstrunk, S.: Frequency-tuned salient region detection. In: Proceedings of the International Conference on Computer Vision and Pattern Recognition, pp. 1597–1604 (2009)

13. Achanta, R., Susstrunk, S.: Saliency detection using maximum symmetric surround. In: Proceedings of the 17th IEEE International Conference on Image Processing (ICIP), pp. 2653–2656 (2010)

14. Bruce, N.D., Tsotsos, J.K.: Saliency based on information maximization. In: Proceedings of NIPS, pp. 155–162 (2005)

15. Harel, J., Koch, C., Perona, P.: Graph-based visual saliency. In: Proceedings of NIPS, pp. 545–552 (2007)

16. Goferman, S., Zelnik-Manor, L., Tal, A.: Context-aware saliency detection. IEEE TPAMI **34**, 1915–1926 (2012)

17. Hou, X., Zhang, L.: Saliency detection: a spectral residual approach. In: Proceddings of the IEEE International Conference on CVPR, pp. 1–8 (2007)

18. Liu, T., Sun, J., Zheng, N., Tang, X., Shum, H.-Y.: Learning to detect a salient object. In: Proceedings of the IEEE International Conference on Computer Vision Pattern Recognition, pp. 1–8 (2007)

An Approach to Improving Single Sample Face Recognition Using High Confident Tracking Trajectories

M. Ali Akber Dewan[1]($\boxtimes$), Dan Qiao[2], Fuhua Lin[1],
Dunwei Wen[1], and Kinshuk[1]

[1] School of Computing and Information Systems,
Athabasca University, Edmonton, Canada
{adewan, oscarl, dunweiw, kinshuk}@athabascau.ca
[2] Department of Mechanical Engineering,
University of Alberta, Edmonton, Canada
dq@ualberta.ca

Abstract. In this paper, single sample face recognition (SSFR) problem is addressed by introducing an adaptive biometric system within a modular architecture where one detector per target individual is proposed. For each detector, a face model is generated with the gallery face image and updated overtime. Sequential Karhunen-Loeve technique is applied to update the face model using representative face captures which are selected from the operational data by using reliable tracking trajectories. This process helps to induce intra-class variation of face appearance and improve representativeness of the face models. The effectiveness of the proposed method is detailed in security surveillance and user authentication using Chokepoint and FIA datasets in SSFR setting.

1 Introduction

In recent years, there has been increasing demand in face recognition, especially for user authentication and security surveillance. In user authentication, verification of persons' identity is performed to provide a customized experience or grant permission to access a specific set of resources or data in a particular system (semi-controlled environment, e.g., inspection lane, chokepoint entry, home or office access control), whereas in security surveillance the detection of individuals of interest is performed for enhanced security and situational awareness in major events (un-controlled environment, e.g., airport or casino) [1]. Though in both applications, many different physiological and behavioral biometrics may be used, the facial biometric has become more popular for its non-intrusive nature, cost effectiveness, and easiness of collecting with the available devices, like cameras. However, after three decades of intense research, the state-of-the-art face recognition approaches can achieve high recognition rate under only controlled settings (verification), whereas the semi- and un-controlled environments still remain challenging [1].

In real-world applications, a more practical scenario may be like this: the watch-list (WL) gallery contains still images of target persons which are usually collected by a

© Springer International Publishing Switzerland 2016
R. Khoury and C. Drummond (Eds.): Canadian AI 2016, LNAI 9673, pp. 115–121, 2016.
DOI: 10.1007/978-3-319-34111-8_16

high quality digital camera under controlled environment, e.g., ID or driver license photo, thus of high resolution, in frontal view, with normal lighting and neutral expression. In contrast, the probes can be a video clip, live or archived, which is taken under semi- and un-constrained environment by webcam or surveillance cameras, thus usually of low image quality, such as low resolution, poor lighting, non-frontal poses, image blur and even serious misalignment. This setting is known as still-to-video face recognition, and widely used in the semi- and un-controlled environments for the user authentication and security surveillance applications. In still-to-video face recognition, most of the reported techniques rely heavily on the size and representativeness of the training data in the WL gallery, and most of them suffer serious performance drop or even fail to work if only one/few training samples per target person is available for generating the *face models*.[1] The most challenging is the Single Sample Face Recognition (SSFR) problem, where a single training face image for a target individual is given for face modeling, and the person is to be identified later in time, in many different and unpredictable poses, lighting, etc. [2].

To enhance the face models in SSFR problem, many different techniques have been proposed, which includes *multiple-representation*, *synthetic generation* and *enlarging the training data* (using some auxiliary set) [2]. However, these techniques may fail to provide more representative face-models since they incorporate limited information on the intra-class variations and uncertainties of a face in the complex operational environment. The solution lies in making the biometric system "adaptive" to the intra-class variations of the operational data [3]. However, the adaptation technique may corrupt the face models if incorrectly updated with the faces from wrong individuals.

In this paper, the SSFR problem is addressed by introducing an adaptive biometric system within a modular architecture, where one detector per target individual is designed. Within each detector, a face model is initialized with the single face image of the target person of the WL gallery, and then the face model is tracked and updated over time. For face model update, representative face captures from the operation data are selected using the reliable parts of the tracking trajectories, where the reliable parts are selected by using *tracking confidence* and *matching confidence*. A particle filter based framework [4] is used for tracking and Sequential Karhunen-Loeve (SKL) technique [5] is used for incremental update of the face models. The strengths of these techniques are their accuracy and computational efficiency, which makes them suitable to use for real-time application of face recognition, especially to improve SSFR problem.

2   Single Sample Face Recognition

The framework of the proposed system is shown in Fig. 1. This framework is used in a modular architecture, where one detector per individual is employed to address the problem with individual specific one class classifier. The convenience of this modular architecture has been widely studied in the literature for setting individual specific

---

[1] Face models refer to one or more facial captures (used for a template matching system) or a set of parameters estimated using the facial captures (used for a pattern classification system) of the target individuals who are enrolled to the watch-list gallery.

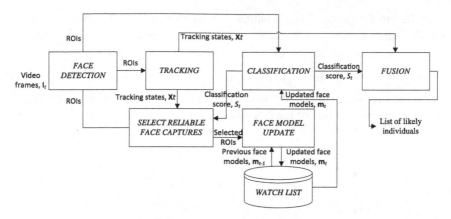

Fig. 1. Framework of the proposed face recognition system

parameters. The proposed framework consists of six main modules: *face detection, tracking, selection of representative face captures, face model update, classification,* and *fusion.*

At a discrete time t, the input for *detection* [6] is a video frame $\mathbf{I}_t$ and output is the set of ROIs delimiting faces in $\mathbf{I}_t$, if any. The *tracking* initiates a new track k once a new face is detected and seeks to estimate its state $\mathbf{X}_t^k \in \{x_t^k, y_t^k, w_t^k, h_t^k\}$ representing the x and y center locations, width, and height of the face ROI in each frame. The features extracted from $\mathbf{X}_t^k$ is represented as $\mathbf{a}_t^k$. During tracking, several states for face k are predicted in each frame using particle filter, and the state $\mathbf{X}_t^k$ that achieves the highest tracking confidence ($S_t^{tracking}$) is finally selected as its actual location

$$S_t^{tracking}\left(\mathbf{a}_{t-1}^k, \mathbf{a}_t^k\right) = 0.5(NCC\left(\mathbf{a}_{t-1}^k, \hat{\mathbf{a}}_t^k\right) + 1) \tag{1}$$

where NCC is a Normalized Correlation Coefficient and $\hat{\mathbf{a}}_t^k$ are the feature vectors extracted from the several predicted states of face k in frame $\mathbf{I}_t$.

The *selection of representative face captures* exploits tracking and selects the representative face captures from the operational data to enhance the face models of the WL gallery. Tracking and matching confidences estimated by (1) and (2), respectively, are used at this stage. If a set of captures for an individual in more than η consecutive frames have the tracking and matching confidences higher than $\tau_{tracking}$ and $\tau_{matching}$, respectively, then the part of the trajectory containing the captures is selected as reliable. Many variations for η, $\tau_{tracking}$, and $\tau_{matching}$, can be used, which we empirically set 10, 0.7, and 0.7, respectively. Figure 2 shows an example of reliable trajectory, where the faces associated with the red dots are selected as the representative captures and the black dots are as the negative samples. It can be observed that the trajectory becomes reliable as soon as it enters into the shaded region where the captures are in more frontal view with minimal changes in appearance.

The *face model update* uses the representative captures to update the face models online. Initially, a single high quality face image is acquired for a target l ($l = 1, \ldots, L$)

and enrolled to the gallery. Then, multiple virtual views are synthesized [7] and constitute the initial training set $\mathbf{B} = \{\mathbf{b}_1^l, \ldots, \mathbf{b}_N^l\}$ with N labelled data for l. The face recognition system then initializes an individual specific detector for person l by generating a face model $\mathbf{m}_{\mathbf{B}}^l = \{\bigcup_{\mathbf{B}}^l, \bar{\mathbf{b}}_{\mathbf{B}}^l, \Sigma_{\mathbf{B}}^l\}$ using $\mathbf{B}$, where $\bigcup_{\mathbf{B}}^l$ is the Eigen vector, $\bar{\mathbf{b}}_{\mathbf{B}}^l$ is the mean vector, and $\Sigma_{\mathbf{B}}^l$ is the covariance matrix computed from the singular value decomposition (SVD) of the centered data matrix of data block, $\mathbf{B}$.

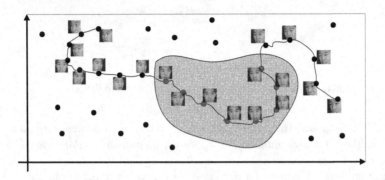

Fig. 2. Selection of reliable part of tracking trajectories

When a person appears in the scene, the person is tracked using particle filter and the representative face captures are selected to form a new data block $\mathbf{A} = \{\mathbf{a}_{n+1}^l, \ldots, \mathbf{a}_{n+q}^l\}$ for person l (with a q additional face captures). Finally, the updated face model $\mathbf{m}_{\mathbf{B}+\mathbf{A}}^l = \{\bigcup_{\mathbf{B}+\mathbf{A}}^l, \bar{\mathbf{b}}_{\mathbf{B}+\mathbf{A}}^l, \Sigma_{\mathbf{B}+\mathbf{A}}^l\}$ is obtained by using the augmented data matrix $[\mathbf{B}\ \mathbf{A}]$ through the SKL algorithm [5] as follows:

Step 1: Compute mean vectors, $\bar{\mathbf{a}}_{\mathbf{A}}^l = 1/m \sum_{i=n+1}^{n+q} \mathbf{a}_i^l$ and $\bar{\mathbf{b}}_{\mathbf{B}+\mathbf{A}}^l = \frac{(f.n)}{(f.n)+q} \bar{\mathbf{b}}_{\mathbf{B}}^l + \frac{q}{(f.n)+q} \bar{\mathbf{a}}_{\mathbf{A}}^l$

Step 2: Form the matrix $\hat{\mathbf{B}} = \left[\left(\mathbf{a}_{n+1}^l - \bar{\mathbf{a}}_{\mathbf{B}}^l\right) \ldots \left(\mathbf{a}_{n+q}^l - \bar{\mathbf{a}}_{\mathbf{B}}^l\right)\sqrt{\{n.q/(n+q)\}} \left(\bar{\mathbf{a}}_{\mathbf{A}}^l - \bar{\mathbf{b}}_{\mathbf{B}}^l\right)\right]$

Step 3: Compute $\tilde{\mathbf{B}} = \mathbf{orth}(\hat{\mathbf{B}} - \mathbf{U}\mathbf{U}^T\hat{\mathbf{B}})$ and $\mathbf{R} = \begin{bmatrix} f\Sigma & \mathbf{U}^T\hat{\mathbf{B}} \\ 0 & \tilde{\mathbf{B}}(\hat{\mathbf{B}} - \mathbf{U}\mathbf{U}^T\hat{\mathbf{B}}) \end{bmatrix}$

Step 4: Compute the SVD of $\mathbf{R} : \mathbf{R} \overset{SVD}{\rightarrow} \tilde{\mathbf{U}}\tilde{\Sigma}\tilde{\mathbf{V}}$

Step 5: Finally, $\mathbf{U}_{\mathbf{B}+\mathbf{A}} = \begin{bmatrix} \mathbf{U} & \tilde{\mathbf{A}} \end{bmatrix}\tilde{\mathbf{U}}$ and $\Sigma_{\mathbf{B}+\mathbf{A}} = \tilde{\Sigma}$

Two key parameters–the forgetting factor f and batch size q–determine the plasticity of the face models over time. $f \in [0\ 1]$ determines the contribution of older observations to be considered and q defines the batch size upon which a face model is updated over time. We empirically set these parameters 0.3 and 10, respectively.

Face Classification compares the face model $\mathbf{m}_t^l = \left\{ \bigcup_t^l, \bar{\mathbf{b}}_t^l, \Sigma_t^l \right\}$ and the input ROI pattern $\mathbf{a}_t^k$ (with tracking identity k) in two steps: *classification* and *fusion*. For each frame $\mathbf{I}_t$, *classification* seeks to measure the similarity between each facial model $\mathbf{m}_t^l$ and the input ROI pattern $\mathbf{a}_t^k$ as follows:

$$S_t^{matching}\left(\mathbf{m}_t^l, \mathbf{a}_t^k\right) = exp\left\{ -\left\| \left(\mathbf{a}_t^k - \bar{\mathbf{b}}_t^l\right) - \mathbf{U}\mathbf{U}^T\left(\mathbf{a}_t^k - \bar{\mathbf{b}}_t^l\right) \right\|^2 \right\} \tag{2}$$

Then the *fusion* accumulates the scores for a target k over the last W ($W = 30$) frames for each trajectory using:

$$acc\_S_t^{matching}\left(\mathbf{m}_t^l, \mathbf{a}_t^k\right) = \frac{1}{W+1} \sum_{i=t-W}^t S_i^{matching}\left(\mathbf{m}_i^l, \mathbf{a}_i^k\right) \tag{3}$$

If the accumulated score for a target surpasses its decision threshold ∂_l, the presence of the individual l is detected. An individual-specific threshold ∂_l is selected using the score distribution obtained by matching the gallery face model $\mathbf{m}_t^l$ to input ROI patterns extracted from video tracks of non-target individuals at a user defined *fpr* of the cumulative probability density function [8].

3   Experiments and Results

To evaluate the performance in security surveillance and user authentication applications, videos from the Chokepoint [9] and FIA [10] datasets are used, respectively. In Chokepoint, 54 videos are recorded, in each 29 individuals walking through a portal are captured with an array of 3 cameras placed above the portal. In FIA, 20-s videos are recorded with an array of 6 cameras positioned at the face level to capture videos from 180 participants mimicking in a passport checking scenario. In the experiments, the ROIs are detected using [6] and scaled into a common size of 48 × 48 pixels. Then, 81-D HOG features are extracted from each ROI and reduced into 32 using PCA. *Correct rejection rate*, *correct recognition rate*, and *accuracy* are used for performance evaluation. The *correct rejection rate* refers to the accuracy in rejecting non-target individuals, whereas the *correct recognition rate* refers to the identification of the targets in the videos. *Accuracy* is the average of the above two parameters, which measures the overall system performance.

In the first experiment, a high quality face image for ID0003 from the Chokepoint is enrolled to the system and compared with 500 face captures from the same individual. Figure 3(a) shows the matching scores produced by the system without considering the face model updates, where 43.7 % scores are above 0.60. In Fig. 3(b), representative face captures are selected using reliable tracking trajectories, and the face model is updated overtime. In this case, 79.8 % of the scores are produced over 0.60, which eventually improves the accuracy of face recognition.

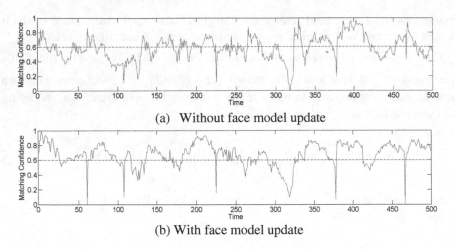

(a) Without face model update

(b) With face model update

Fig. 3. Analysis of matching confidence

The second experiment is conducted on FIA and Chokepoint datasets considering the user authentication and the security surveillance applications, where the WL sizes are varied from 5 to 20 and 3 to 9, respectively. Considering the application type, higher skewness (≥ 0.3) is allowed in probe for the user authentication than the security surveillance (< 0.3). This causes more frequent presence of the targets in the probe and eventually more frequent updates of the face models in user authentication than the other. The averages of 50 randomized runs of the above experiment are listed in Table 1. Because of less frequent updates of the face models in the security surveillance, lower performance is achieved than in the user authentication application.

Table 1. System performance in user authentication and security surveillance

User authentication				Security surveillance			
WL size	Average correct rejection	Average correct recognition	Average accuracy	WL size	Average correct rejection	Average correct recognition	Average accuracy
5	93.23 (± 0.09)	94.43 (± 0.07)	93.83 (± 0.08)	3	85.53 (± 0.07)	86.22 (± 0.05)	85.87 (± 0.06)
10	91.44 (± 0.05)	93.75 (± 0.08)	92.59 (± 0.06)	5	84.63 (± 0.06)	88.52 (± 0.03)	86.57 (± 0.04)
15	89.46 (± 0.07)	92.19 (± 0.02)	93.83 (± 0.04)	7	87.99 (± 0.05)	91.45 (± 0.06)	89.72 (± 0.05)
20	90.53 (± 0.03)	91.62 (± 0.07)	93.83 (± 0.05)	9	83.23 (± 0.06)	84.32 (± 0.07)	83.77 (± 0.06)

4 Conclusion

In this paper, a novel approach has been proposed addressing the Single Sample Face Recognition problem. To overcome the challenge of limited representativeness of the face models, the face images for the target individuals are captured from the operational data by using reliable tracking trajectories and the face models are updated over time. This improves intra-class variation of faces in the face models as well as the performance of the face recognition system. Experiments on security surveillance and user authentication demonstrate the effectiveness of the proposed approach.

Acknowledgement. This research is supported by Academic Research Fund and Research Incentive Grant, Athabasca University, and NSERC, Canada.

References

1. Puvvadi, U.L.N., Benedetto, K.D., Patil, A., Kang, K.-D., Park, Y.: Cost-effective security support in real-time video surveillance. IEEE Trans. Industr. Inf. **11**(6), 1457–1465 (2015)
2. Tan, X., Chen, S., Zhou, Z.-H., Zhang, F.: Face recognition from a single image per person: a survey. Pattern Recogn. **39**(9), 1725–1745 (2006)
3. Dewan, A., Granger, E., Roli, F., Sabourin, R., Marcialis, G.: Adaptive appearance model tracking for still-to-video face recognition. Pattern Recogn. **49**, 129–151 (2016)
4. Ross, D.A., Lim, J., Lin, R.-S., Yang, M.-H.: Incremental learning for robust visual tracking. Int. J. Comput. Vis. **77**(1), 125–141 (2008)
5. Levy, A., Lindenbaum, M.: Sequential Karhunen–Loeve basis extraction and its application to images. IEEE Trans. Image Process. **9**(8), 1371–1374 (2000)
6. Viola, P., Jones, M.J.: Robust real-time face detection. Int. J. Comput. Vis. **57**(2), 137–154 (2004)
7. Zhang, T., Li, X., Guo, R.-Z.: Producing virtual face images for single sample face recognition. Optik-Int. J. Light Electron Opt. **125**(17), 5017–5024 (2014)
8. Bashbagi, S., Granger, E., Sabourin, R., Bilodeau, G.-A.: Watch-list screening using ensembles based on multiple face representations. In: ICPR, Stockholm, Sweden (2014)
9. Wong, Y., Chen, S., Mau, S., Sanderson, C., Lovell, B.C.: Patch-based probabilistic image quality assessment for face selection and improved video-based face recognition. In: CVPRW, Colorado, USA (2011)
10. Goh, R., Liu, L., Liu, X., Chen, T.: The CMU face in action database. In: AMFG, Beijing, China (2005)

Neural Network-POMDP-Based Traffic Sign Classification Under Weather Conditions

Shervin Shahryari[✉] and Cameron Hamilton

Institute for Artificial Intelligence, University of Georgia, Athens, GA, USA
{shervin,cameron}@cs.uga.edu

Abstract. Despite their initial success in operating autonomously, self-driving cars are still unable to navigate under severe weather conditions. In the proposed system, a multi-layer perceptron (MLP) performs initial traffic sign classification. The classification is input as an observation into a partially observable Markov decision process (POMDP) in order to determine whether taking another picture of the sign, or accepting the classification determined by the MLP is the more optimal action. The synergistic combination of the MLP with the POMDP was shown to have a greater functionality than the sum of the MLP and POMDP operating in isolation. The results demonstrate the MLP-POMDP system is capable of training faster and more accurately classifying traffic sign images obscured by fog than a MLP. With further development of this model, one of the greatest shortcomings of autonomous driving may be overcome by accurately classifying signs despite obstruction by weather.

Keywords: Neural network · Convolution neural network · POMDP · Decision making · Autonomous vehicles

1 Introduction

With the advent of autonomous vehicles, particularly self-driving cars, there has been a steady push to develop technologies that will ensure these vehicles are safe and reliable. Although Google's self-driving cars have logged nearly 700,000 autonomous miles total [14], there still remain certain limitations that prevent these vehicles from operating effectively under conditions that drivers regularly face. One of the greatest limitations is that these vehicles cannot "handle heavy rain and snow-covered roads" [7, 13]. As it is vital to overcome this limitation before self-driving cars and other like vehicles are ready for consumers, the current study aimed to develop a traffic sign classification system that is robust enough to handle obstruction of signs due to weather conditions.

2 Methods

2.1 Traffic Sign Selection

Seven traffic signs were selected for the classifier system to learn: Stop, Yield, Do Not Enter, No Right Turn, No Left Turn, Road Work Ahead, and Speed Limit 40. The first

© Springer International Publishing Switzerland 2016
R. Khoury and C. Drummond (Eds.): Canadian AI 2016, LNAI 9673, pp. 122–127, 2016.
DOI: 10.1007/978-3-319-34111-8_17

six signs that were selected are United States traffic signs, while the last is a German traffic sign. This German sign was selected over the United States' speed limit sign because of the German sign's resemblance to the United States' No Right Turn and No Left Turn signs as shown in Fig. 1. The idea of training the classifier on signs with similar features was to ensure the classifier is robust enough to form accurate distinctions between signs despite their similarities.

Fig. 1. Traffic signs used for the training set and for generating testing set images scaled to 64 × 64 pixels

2.2 Construction of the Multi-layer Perceptron (MLP)

A multi-layer perceptron (MLP) was implemented to demonstrate how the MLP-POMDP system can significantly improve performance over the stand-alone MLP. The input into the neural network was simply the RGB intensity values for each pixel: a node was dedicated to each pixel's red, green, and blue intensity values respectively. Each traffic sign image was scaled to 32 × 32 pixels. Thus, the input layer of the neural network consisted of 3072 nodes. A four hidden layer architecture with 300 nodes in each layer also provided the most accurate classifications within this approach. This architecture was determined through a genetic algorithm search strategy.

The 3072-300-300-300-300-7 MLP was trained on the seven noise-free traffic sign images for 3000–5000 iterations using the backpropagation algorithm. The training set was limited to noise-free images, such that the network would not overfit to particular patterns of noise obscuring the signs, but would rather learn the features of the signs apparent in the noise-free images. The POMDP compensates for the fact that the MLP was trained only on noise-free images by maintaining a probability distribution over all the possible signs (i.e. a belief vector) that is continually updated with each successive action (take another picture or classify as sign s) and observation (i.e. each picture taken). Thus, the MLP-POMDP system continues to take additional pictures until enough features of the sign have been recognized for the system to have a sufficiently strong belief that sign s is the proper classification (see Fig. 2).

Fig. 2. Sample of testing set traffic sign images (32 × 32 pixels) with fog filter opacity in range 50–90 %

2.3 A Formalization of the MLP-POMDP System

The POMDP implemented within the classifier system is comprised of the following tuple <S, A, Z, T, O, R>:

- *S:* States – the set of possible traffic signs {Stop, Yield, No Right Turn, No Left Turn, Speed Limit 40, Road Work Ahead, Do Not Enter}
- *A:* Actions – the set of available actions for the classifier {take another image at next time step, classify image as Sign 1, classify image as Sign 2,..., classify image as Sign 7}. It is important to note that once a classification operation is performed, the system resets itself such that the probability distribution for the set of possible signs (i.e. the belief vector) becomes uniform. The reset operation is based on the assumption that once a classification has been decided upon, the vehicle has already traveled beyond that sign and the cameras on the vehicle are now taking pictures of a new sign.
- *Z:* Observations – Classification output from the neural network as a confidence between 0 and 1.
- *T: SxA* $\longrightarrow$ *S':* State Transition Function – as the sign will remain the same over time, the transition function will simply be an identity matrix for the "take another image" action. However, once one of the "classify image" actions is taken, it is assumed that a new sign is now presented to the classifier system. As there is no means for determining what the sign is *a priori*, each state (i.e. each possible sign) is assumed to have an equal probability of occurrence.
- *O: SxA* $\longrightarrow$ *Z:* Observation Function – defines the probability that an observation of classification confidence z will occur given the actual traffic sign contained in the image s and action a. Thus, the observation function of this model is the confusion matrix given by the neural network for the "take another image" action. The confusion matrix is determined by training the MLP prior to its integration into the MLP-POMDP and the resulting observation function it derives remains fixed. After the "classify image" action has been performed, there is no indication of what the new sign is, and thus each observation is assumed to have a uniform probability of occurrence.
- *R: SxA* $\longrightarrow$ *R:* Reward Function – In general, actions where the assigned classification matches the actual sign, the reward is given a positive value. For actions where the assigned classification does not match the actual sign, a negative reward is provided. For taking another picture, a small negative reward is provided.

3   Results

The MLP was able to classify three of the seven with over 99 % accuracy, despite their occlusion with randomized fog filters. The accuracy attained on the remaining signs was below 80 % for each, as shown in the confusion matrix represented in Fig. 3. A Chi-Squared test was used in order to determine whether the hybridization of the MLP with the POMDP significantly improved the performance of the classifier system in comparison to the performance of the MLP by itself. As shown in Fig. 4, the MLP-POMDP hybrid classifier system was significantly more accurate than the stand-alone MLP (i.e. a $p < 0.001$ was found for each traffic sign). This improvement was most drastic for the "Do Not Enter," "No Right Turn," and "Speed Limit 40" signs. The MLP-POMDP system achieved an overall accuracy of 92.9 % (see Tables 1 and 2).

Table 1. Confusion matrix of the multilayer perceptron on the testing set images.

Traffic sign	Road work	Do not enter	No left turn	No right turn	Stop	Speed limit	Yield
Road work	0.999	0	<0.001	<0.001	<0.001	0	<0.001
Do not enter	0	0.629	0.37	0	<0.001	0	<0.001
No left turn	0	0	0.996	0	0.002	0	0.002
No right turn	0.021	0	0.224	0.209	0.546	0	0
Stop	0	0	0	0	1	0	0
Speed limit	<0.001	0	0.225	0	0.16	0.615	<0.001
Yield	0.003	0	0.211	0	0.014	0.001	0.771

Table 2. Comparison of the accuracy of the stand-alone MLP versus the accuracy of the hybrid MLP-POMDP system.

	MLP	MLP-POMDP
Road work ahead	0.999	1.0
Do not enter	0.629	0.824
No left turn	0.996	1.0
No right turn	0.209	0.881
Stop	1.0	1.0
Speed limit 40	0.615	0.998
Yield	0.771	0.8

4   Conclusion

4.1   Discussion of the Results

The hybrid MLP-POMDP classifier system was shown overcome the MLP's inability to extract features from the obscured sign images and to provide significantly more accurate classifications of the traffic sign images filtered with random fog masks than the stand-alone MLP. This finding suggests that it possible to accurately classify traffic signs under weather conditions by using a history of image observations and classifications as input to a POMDP. The confusion matrix for the neural network or other base classifier should also be specified as the POMDP's observation model, such that the reward for each given sign can be defined in proportion to the base classifiers accuracy on that sign. In addition, the training time for the hybrid MLP-POMDP system was less than five minutes, while other approaches like multi-column deep neural networks (MCDNN) can take days to train [2] Through the use of hybrid neural network and POMDP classifier systems, self-driving cars may be able to drive more effectively under a variety of weather conditions, given this system allows the vehicle to accurately classify traffic signs with as fewest images as possible.

4.2   Limitations and Future Directions

A number of important traffic signs were excluded from the system's training set in the interest of first demonstrating how this novel system can more accurately classify occluded traffic signs than a MLP, with the idea that a larger traffic sign database could be utilized in subsequent studies. Second, training and testing instances utilized the same seven traffic sign images that were forward-facing, rather than at an angle and/or partially occluded as is often the case when encountering signs on the road. Furthermore, only one type of weather condition was considered, and that condition was simulated using fog images. This limitation was due to the lack of available US traffic sign data with these qualities. However, overcoming these shortcomings is straightforward, as real images of traffic signs under varying weather conditions can be taken by a vehicle-mounted camera within future studies. Since the POMDP decides upon a final classification based on the history of observations, actions, and its observation model, the hybrid-classifier system should be able to handle multiple weather conditions, partial occlusion and different sign angles with comparable efficiency to the testing set images used in this study. Third, no convolutional neural networks or multi-column deep neural networks were implemented as a benchmark for comparison with the MLP-POMDP system, despite that these networks represent the current state-of-the-art in image classification. Furthermore, a multi-column deep neural network or a stand-alone convolutional neural network could be used in place of an MLP within the MLP-POMDP, in order to further improve overall classification accuracy through hierarchical feature extraction. A MLP was used in this primary study to demonstrate how the proposed system/approach can drastically improve a under-performing model, but in order demonstrate how the current system compares with the state-of-the-art in image classification, the system will have to be compared with convolutional neural networks and other deep neural networks in future studies. Although these sorts of neural networks take considerable time to train, if these networks can be trained offline, then the boost in accuracy they afford is worth the training time.

4.3 Final Remarks

In sum, the hybrid neural network and POMDP classifier system developed in this study can be used as a framework for overcoming one of the greatest weakness currently facing autonomous driving - operating under heavy weather conditions other types of interference with the sign image. With this system in place, self-driving cars loom closer to becoming a reality.

References

1. Baró, X., Escalera, S., Vitrià, J., Pujol, O., Radeva, P.: Traffic sign recognition using evolutionary adaboost detection and forest-ECOC classification. IEEE Trans. Intell. Transp. Syst. **10**(1), 113–126 (2009)
2. Cireşan, D., Meier, U., Masci, J., Schmidhuber, J.: Multi-column deep neural network for traffic sign classification. Neural Netw. **32**, 333–338 (2012)
3. Daugman, J.G.: Uncertainty relation for resolution in space, spatial frequency, and orientation optimized by two-dimensional visual cortical filters. J. Opt. Soc. Am. **2**(7), 1160–1169 (1985)

4. De la Escalera, A., Armingol, J.M., Mata, M.: Traffic sign recognition and analysis for intelligent vehicles. Image Vis. Comput. **21**, 247–258 (2003)

5. La Escalera, D., Moreno, L.E., Salichs, M.A., Armingol, J.M.: Road traffic sign detection and classification. IEEE Trans. Ind. Electron **44**(6), 848–859 (1997)

6. Egmont-Petersen, M., de Ridder, D., Handels, H.: Image processing with neural networks—a review. Pattern Recogn. **35**(10), 2279–2301 (2002)

7. Gomes, L.: Hidden Obstacles for Google's Self-Driving Cars, 28 August 2014. http://www.technologyreview.com/news/530276/hidden-obstacles-for-googles-self-driving-cars/. Accessed 4 May 2015

8. Lawrence, S., Giles, C.L., Tsoi, A.C., Back, A.D.: Face recognition: a convolutional neural-network approach. IEEE Trans. Neural Netw. **8**(1), 98–113 (1997)

9. LeCun, Y., Bengio, Y.: Convolutional networks for images, speech, and time series. In: The handbook of Brain Theory and Neural Networks, 3361, 310 (1995)

10. Marčelja, S.: Mathematical description of the responses of simple cortical cells. J. Opt. Soc. Am. **70**(11), 1297–1300 (1980)

11. Paclık, P., Novovičová, J., Pudil, P., Somol, P.: Road sign classification using Laplace kernel classifier. Pattern Recogn. Lett. **21**(13), 1165–1173 (2000)

12. Russell, S., Norvig, P.: Artificial Intelligence: A Modern Approach. Pearson, New York (2010)

13. Urmson, C.: The latest chapter for the self-driving car: Mastering city street driving, 28 April 2008. http://googleblog.blogspot.com/2014/04/the-latest-chapter-for-self-driving-car.html. Accessed 4 May 2015

14. Zaklouta, F., Stanciulescu, B., Hamdoun, O.: Traffic sign classification using K-d trees and random forests. In: The 2011 International Joint Conference on Neural Networks (IJCNN), pp. 2151–2155. IEEE, July 2011

Natural Language Processing

Poetry Chronological Classification: Hafez

Arya Rahgozar[✉] and Diana Inkpen

School of Electrical Engineering and Computer Science, University of Ottawa,
Ottawa, ON K1N 6N5, Canada
{arahgoza,diana.inkpen}@uottawa.ca

Abstract. We propose a novel task, the chronological classification of poems. Houman's research methodology, which we have adopted here, has provided us with labels for the ghazals that we used as training data in order to train automatic classifiers to classify the rest of the ghazals. To achieve our objective, we have prepared a corpus of Hafez poems in digital form with consistent idiosyncratic linguistic behaviors. Our classification framework uses a combination of unsupervised and supervised methods: a Support Vector Machine (SVM) classifier that uses Latent Dirichlet Allocation (LDA)-based similarity features.

Keywords: Poetry · Chronological classification · LDA · SVM · Similarity · Prototype

1 Introduction

The purpose of our automatic classification of Hafez's ghazals chronologically is to establish the relative timing of any poem, with reference to Hafez's lifetime and hence to help understand his poetry better while applying a semantic approach in machine learning (ML) techniques.

Dr. Mahmoud Houman's work inspired us. In his book about Hafez [1], Houman has done this chronological classification by hand, about 80 years ago. Although he only classified about half of the most straightforward poems, it is our understanding that he mainly meant to provide us with a novel and pragmatic perspective for the classification in which he employed semantic analysis as opposed to subjective and intuitive speculations.

Houman provides a psychological and personality growth perspective on the poet Hafez. This perspective plays an integral role in the interpretation of the poems and their chronological classification shown in Fig. 1. Our research is, therefore, a continuation of Houman's research, but with the use of NLP and ML techniques.

From the very beginning, we realized the great challenges involved. Most importantly, there was no large and reliable corpus of Hafez poems available in electronic form. On the other hand, nearly all classification tasks in the literature were using very large corpora as opposed to our case: 468 ghazals each about 10 lines.

Very few or none of the datasets and resources for text classification were on poems and in Persian. Persian is Right-to-Left, making heavy use of dots, oblique strokes, with hidden vowels, highly cursive and with many position-driven words. Therefore, we not

© Springer International Publishing Switzerland 2016
R. Khoury and C. Drummond (Eds.): Canadian AI 2016, LNAI 9673, pp. 131–136, 2016.
DOI: 10.1007/978-3-319-34111-8_18

only had to change the available software libraries to work with Persian, since classification researchers had mostly designed software libraries to classify English text, but we also had to adopt an efficient and accurate classification methodology applicable to our classification task with such a small corpus.

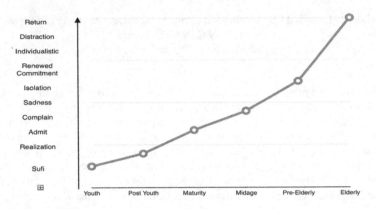

Fig. 1. Hafez's evolutionary growth curve

2   Hafez Corpus

We used Houman's Hafez[1] and followed his approach. While typing in the poems, we applied predefined linguistic rules to all poems. In other words, we ensured consistency while creating our Hafez corpus. We have used multiple types of white spaces to separate or join the one-word terms that we write as counter-intuitive in Persian[2]. In places of potential confusion, we have specified the otherwise unwritten vowels and diacritics inline.

As we see in Fig. 1, in the chronological and conceptual poem chart, each poem essentially would reside on a specific curve point depending on its determined point in time and based on its semantic elements, theme and attributes that Houman detected in the poems. Our corpus complies with Houman's order of ghazals; that is, the timing annotation is, in fact, the actual location of the ghazal with respect to others; this method was the easiest way to record Houman's classification and set the timing attribute of the poem during the preparation of our Hafez corpus.

There are 468 ghazals, out of which Houman labeled only 249 with time information; we have consolidated the 6 classes of chronological pairs into 3 to facilitate classification experiments $\alpha + b = \alpha'$, $c + d = b'$, $e + f = c'$ as shown in Table 1.

---

[1]   Mohammad Ghazvini, 1874-1949 was an Iranian scholar who corrected and prepared today's most reliable prints of Hafez ghazals.
[2]   For example, *dãnəʃ-ãmuz* means student and it is one word but we write it as two in Persian.

Table 1. Corpus training labels

6 Classes		3 Classes
Youth = 38	a	a'
After youth = 25	b	
Maturity = 79	c	b'
Middle age = 66	d	
Before elderly = 28	e	c'
Elderly = 13	f	

3   Related Work

To approximate publication time of the lyrics and detection of the genre [2] used features such as vocabulary, style, semantics, orientation and structure of the song. The authors classified the genres by using SVM.

There is a comparison of semantic classification methods of LSI features for SVM and Vector Space Model (VSM) features for SVM with LDA features for SVM, which turns to be the final top performer [3]. In addition, LDA-SVM is the best performing classifier in finding main subject heading of Arabic texts; they compared this top performer with Naive Bayes, SVM and kNN classifiers [4].

In search of an efficient text classification method, we decided to use SVM [5], because it is a state-of-the-art classification algorithm [6]. Apart from the need for a reliable and consistent corpus, effective feature engineering for SVM is of importance. LDA is used in many industrial areas as an effective tool for a while now [7, 8]. In semantic text classification, many researchers apply LSA or LDA as feature engineering mechanisms for SVM classifiers. For example, [9] uses SVM with multilevel LDA features to classify social media messages and newsgroup texts.

Orthographic, syntactic and phonemic features were used to classify poems by style [10]. In analyzing poems and their aesthetics and in order to reach semantics of imagery, other researchers employ sound devices such as alliteration, consonance, and rhyme [11]. In the case of Hafez's ghazals, such features are consistent across all classes. More work uses NER and part-of-speech taggers to create features to classify poems by style [12]. [18] classified poems into multiple subjects.

Unlike the related work on poetry classification, we classify the poems by one poet alone (Hafez) in chronological order and the poems contain many symbols and hidden semantics that we captured by LDA-driven cosine *similarities* in vector space.

4   Proposed Methodology

As shown in Fig. 2, we used feature-engineering techniques based on layers of Bag-Of-Words[3], Term Frequency-Inverse Document Frequency (TF-IDF) that are transformed into the vector space of LSI or LDA. We then used that for training the SVM classifier

---

[3] Frequency of each word used as feature, irrespective of grammar, order or semantic relations.

model. To achieve our top performing SVM classifier, we have applied our LDA-based cosine similarity features. The dictionary maps poems' normalized words to their index.

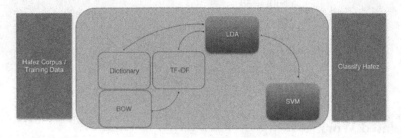

Fig. 2. Architecture of our Hafez classification system

Similar to [16] that adopted from [17], we have averaged *Prototype* similarity vectors for each class. That is, each poem has three *Prototype* features for SVM. Each of the three features is ghazal's average LDA driven *cosine similarity* to all other poems of the class to capture their probabilistic semantic relatedness.

We have used the highest probability terms among all six topics for each class *Youth, Maturity* and *Elderly*. Technically, we use the GENSIM library created by [13] to develop the features; the similarity features in GENSIM[4] and its indexing mechanism by LSI concepts are based on [14]. Then we used WEKA [15] to train the SVM classifier. We grouped the six classes of Houman into pairs, for performance reasons.

5   Experiments and Results

The baseline[5] accuracy for the classification of the three amalgamated classes is 58.2 %.

In Table 2, we are showing the accuracy results of the 10-fold evaluation method for our SVM classifiers with different sets of features. We show evaluation measures for the weighted average of three classes, along with the experiments using Weka.

Table 2. SVM classification evaluation results for 3 classes

Feature	Accuracy	Precision	Recall	F-score
BOW	61.0 %	0.605	0.61	0.50
LDA	58.2 %	0.36	0.56	0.43
BOW + LSI	61.4 %	0.61	0.62	0.50
BOW + LDA	61.8 %	0.62	0.62	0.50
LDA similarity	79.5 %	0.67	0.795	0.72

---

[4] Please refer to https://radimrehurek.com/gensim/ for a more detailed description of transformations and *Similarity* interface.

[5] Baseline is a classifier that always chooses the most frequent class, b', out of the three.

In our first experiment, we created the BOW training data as input to the SVM classifier and increased the accuracy to 61 % compared to the baseline. The LDA factors standalone did not improve over the baseline of 58 %. However, keeping the BOW and adding the LSI or LDA factors on top, only slightly improved accuracy over the BOW alone.

Before arriving at our best result, we tried the crude version of LDA-cosine similarity factors on top of other features, not shown here for brevity reasons. At that point, we hypothesized that the LSI or the LDA driven *similarity* factors alone should provide us with strong enough training features. Therefore, in the next experiments, we went back and created the SVM training data only with normalized similarity factors, once with LSI and then once with LDA. LDA driven similarity factors proved stronger than those of LSI did. That is, as we observed the remarkable strength of these features, we only kept the BOW and LDA factors in the similarity factor calculations, in the final SVM training data. Yet this method brought the accuracy of the classifier to our best result of 79.5 % using our Persian training dataset (all the reported results are via 10-fold cross-validation on the training data). We also ran our model on the unlabeled ghazals and had an expert check a few of the predictions, for a preliminary inspection.

6   Conclusion and Future Work

We have developed the first automatic classifier of the Hafez ghazals. This main task required us to first prepare the Hafez's corpus that retains a consistent idiosyncratic behavior in its latest written Persian version. We utilized NLP, ML, feature engineering and classification methods such as LDA and SVM that were the most appropriate to the detection and semantic interpretation of various nuances and historical mentions in Hafez's ghazals. In the future work, we will make efforts towards the generation of more intuitive clusters of important terms. We also plan to use translations of Hafez poems, in order to do a multilingual classification.

Acknowledgements. We would like to thank Mr. Mehran Rahgozar for his invaluable linguistic insights, expert evaluations and remarkable support especially in the preparation of the Hafez Corpus. Our gratitude also goes to Mr. Mehran Raad who helped with evaluating the results.

References

1. Houman, M.: Hafez. In: khoi, E. (ed.) Published by Tahouri Library and Printed by Sepehr printing office, Tehran, Iran (1938)
2. Fell, M.: Lyrics classification. Master thesis, Saarland University (2014)
3. Liu, Y., Ponraj, M., Liu, Z., Li, M.: Performance evaluation of Latent Dirichlet Allocation in text mining. In: Proceedings of the 8th International Conference on Fuzzy Systems and Knowledge Discovery (2011)
4. Mars, M., Maraoui, M., Zrigui, M., Ayadi, R.: Arabic text classification framework based on Latent Dirichlet Allocation. J. Comput. Inf. Technol. **20**(2), 125–140 (2012)
5. Vapnik, V.N.: The Nature of Statistical Learning Theory. Springer, New York (1995)

6. Joachims, T.: Text categorization with support vector machines: learning with many relevant features. In: Nédellec, C., Rouveirol, C. (eds.) ECML 1998. LNCS, vol. 1398, pp. 137–142. Springer, Heidelberg (1998)
7. Marcialis, F., Roli, G.L.: Fusion of LDA and PCA for face recognition. In: Proceedings of the Workshop on Machine Vision and Perception. Meeting of the Italian Association for Artificial Intelligence (AI*IA) (2002)
8. Martinez, A.C., Kak, A.M.: PCA versus LDA. IEEE Trans. Pattern Anal. Mach. Intell. **23**(2), 228–233 (2001)
9. Inkpen, D., Razavi, A.H.: Topic classification using Latent Dirichlet Allocation at multiple levels. In: Proceedings of CICLing 2014, Kathmandu, Nepal, April 2014
10. Kaplan, D.M., Blei, D.M: A computational approach to style in American poetry. In: Seventh IEEE International Conference on Data Mining, vol. 1550-4786/07 (2007)
11. Kao, J., Jurafsky, D.: A computational analysis of style affects, and imagery in contemporary poetry. Association for Computational Linguistics (2012)
12. Delmonte, R.: Computing poetry style. In: Proceedings for Scientific Workshops, CEUR Workshop, Emotion, and Sentiment in Social and Expressive Media, vol. 1096 (2013)
13. Rehurek, R., Sojka, P.: A software framework for topic modeling with large corpora. In: Proceedings of the LREC 2010 Workshop on New Challenges for NLP Frameworks, pp. 45–50, Valletta, Malta. ELRA, May 2010. http://is.muni.cz/publication/884893/en
14. Deerwester, S., Dumais, G.F., Landauer, S.: Knowledge and natural language processing. (KBNL knowledge-based natural language). Commun. ACM, 33(8), 50–71 (1990). Programs with Common Sense
15. Hall, M., Frank, E., Holmes, G., Pfahringer, B., Reutemann, P., Witten, I.H.: The Weka data mining software: an update. SIGKDD Explor. **11**(1), 10–18 (2009)
16. Bezdek, J.C., Reichherzer, T.R., Lim, G.S., Attikiouzel, Y.: Multiple-prototype classifier design. IEEE Trans. Syst. Man Cybern. **28**(1), 67–79 (1998)
17. Chang, C.L.: Finding prototypes for nearest neighbor classification. IEEE Trans. Comput. **23**, 1179–1184 (1974)
18. Lou, A., Inkpen, D., Tănăsescu, C.: Multi-label subject-based classification of poetry. In: Proceedings of FLAIRS. Association for the Advancement of Artificial Intelligence (2015)

f: Phrase Relatedness Function Using Overlapping Bi-gram Context

Md. Rashadul Hasan Rakib[(⊠)], Aminul Islam, and Evangelos Milios

Faculty of Computer Science, Dalhousie University, Halifax, Canada
{rakib,islam,eem}@cs.dal.ca

Abstract. We present an unsupervised phrase relatedness function (f) that has been applied in a Semantic Textual Similarity system ($TrWP$) of SemEval-2015. The best run of $TrWP$ was ranked 33 among 73 runs. f finds the relatedness strength between two phrases using overlapping bi-gram context extracted from the Google-n-gram corpus. The relatedness strength is the strength of association capturing how similar or dissimilar two phrases are. In order to find the relatedness strength, f applies a sum-ratio (SR) technique based on the statistics of the overlapping n-grams associated with two input phrases. The experimental result from f demonstrates improvement over existing phrase relatedness methods on two standard datasets of 216 phrase-pairs. f does not require any human annotated resource and is independent of the syntactic structure of phrases.

Keywords: Phrase relatedness · Google-n-gram · Unsupervised · Overlapping bi-gram context

1 Introduction

Generally, a phrase is an ordered sequence of multiple words that all together refer to a particular meaning [1,2]. Bi-grams and Tri-grams are commonly used methods to extract and identify meaningful phrases in statistical natural language processing [2,3]. Phrase relatedness plays an important role in different Text Mining tasks; for instance, document similarity[1], classification and clustering are performed on the documents composed of phrases. Several document clustering methods use phrase similarity to determine the similarity between documents so as to improve the clustering result [2,4]. SpamED [5] uses the bi-gram and tri-gram phrase similarity between an incoming e-mail message and a previously marked spam to enhance the accuracy of spam detection.

Some methods for computing phrase relatedness adopt compositional distributional semantics (CDS) [6–8], some use the syntactic structure of the phrases in CDS [9], some use different tools and knowledge-based resources [10,11] and some use the co-occurrence statistics (counts) from the web corpus [12–16].

---

[1] We use 'relatedness' and 'similarity' interchangeably in our paper, albeit 'similarity' is a special case or a subset of 'relatedness'.

© Springer International Publishing Switzerland 2016
R. Khoury and C. Drummond (Eds.): Canadian AI 2016, LNAI 9673, pp. 137–149, 2016.
DOI: 10.1007/978-3-319-34111-8_19

f computes relatedness strength between two phrases by applying a Sum-Ratio (SR) technique[2] on the counts of any two Google-n-grams that contain two target phrases and an overlapping bi-gram context following the approach in $TrWP$ [17]. Relatedness strength is normalized by the normalization technique used in Normalized Google Distance (NGD) [13].

f uses overlapping bi-gram context because the co-occurrences (e.g., "large number and vast amount") of two phrases (e.g., "large number", "vast amount") in the Google-n-gram corpus are too rare to be usable. Consider two Google-n-grams "large number of death" and "vast amount of death" where "large number" and "vast amount" are the target phrases and "of death" is an overlapping bi-gram context.

f considers the whole phrase as a single unit because splitting phrases into words ignores the word order that might change the meaning of phrases leading to inconsistent phrase relatedness score [18]. For example, if we split the phrases "boat house" and "house boat" into words, we get the relatedness score one, nonetheless as a whole unit, these two phrases do not refer to exactly the same meaning [18]. "boat house" means a house for sheltering boats whereas "house boat" means a boat that serves as a house.

The relatedness measures such as Jaccard [15], Simpson [12], Dice [19], PMI [16] and NGD [13] can be applied to phrase relatedness task using web corpus where the phrases are considered as single units. We treat these measures as web-based phrase relatedness methods. However these web-based methods are infeasible to use for document classification or clustering tasks since these tasks require a locally computable relatedness method which is executed in their inner loop without any need to query a search engine with associated network communication overhead.

The reason for using Google-n-gram [20] corpus is that it covers all the topics, words and phrases with their statistics (counts) that are used in the real world. Moreover, to generate n-grams (e.g., 3-grams, 4-grams, 5-grams) and their corresponding frequencies from a corpus is computationally expensive.

We summarize the features of f as follows: (i) f shows a new approach (Sum-Ratio technique) to compute phrase relatedness using overlapping context. (ii) f is a generic corpus based unsupervised phrase relatedness function. (iii) f is independent of the syntactic structure of phrases and does not take into account the types of words within phrases. (iv) f does not split a phrase into words and thus preserves the internal semantics of a phrase. (v) f requires neither any human annotated resource (e.g., WordNet) nor any external Natural Language Processing tools (e.g., lemmatizer, POS tagger).

2   Related Work

Most of the phrase relatedness tasks [6,7,9,21] use compositional distributional semantic (CDS) models where the phrases are split into words and each word is

---

[2] We use the term Sum-Ratio as the weighted mean of two numbers.

represented by a vector in Vector Space Model (VSM), constituted by the co-occurrences of that word with different contexts (attributes). The phrase meaning is obtained by combining the vectors of its individual words using different composition functions (e.g., additive, multiplicative).

Mitchell and Lapata [7] represent a two-word phrase by two vectors with attributes selected from the BNC corpus using two different techniques: semantic space [22] and LDA topic model [23]. Several composition functions (e.g., additive, multiplicative) are applied to the vectors for a particular phrase in order to produce a resultant vector; then the similarity between two phrases is computed using two resultant vectors by the widely used cosine measure. Hartung and Frank [6] also represent phrases by vectors to compute similarity between them. They apply Controlled LDA [6] to select attributes of a word within a phrase. Both Mitchell and Lapata [7] as well as Hartung and Frank [6] need to train an LDA topic model in order to extract attributes.

Annesi et al. [9] considers each of the given phrases has same syntactic structure, for instance, first and second word of each phrase are adjective and noun respectively e.g., *phrase1 = adjective1-noun1, phrase2 = adjective2-noun2*. The similarity between two phrases is calculated by aggregating the similarity between *adjective1* and *adjective2*, and the similarity between *noun1* and *noun2*.

UMBC [10] and OMIOTIS [11] have phrase relatedness web services built on a human annotated knowledge base (e.g., WordNet). They use word-pair relatedness from WordNet to compute the phrase-pair relatedness irrespective of the word order within phrases. A dynamic programming based algorithm [24] measures phrase similarity through integrating both edit distance between the parse trees of the phrases and word-pair similarity.

The web-based phrase relatedness methods [25]: Jaccard [15], Simpson [12], Dice [19], PMI [16] and NGD [13] use the co-occurrence counts of two phrases $P1$ and $P2$ from the web corpus, defined in Eqs. 1–5 respectively. The co-occurrence is measured based on the number of web documents in which $P1$ and $P2$ appear together denoted by $D(P_1, P_2)$. $D(P_1, P_2)$ is normalized with respect to the independent occurrences of P_1 and P_2 represented as $D(P_1)$ and $D(P_2)$ respectively. $D(P_1)$ = number of web documents that contain P_1. $D(P_2)$ is computed similarly. We submit the queries: **"P1"**, **"P2"** and **"P1" and "P2"** to Google in order to obtain the number of independent occurrences and co-occurrences of the phrases $P1$ and $P2$ respectively. T is the total number of web documents. The value of NGD is unbounded, ranging from 0 to ∞. Therefore, the value is bounded between 0 and 1 using the variation of NGD defined in [26].

$$Jaccard(P_1, P_2) = \frac{D(P_1, P_2)}{D(P_1) + D(P_2) - D(P_1, P_2)} \tag{1}$$

$$Simpson(P_1, P_2) = \frac{D(P_1, P_2)}{min(D(P_1), D(P_2))} \tag{2}$$

$$Dice(P_1, P_2) = \frac{2 \times D(P_1, P_2)}{D(P_1) + D(P_2)} \tag{3}$$

$$PMI(P_1, P_2) = log_2 \left(\frac{\frac{D(P_1,P_2)}{T}}{\frac{D(P_1)}{T} \quad \frac{D(P_2)}{T}} \right) \qquad (4)$$

$$NGD(P_1, P_2) = e^{-2 \times \frac{max(log\, D(P_1),\, log\, D(P_2)) - log\, D(P_1,P_2)}{log\, T - min(log\, D(P_1),\, log\, D(P_2))}}. \qquad (5)$$

3 Terminology Used in Phrase Relatedness

The terminologies used in measuring phrase relatedness are described below.

Bi-gram Context: Bi-gram context is a bi-gram, extracted by placing a phrase in the left most, middle and right position within the Google-n-grams. Sample bi-gram contexts for the phrase "large number" are shown in Table 1.

Table 1. Positions of the phrase ("large number") in Google-n-grams and corresponding bi-gram contexts marked bold.

Phrase position	Google-n-grams	Bi-gram context
Left most	large number **of files**	of files
Middle	**very** large number **generator**	very.. generator
Right most	**multiply a** large number	multiply a

Overlapping Bi-gram Context: The overlapping bi-gram context is a bi-gram which is overlapped between two Google-n-grams that contain two target phrases at the same position.

Sum-Ratio (SR): Sum-Ratio refers to the product of the sum and ratio between the minimum (min) and maximum (max) of two numbers. Given two numbers a and b, the Sum-Ratio of a and b is defined as follows.

$$Sum(a, b) = a + b \qquad \text{e.g., Total strength}$$
$$Ratio(a, b) = min(a, b)/max(a, b) \qquad \text{e.g., Relative strength}$$
$$Sum\text{-}Ratio(a, b) = Sum \times Ratio \qquad \text{e.g., Quantified Total strength}$$

The Sum-Ratio of two numbers indicates the strength of association (association strength) between them by maximizing the sum of two numbers with respect to their ratio. For instance, the strength of association between 1 and 1000 is $(1 + 1000) \times (min(1, 1000)/max(1, 1000)) = 1.001$ and between 500 and 501 is $(500 + 501) \times (500/501) = 999.002$. The sum of the first and second pair of two numbers are same (e.g., 1001), but the association strengths are different due to having different ratios $(1/1000) = 0.001$ and $(500/501) = 0.998$, respectively. The objective of Sum-Ratio is to capture the strength of association between the counts of two overlapping Google-n-grams.

Relatedness Strength: Relatedness strength is the strength of association between two phrases P_1 and P_2, which is computed using the Sum-Ratio values obtained from the counts of any two Google-n-grams that contain P_1 and P_2 respectively and an overlapping bi-gram context.

Given a pair of phrases P_1 and P_2, any n-gram pair that contains P_1 and P_2 and an overlapping bi-gram context, represents an association between P_1 and P_2. In order to measure the strength of association between two phrases, we use the counts of Google-n-grams in which they appear. To measure the strength of association between two counts, the phrase relatedness function (f) uses the Sum-Ratio technique.

4   Computing Phrase Relatedness

To compute phrase relatedness we present an unsupervised function f that computes relatedness strength between two phrases P_1 and P_2 using the Google-n-gram corpus [20]. The relatedness strength is normalized between 0 and 1 using NGD [13] in conjunction with NGD' [26]. The steps of computing phrase relatedness are depicted in Fig. 1.

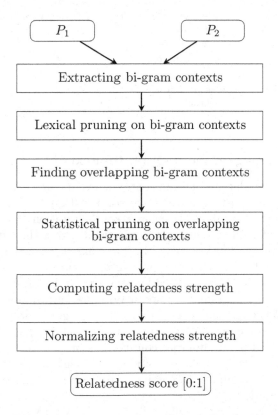

Fig. 1. Steps of computing phrase relatedness.

4.1 Extracting Bi-gram Contexts

The relatedness between two phrases is computed based on the bi-gram contexts extracted from Google-n-grams. Therefore f needs to extract all possible bi-gram contexts from Google-n-grams in which the target phrases appear as shown in Table 1. Sample extracted (non) overlapping bi-gram contexts for the phrases $P_1 =$ "large number" and $P_2 =$ "vast amount" and their associated frequencies (counts) are shown in Table 2.

Table 2. Sample (non) Overlapping Google-n-grams for phrases "large number" and "vast amount" along with artificial counts.

Large number, count $= 1000$		Vast amount, count $= 2000$		Bi-gram context	
Google-n-gram	Count	Google-n-gram	Count	Overlap	Non-overlap
large number of death	100	vast amount of death	200	of death	
consider the large number	300	consider the vast amount	400	consider the	
very large number generator	150	very vast amount generator	300	very.. generator	
large number of data	200	vast amount of data	300	of data	
large number of items	100	vast amount of land	200		of items, of land

4.2 Lexical Pruning on the Bi-gram Contexts

Some phrases along with their bi-gram contexts do not convey meaningful insight due to the improper positioning of stop-words within the bi-gram contexts. Therefore we perform lexical pruning[3] based on the position of stop-words inside the bi-gram contexts. When the target phrase is placed at the left most position within a Google-n-gram followed by a bi-gram context, then the Google-n-gram is pruned if the right most word of that bi-gram context is a stop-word. When the phrase is in the middle surrounded by two context words within a Google-n-gram, then the Google-n-gram is pruned if both the surrounding context words are stop-words. When the phrase is situated at the right most position within a Google-n-gram preceded by a bi-gram context, then the Google-n-gram is pruned if the left most word of the bi-gram context is a stop-word. After performing lexical pruning, we have two sets of non-pruned Google-n-grams containing the bi-gram contexts of two phrases respectively. The pruning conditions with examples are given in Table 3.

Consider the phrase "large number" and the container Google-n-gram "large number of death" given in Table 3. The phrase "large number" and the bi-gram

---

[3] Perform pruning on the bi-gram contexts implies to the pruning of the Google-n-grams from which those contexts are extracted.

context "of death" all together refer to the amount of death. Hence "of death" is kept. In contrast, the phrase "large number" and the bi-gram context "of the" all together do not express any significant meaning, therefore "of the" has been pruned.

Table 3. Conditions for the lexical pruning on bi-gram contexts.

Phrase position	Phrase	Sample bi-gram contexts		Pruning conditions on bi-gram contexts
		Meaningful (Kept)	Meaningless (Pruned)	
Left most	large number	large number **of death**	large number **of the**	right most uni-gram is a stop-word
Middle	different parts	**combine** different parts **carefully**	**in** different parts **to**	Both uni-grams are stop-words
Right most	certain circumstance	**state under** certain circumstance	**of under** certain circumstance	Left most uni-gram is a stop-word

4.3   Finding Overlapping Bi-gram Contexts

In this step, we find the overlapping bi-gram contexts between two sets of non-pruned Google-n-grams extracted for two phrases. The Google-n-grams having overlapping bi-gram contexts are separated from the Google-n-grams that have no overlapping contexts. For each overlapping Google-n-gram pair, the strength of association is calculated to ultimately compute the relatedness strength between the phrases. The set of overlapping bi-gram contexts for the phrases "large number" and "vast amount" is { "of death", "consider the", "very.. generator", "of data"} as shown in Table 2.

4.4   Statistical Pruning on the Overlapping Bi-gram Contexts

Each Google-n-gram pair with overlapping bi-gram context possesses a strength of association (association strength). We presume that if most of the Google-n-gram pairs have higher association strengths, the relatedness score between two phrases tends to be higher and vice versa. Now the question arises, should we consider all the overlapping bi-gram contexts of a particular phrase pair. The answer is no because all of them are not reliable. Therefore we apply statistical pruning using Normal Distribution [27] to discard the overlapping bi-gram contexts whose association strengths are not reliable enough. The reliable association strengths tend to be closer to their center (e.g., they lay within their mean ± standard deviation). In Normal Distribution, most of the samples tend to be around their center (mean) (e.g., most of them exist within their mean ± standard deviation).

In order to perform statistical pruning, we divide each Google-n-gram count (frequency) within a pair by the count of its corresponding phrase, resulting a normalized count. For each Google-n-gram pair, the minimum and maximum among the two normalized counts are determined. After that we calculate the ratio (e.g., minimum/maximum) between them. Following that, for each Google-n-gram pair, we multiply the ratio by the sum of two Google-n-gram counts, which produces a resultant product (e.g., strength of association). Later on we compute the mean (u_{sr}) and standard deviation (sd_{sr}) from the association strengths of Google-n-gram pairs. If the association strength is within $u_{sr} \pm sd_{sr}$, it is kept otherwise pruned.

How the statistical pruning is being performed, is illustrated using the counts of the phrases and Google-n-grams from Table 2. Consider the Google-n-gram pair ("large number of data", "vast amount of data") where the target phrases are "large number" and "vast amount" respectively. The counts of the phrases "large number" and "vast amount" are 1000 and 2000 correspondingly; and the counts of the n-grams "large number of data" and "vast amount of data" are 200 and 300 respectively. The normalized counts of the n-grams "large number of data" and "vast amount of data" are $200/1000 = 0.2$ and $300/2000 = 0.15$ respectively and the ratio between them is $min(0.2, 0.15) \div max(0.2, 0.15) = 0.15/0.2 = 0.75$ which is multiplied by the sum of these n-gram counts (e.g., $200 + 300 = 500$) produces the resultant product $500 \times 0.75 = 375$ (e.g., strength of association). The association strengths (e.g., 300, 466.67, 450) for other Google-n-gram pairs are calculated similarly. The u_{sr} and sd_{sr} from these association strengths are 397.92 and 76.49 respectively. The $u_{sr} + sd_{sr} = 474.41$ and $u_{sr} - sd_{sr} = 321.43$ are used to identify the reliable association strengths which are in between $u_{sr} \pm sd_{sr}$. So the association strengths 375, 450 and 466.67 along with their Google-n-gram pairs are kept and the association strength 300 along with the n-gram pair ("large number of death", "vast amount of death") is pruned.

4.5 Computing Relatedness Strength

We compute the relatedness strength between two phrases P_1 and P_2 by combining the relatedness strengths obtained from overlapping and all bi-gram contexts.

Using Overlapping Bi-gram Contexts: For each non-pruned Google-n-gram pair having overlapping bi-gram context, the strength of association is calculated following the Sum-Ratio technique. We sum the two Google-n-gram counts and find the minimum and maximum among them. After that we calculate the ratio (e.g., minimum/maximum) between them. Then the Sum-Ratio value is calculated by multiplying the sum with ratio which signifies the strength of association for a Google-n-gram pair. By summing up the strength of association of each Google-n-gram pair, we get the relatedness strength between the phrases P_1 and P_2 denoted by $RSOB(P_1, P_2)$ as shown in Eq. 6. GP_1 and GP_2 are the Google-n-grams that contain P_1 and P_2, respectively and a overlapping bi-gram context. $C(GP_1)$ and $C(GP_2)$ are the counts of GP_1 and GP_2 respectively. $C(P)$ is the count of phrase P. k is the number of non-pruned Google-n-gram pairs.

$$RSOB(P_1, P_2) = \sum^{k} \frac{min(C(GP_1), C(GP_2))}{max(C(GP_1), C(GP_2))} \times sum(C(GP_1), C(GP_2)). \quad (6)$$

Using All Bi-gram Contexts: All bi-gram contexts of a phrase P_1 include both the non-pruned overlapping and non-overlapping bi-gram contexts that are extracted from the Google-n-grams where P_1 appears. Two vectors V_1 and V_2 in Vector Space Model are constructed for the phrases $P_1 =$ "large number" and $P_2 =$ "vast amount", respectively using their corresponding all bi-gram Contexts from Table 2, as shown in Table 4. The elements of V_1 and V_2 are binary and reflect the presence or absence of a bi-gram context belonging to the phrases P_1 and P_2, respectively. The relatedness strength between P_1 and P_2 using all bi-gram contexts is designated as $cosSim(V_1, V_2)$, and computed by the cosine similarity between the vectors V_1 and V_2, defined in Eq. 7. The similarity score refers to the number of commonalities between the vectors in terms of the bi-gram contexts of P_1 and P_2 respectively.

Table 4. Vectors V_1 and V_2 from non-pruned (non) overlapping bi-gram contexts for the phrases "large number" and "vast amount", respectively.

	consider the	very.. generator	of data	of items	of land
V_1	1	1	1	1	0
V_2	1	1	1	0	1

$$cosSim(V_1, V_2) = \frac{V_1.V_2}{||V_1|| \; ||V_2||}. \quad (7)$$

Combining Relatedness Strengths from Overlapping and All Bi-gram Contexts: We combine the relatedness strengths obtained from overlapping and all bi-gram contexts to compute the overall relatedness strength $f(P_1, P_2)$ between the phrases P_1 and P_2, by quantifying $RSOB(P_1, P_2)$ with respect to $cosSim(V_1, V_2)$, as defined in Eq. 8. In order to quantify, $RSOB(P_1, P_2)$ is multiplied by $cosSim(V_1, V_2)$. To illustrate the reason, why we multiply $RSOB(P_1, P_2)$ by $cosSim(V_1, V_2)$, let us consider the Google-n-grams given in Table 2. The $RSOB$ (P_1, P_2) and $cosSim(V_1, V_2)$ between the phrases "large number" and "vast amount" are 1083.33 and 0.75, respectively. After multiplying them in Eq. 8, we get $f(P_1, P_2) = 812.49$ which is less than 1083.33 because

$cosSim(V_1, V_2) = 0.75$. If all the bi-gram contexts of two phrases are overlapped, then we get the maximum value for $f(P_1, P_2)$ since $cosSim(V_1, V_2)$ is 1, and vice versa.

$$f(P_1, P_2) = RSOB(P_1, P_2) \times cosSim(V_1, V_2). \tag{8}$$

4.6 Normalizing Overall Relatedness Strength

The overall relatedness strength $f(P_1, P_2)$, between the phrases P_1 and P_2 is normalized by the normalization approach used in NGD [13] as defined in Eq. 9. N is the total number of web documents used in the Google-n-gram corpus. The value of NGD is unbounded, ranging from 0 to ∞. Therefore $NGDf$ adopts NGD' [26], a variation of NGD, in order to bound the normalized value between 0 and 1 which is the ultimate relatedness score between the phrases P_1 and P_2.

$$NGDf(P_1, P_2) = e^{-2 \times \frac{max(log\, C(P_1),\, log\, C(P_2)) - log\, f(P_1, P_2)}{log\, N - min(log\, C(P_1),\, log\, C(P_2))}}. \tag{9}$$

5 Evaluation

To evaluate the phrase relatedness method ($NGDf$), the adjective-noun (AdjN) and noun-noun (NN) section from Mitchell and Lapata's [7] dataset have been used. Each section contains 108 phrase-pairs with human similarity ratings between 1 to 7. Hartung and Frank [6] and Reddy et al. [21] evaluate their phrase relatedness methods using AdjN and NN phrase-pairs from Mitchell and Lapata's [7] dataset respectively.

We evaluate $NGDf$ via correlation analysis. In particular, we calculate the Pearson correlation[4] r between the relatedness scores generated by different phrase relatedness methods[5] and human similarity ratings as shown in Table 5. $NGDf$ outperforms all the other measures by achieving the highest Pearson's r for combined 216 phrase-pairs.

To find whether the difference between two correlations is statistically significant, we use the procedure in [29]. For 108 AdjN phrase-pairs, the correlation differences between $NGDf$ and the rest except $cosSim$ are statistically significant at 0.05 level. For 108 NN phrase-pairs, the correlation differences between $NGDf$ and the rest except Annesi et al. [9] are statistically significant at 0.05 level. For combined 216 phrase-pairs, the correlation differences between $NGDf$ and the rest except Annesi et al. [9] are statistically significant at 0.05 level.

OMIOTIS [11] and UMBC [10] have phrase relatedness services that split phrases into words and use word-pair relatedness obtained from WordNet to

---

[4] We prefer Pearson's r to Spearman's ρ because Agirre et al. [28] stated that Pearson's r is more informative than Spearman's ρ. Spearman's ρ considers the rank differences while Pearson's r takes into account the value differences. Moreover, SemEval-2013 [28] used Pearson's r for evaluation task.

[5] Pearson's r is not computed using Mitchell and Lapata's [7] system due to the unavailability of their individual phrase-pair score. Moreover, in an attempt to reproduce Mitchell and Lapata's [7] method, Hartung and Frank [6] get Spearman's $\rho = 0.34$ instead of $\rho = 0.46$ on 108 adjective-noun pairs.

Table 5. Pearson's r on 108 AdjN, 108 NN and 216 combined phrase-pairs. For a specific dataset, the highest correlation (r) is marked bold. The correlation (r) of $NGDf$ is statistically significant at 0.05 level marked (∗).

Method type	Phrase rel. method	AdjN	NN	Combined
Knowledge based	OMIOTIS [11]	0.491*	0.352*	0.399*
	UMBC [10]	0.550*	0.357*	0.436*
Computational Distributional Semantics (CDS) [8]	Hartung and Frank [6]	0.502*	×	×
Syntactic structure in CDS	Annesi et al. [9]	0.602*	**0.703**	0.639
Web-based	Jaccard [15]	0.324*	0.007*	0.163*
	Simpson [12]	0.360*	−0.003*	0.184*
	Dice [19]	0.332*	0.031*	0.169*
	PMI [16]	0.223*	0.252*	0.186*
	NGD [13]	0.247*	0.176*	0.160*
Google-n-gram based	cosSim	0.722	0.536*	0.589*
	NGDf	**0.743**	0.636	**0.656**

calculate the phrase-pair relatedness. Due to the lack of entry in WordNet, we get more zero (or near to zero) relatedness scores by OMIOTIS and UMBC than other methods. For example, UMBC gives zero relatedness score for 29 among 216 phrase-pairs. This reduces the overall performance of OMIOTIS and UMBC.

Annesi et al. [9] considers the syntactic structure of the phrases; because of this, the correlation r on NN phrase-pairs is higher than that of AdjN pairs. Consider the syntactic structures of two AdjN phrases, $phrase1 = adj1$-$noun1$, $phrase2 = adj2$-$noun2$. $rel(phrase1, phrase2) = rel(adj1, adj2)$ (× or +) $rel(noun1, noun2)$ where rel refers to the relatedness function. Since they multiply or add two relatedness scores generated from two different types of word-pairs e.g., $(adj1, adj2)$ and $(noun1, noun2)$, the correlation r for AdjN phrase-pairs is lower than that of NN pairs. For NN pairs any operation (× or +) is performed on the relatedness scores, obtained from the same type of word-pairs.

The web-based methods [25]: Jaccard [15], Simpson [12], Dice [19], PMI [16] and NGD [13] use the co-occurrence counts of two phrases from the web documents. Their results vary due to normalizing the co-occurrence count by different relatedness measures.

We consider $cosSim(V_1, V_2)$ as a baseline method that uses simple Cosine similarity to compute the similarity between two phrases where the vectors V_1 and V_2 are constructed for the phrases P_1 and P_2 respectively using the bi-gram contexts extracted from Google-n-grams in which the phrases appear. For combined 216 phrase-pairs the correlation r from $cosSim(V_1, V_2)$ is lower than that of $NGDf$ since $cosSim(V_1, V_2)$ does not take into account the relatedness strengths between the counts of the Google-n-grams having overlapping bi-gram contexts.

6   Conclusion and Future Work

The phrase relatedness function f computes relatedness strength between two phrases using the sum-ratio technique in conjunction with cosine similarity. $NGDf$ normalizes the relatedness strength obtained from f and outperforms other existing phrase relatedness methods. f is unsupervised, does not require any human annotated resource and is independent of the syntactic structure of phrases. Unlike other phrase relatedness methods based on word relatedness, f considers the whole phrase as a single unit without losing the inner semantic meaning of a phrase. In the future, we plan to study how the use of phrase relatedness along with word relatedness affects the accuracy of document relatedness, classification and clustering.

References

1. Zamir, O., Etzioni, O.: Grouper: a dynamic clustering interface to web search results. In: Proceedings of the Eighth International Conference on World Wide Web, WWW 1999, New York, USA, pp. 1361–1374 (1999)
2. Chim, H., Deng, X.: Efficient phrase-based document similarity for clustering. IEEE Trans. Knowl. Data Eng. **20**(9), 1217–1229 (2008)
3. Charniak, E.: Statistical Language Learning. MIT Press, Cambridge (1993)
4. Hammouda, K., Kamel, M.: Efficient phrase-based document indexing for web document clustering. IEEE Trans. Knowl. Data Eng. **16**(10), 1279–1296 (2004)
5. Pera, M.S., Ng, Y.K.: Spamed: a spam e-mail detection approach based on phrase similarity. J. Am. Soc. Inf. Sci. Technol. **60**(2), 393–409 (2009)
6. Hartung, M., Frank, A.: Assessing interpretable, attribute-related meaning representations for adjective-noun phrases in a similarity prediction task. In: Proceedings of the GEMS 2011 Workshop, Stroudsburg, PA, USA, pp. 52–61(2011)
7. Mitchell, J., Lapata, M.: Composition in distributional models of semantics. Cogn. Sci. **34**(8), 1388–1429 (2010)
8. Baroni, M.: Composition in distributional semantics. Lang. Linguist. Compass **7**(10), 511–522 (2013)
9. Annesi, P., Storch, V., Basili, R.: Space projections as distributional models for semantic composition. In: Gelbukh, A. (ed.) CICLing 2012, Part I. LNCS, vol. 7181, pp. 323–335. Springer, Heidelberg (2012)
10. Han, L., Kashyap, A.L., Finin, T., Mayfield, J., Weese, J.: UMBC_EBIQUITY-CORE: semantic textual similarity systems. In: Proceedings of the Second Joint Conference on Lexical and Computational Semantics, June 2013
11. Tsatsaronis, G., Varlamis, I., Vazirgiannis, M., Nørvåg, K.: Omiotis: a thesaurus-based measure of text relatedness. In: Buntine, W., Grobelnik, M., Mladenić, D., Shawe-Taylor, J. (eds.) ECML PKDD 2009, Part II. LNCS, vol. 5782, pp. 742–745. Springer, Heidelberg (2009)
12. Bollegala, D., Matsuo, Y., Ishizuka, M.: A web search engine-based approach to measure semantic similarity between words. IEEE Trans. Knowl. Data Eng. **23**(7), 977–990 (2011)
13. Cilibrasi, R.L., Vitanyi, P.M.B.: The google similarity distance. IEEE Trans. Knowl. Data Eng. **19**(3), 370–383 (2007)
14. Lin, D.: An information-theoretic definition of similarity. In: Proceedings of the Fifteenth ICML, ICML 1998, San Francisco, CA, USA, pp. 296–304 (1998)

15. Salton, G., McGill, M.J.: Introduction to Modern Information Retrieval. McGraw-Hill Inc., New York (1986)
16. Turney, P.D.: Mining the web for synonyms: PMI-IR versus LSA on TOEFL. In: Flach, P.A., De Raedt, L. (eds.) ECML 2001. LNCS (LNAI), vol. 2167, p. 491. Springer, Heidelberg (2001)
17. Rakib, M.R.H., Islam, A., Milios, E.: TrWP: text relatedness using word and phrase relatedness. In: Proceedings of the SemEval 2015, Colorado, pp. 90–95 (2015)
18. Turney, P.D., Pantel, P.: From frequency to meaning: vector space models of semantics. J. Artif. Int. Res. **37**(1), 141–188 (2010)
19. Lin, D.: Automatic retrieval and clustering of similar words. In: Proceedings of the 36th Annual Meeting of the Association for Computational Linguistics, ACL 1998, pp. 768–774 (1998)
20. Brants, T., Franz, A.: Web 1T 5-gram corpus version 1.1. Linguistic Data Consortium (2006)
21. Reddy, S., Klapaftis, I., McCarthy, D., Manandhar, S.: Dynamic and static prototype vectors for semantic composition. In: Proceedings of the 5th International Joint Conference on Natural Language Processing, Thailand, pp. 705–713, November 2011
22. Lund, K., Burgess, C.: Producing high-dimensional semantic spaces from lexical co-occurrence. Behav. Res. Methods Instrum. Comput. **28**(2), 203–208 (1996)
23. Blei, D.M., Ng, A.Y., Jordan, M.I.: Latent Dirichlet allocation. J. Mach. Learn. Res. **3**, 993–1022 (2003)
24. Vilares, M., Ribadas, F.J., Vilares, J.: Phrase similarity through the edit distance. In: Galindo, F., Takizawa, M., Traunmüller, R. (eds.) DEXA 2004. LNCS, vol. 3180, pp. 306–317. Springer, Heidelberg (2004)
25. Islam, A., Milios, E., Kešelj, V.: Comparing word relatedness measures based on google-n-grams. In: COLING (Posters), pp. 495–506 (2012)
26. Gracia, J., Trillo, R., Espinoza, M., Mena, E.: Querying the web: a multiontology disambiguation method. In: Proceedings of the 6th International Conference on Web Engineering, ICWE 2006, pp. 241–248. ACM, New York (2006)
27. Bohm, G., Zech, G.: Introduction to statistics and data analysis for physicists. DESY (2010)
28. Agirre, E., Cer, D., Diab, M., Gonzalez-Agirre, A., Guo, W.: *SEM 2013 shared task: semantic textual similarity. In: Second Joint Conference on Lexical and Computational Semantics, Atlanta, Georgia, USA, pp. 32–43, June 2013
29. Zou, G.Y.: Toward using confidence intervals to compare correlations. Psychol. Methods **12**(4), 399–413 (2007)

Harnessing Open Information Extraction
for Entity Classification in a French Corpus

Fabrizio Gotti[✉] and Philippe Langlais

RALI, Université de Montréal, CP 6128 Succursale Centre-Ville,
Montréal H3C 3J7, Canada
{gottif,langlais}@iro.umontreal.ca

Abstract. We describe a recall-oriented open information extraction system designed to extract knowledge from French corpora. We put it to the test by showing that general domain information triples (extracted from French Wikipedia) can be used for deriving new knowledge from domain-specific documents unrelated to Wikipedia. Specifically, we can label entity instances extracted in one corpus with the entity types identified in the other, with little supervision. We believe that the present study is the first one that focusses on such a cross-domain, recall-oriented approach in open information extraction.

Keywords: Natural language processing · Open information extraction · Named entities · Entity classification

1 Introduction

Extracting knowledge from a large set of mostly unstructured documents (such as the Web) and organizing it into a knowledge base (KB) is a key challenge in artificial intelligence [1]. Intuitively, such KBs should directly impact the quality of many NLP applications such as question answering or information retrieval. Open information extraction (OIE), the task of extracting knowledge from texts without much supervision (especially not a prescription of the kind of information to mine), has brought new hope for such an endeavour. It has given rise to a number of exciting realizations, many fostered by major search engine companies. One of the most striking projects is IBM's Watson question answering system [2], which exploits the information extracted from over 200 million web pages, and went on to win a 2011 *Jeopardy!* television game show against two (human) champions. The work of Microsoft on the Literome project [3] is another impressive realization, where information mined from scientific articles available in PubMed[1] has been exploited for assisting medical researchers.

Many initiatives have been launched for acquiring large repositories of structured semantic knowledge about our world, including FreeBase [4], Yago [5], DBpedia [6] or more generally the Linked Open Data [7]. Many such repositories are often collaborative. For instance, DBpedia is built automatically from Wikipedia, which is

---

[1] http://www.ncbi.nlm.nih.gov/pubmed.

© Springer International Publishing Switzerland 2016
R. Khoury and C. Drummond (Eds.): Canadian AI 2016, LNAI 9673, pp. 150–161, 2016.
DOI: 10.1007/978-3-319-34111-8_20

definitely a collaborative effort. A few initiatives are (almost) unsupervised, such as the NELL system [8], which continuously learns to extract knowledge from web pages.[2]

While these repositories are continuously growing, they still suffer from two main shortcomings. First, they lack coverage for specialized domains. There does not seem to be many repositories that would be useful for, say, developing a system to answer questions on network protocols. Second, they are mainly English-centric. One might argue that this is not an issue since semantics are not language-specific, but this amounts to an oversimplification. Texts in a given language could very well yield some useful information (or points of view) that are glossed over or simply absent from English documents. More practically, concepts in KBs are associated with English strings (e.g. `ref:label` in DBpedia) that systems can locate in texts, which limits portability to other languages. We are aware of multilingual initiatives, such as BabelNet [9], but their coverage is poor.

This study attempts to tackle some of these shortcomings. We describe a recall-oriented OIE system designed to extract knowledge (triples) from French corpora. We then put it to the test by assessing how general domain knowledge (from French Wikipedia) can be used for deriving new information in domain-specific documents (Érudit, a collection of scholarly papers in the humanities). We believe that the present study is the first one that focusses on such a cross-domain use in OIE. Although a few domain-specific OIE systems have been designed, such as the Literome project aforementioned, they mostly rely on a huge collection of domain-specific texts.

We start by discussing related work in Sect. 2. We describe in Sect. 3 our effort to develop an OIE system for the French language. In Sects. 4 and 5, we show how it is possible to derive and characterize entity types from triples extracted in Wikipedia and then use these types to classify entity instances found in triples obtained from another corpus. We conclude in Sect. 6.

2   Related Work

Since the seminal work conducted on TextRunner [10], several toolkits have emerged to allow Open Information Extraction (OIE) in NLP applications. For instance, REVERB [11] relies on part-of-speech tagging and is available for analyzing English text. The WOE extractor [12] was one of the first to propose distant supervision (in their case, they used arguments in Wikipedia infoboxes) in order to mine patterns of interest. Other extractors, such as OLLIE [13], exploit the dependency parse of sentences in order to extract more precise and more diversified relations.

Dealing with facts acquired by such tools on a large corpus is a daunting task. See for instance the impressive effort made at Google [14] for building a huge collection of facts from as many sources as possible (including manually curated databases). Structuring those facts into a useful KB is an even more challenging endeavour, one that has benefited from exciting developments in the recent years.

---

[2] Some supervision was provided at the very beginning of the project in order to identify a number of interesting relations, and there is also human feedback after each iteration of the system in the form of ratings on some newly extracted facts.

Some systems have been designed to learn to structure extraction patterns. For instance, PATTY [15] exploits types (*person*, *location*, etc.) defined for some named entities in manually curated repositories such as YAGO to learn that, for instance, the extraction pattern *<person> winner of <award>* implies the pattern *<person> nominated for <award>*. In other systems, the types of entities or relations in texts emerge from the acquired facts thanks to clustering. A convincing example of this is the WEBRE system [16] in which *itemsets* emerge from a collection of untyped facts, such as {*marijuana, caffeine, ...*}, {*cause, result in, ...*}, {*insomnia, emphysema, ...*}.

Such realizations lead to interesting new applications. For instance, in [17], the authors describe a tool primarily aimed at data scientists, allowing them to explore a large collection of documents, thanks to structured extraction patterns. This is useful for rapidly designing a specific extractor, e.g. a tool for mining architects in texts.

Despite these developments, there are still a number of issues that point to a need for better technology, as we show later in this study. First, extraction toolkits are prone to errors, and better extractors must be learned (see for instance [18] for some possible directions). Second, many (if not most) facts acquired are either uninformative and/or anaphoric (e.g. (*she, continues, her study*)). While anaphoric facts may be partially sanitized by coreference resolution, measuring the informativeness of a fact is still an unresolved issue.

3 Partial Adaptation of REVERB to French

The original REVERB Open IE system [11] proceeds in two phases, both of which are language-dependent.

1. The first stage reads POS-tagged and NP-chunked sentences and produces a (possibly empty) set of extraction triples (*arg1, r, arg2*), e.g. (*President Obama, gave a talk at, the White House*). To be extracted, a triple must satisfy a few constraints. A syntactic constraint stipulates that a relation *r* must match a regular expression based on parts of speech. A lexical constraint learned on a large corpus removes overspecified relations (e.g. *are only interested in part of the solution for*). If a relation is located, noun phrases to its right and left must also be present and valid.
2. A second stage uses a logistic regression classifier in order to filter out dubious or uninformative triples. REVERB's authors selected 19 features that are used to build this confidence function crucial to weed out uninformative and incoherent extractions (at the cost of recall).

We wanted to take advantage of the extraction engine already provided by REVERB for our study. In our case, however, we did not implement the French equivalents of the lexical constraint and the classifier. We conjectured that the filtering nature of many of the manipulations made on these French triples in this experiment would naturally eliminate some of the noise.

Thanks to the high quality of its API, REVERB lends itself very well to a "translation" in another language. For French, we had to make the following modifications.

- The preprocessing steps relying on Apache OpenNLP were adapted to use French statistical models for sentence segmentation, word tokenization, part-of-speech tagging and noun-phrase chunking. These models are trained on a large generic corpus [19] and are freely available on the web[3] for the OpenNLP framework.
- The empirical POS-based regular expression at the heart of relation extraction was changed to the one shown in Fig. 1, which is the result of our own attempts to capture as many relations as possible on a small development set. Aside from the fact that the French POS tagset differs slightly from the English one, a noteworthy difference is the presence of clitics in this pattern. Indeed, clitics frequently occur in French verb phrases, e.g. *ils s'y sont revus* (*they saw each other there*), whose POS tags are CLI (*ils*) CLI (*s'*) CLI (*y*) V (*sont*) PP (*revus*). Another distinction is the absence of nouns and determinants in the expression, which greatly reduces the need for the afore-mentioned lexical constraint present in the English REVERB, at the cost of missing some French verb phrases containing these elements (e.g. *faire partie de – be part of*). We felt that the added precision was worth this alteration (Fig. 2).

ADV? CLI* V PP? ADV? PP? INF? ADV? PREP?
INF = (VINF | PREP VINF)
ADV = *adverb*, CLI = *clitic*, V = *verb*, PP = *past participle*, VINF = *infinitive*

Fig. 1. Regular expression used in the REVERB extraction engine for French. The special symbols ? and * indicate respectively "once or not at all" and "zero or more times".

Dans la première partie de Surveiller et punir, Michel Foucault décrit le rôle politique et social du supplice durant l'époque qui précède les réformes pénales de l'âge classique.
(*Michel Foucault, décrit, le rôle politique*)
(*l'époque, précède, les réformes pénales de l'âge classique*)

Fig. 2. A sample sentence extracted from the `erudit` corpus yields two extracted triples.

4   Triple Extraction

4.1   Corpora

We extracted triples from two distinct French corpora, called `wiki` and `erudit`. The `wiki` corpus is a text serialization of all French Wikipedia articles as of June 2014 – 1.5 million articles, in total[4]. The `erudit` corpus is derived from the online collection of scholarly and cultural journals curated by the Érudit Consortium, which consists of 158 journals, mostly in the humanities[5]. We extracted the raw text from 19k XML documents in French, a task made easy by the principled tagging effort carried out on these documents by the Érudit team.

---

[3] sites.google.com/site/nicolashernandez/resources/opennlp.

[4] http://rali.iro.umontreal.ca/rali/en/wikipedia-dump.

[5] http://erudit.org/revue/.

The statistics for these corpora are shown in Table 1. It is worth noting that, beside their dissimilar domains, the two corpora also differ sharply in their nature by their average sentence length, most likely due to the comparative verbosity of scholarly papers that make up the `erudit` corpus.

Table 1. Corpus statistics for the Wikipedia corpus and the corpus derived from Érudit.

Corpus	Domain	Docs	Sentences	Tokens	Forms	Tokens/Sent
`wiki`	Generic	1.5 M	31.1 M	668 M	28.7 M	20
`erudit`	Humanities	19 k	2.8 M	96 M	2.8 M	34

4.2   Extraction of Triples

We carried out the extraction of triples (*arg1*, *r*, *arg2*) using the modified version of ReVerb described in Sect. 3. We completed the extraction process by lemmatizing the verbs present in *r*. Therefore, the following discussion only concerns verbs in their infinitive form. Table 2 shows the extraction statistics.

Table 2. Extraction statistics for corpora `wiki` and `erudit`. We only use triples without pronouns in this study, losing about a third of the original triples. The remaining statistics indicate the number of different relations, arg1s and arg2s found in these filtered triplets.

Corpus	Raw triples	Triples w/o pronouns	Relations	arg1s	arg2s
`wiki`	30.4 M	20.8 M	1.2 M	7.2 M	7.4 M
`erudit`	4.7 M	3.1 M	0.4 M	1.3 M	1.4 M

The triples in both corpora follow a Zipfian distribution. For `wiki`, the frequency of a triple can be approximated by $freq = 10453 \times rank^{-0.716}$, with $R^2 = 0.997$, although the three most frequent triples are overrepresented. They have a frequency of about 30k each and are the product of a Wikipedia template on demographics. The most frequent reads (*The evolution in population, is known, throughout*).

The most frequently occurring relations in corpus `erudit` are *être, avoir, faire, devenir* (*be, have, do, become*), involved in 17 % of all triples extracted. The arguments *arg1* and *arg2* are dominated by pronouns. For both corpora, the ten most frequent *arg1*s are all pronouns. Since these anaphora render their triplets uninformative, we filtered them using a simple blacklist, losing a third of the raw triples.

5   Classification of Entity Instances

In this study, we attempt to classify **entity instances** found in extracted triples (e.g. *Michel Foucault*) into **entity types** (e.g. *auteur – author*). Entity instances are not limited to traditional named entities. For example, we would also like to classify an *article* as an *étude* (*scientific study*). An additional challenge stems from the fact that we use the large `wiki` corpus to define the entity types and then proceed to classify instances found in triples extracted from the `erudit` corpus.

5.1 Selection of Entity Types

We start by defining a set of entity types in a loosely supervised way. We filter the triples (*arg1*, *r*, *arg2*) from `wiki` by keeping only those that satisfy the constraints that *r* must be the verb *être* (*to be*) and *arg2* must contain a common noun *t*. All such elements *t* whose frequency is greater than an empirically devised threshold of 1000 are kept and constitute the set *T* of entity types considered in this study.

The set *T* contains 358 types. The five most frequent types are *commune* (*municipality*), *espèce* (*species*), *village*, *film* and *ville* (*city*), presumably reflecting the relatively large number of Wikipedia articles devoted to these topics. Some topics have overlapping meanings, like *commune* and *ville*. Other such overlapping topics comprise *philosophe* (*philosopher*), *auteur* (*author*) and *écrivain* (*writer*). We did not regroup or filter out these overlapping categories. On a related note, we consider entity types as non-hierarchical: even when a type t_1 semantically subsumes a type t_2, they are considered completely distinct.

5.2 Characterizing Entity Types by Their Relations

Once the set *T* of entity types is built, we characterize each $t \in T$ by the relations where *t* is involved, reasoning that different instances of a given type *t* must have similar relations, and that these relations should differ from those involved with instances of a different type.

We therefore seek to build a **relation profile** P_t for each *t*. At its simplest, this profile will include relations (i.e. verbs) and their associated count. We find the relations of interest in the triples extracted from the `wiki` corpus.

We start by identifying a set of instances for each *t*. We gather all *arg1*s from triples (*arg1*, *r*, *arg2*) where *r* is the verb *être*, *arg2* contains the common noun *t*, and *arg2* is not a hapax. This works reasonably well, but instances are often contaminated with ubiquitous instances. For example, a third of all 358 topics contain the instance *son père* (*his father*). Manual examination revealed that anaphora is to blame for this phenomenon, since these pervasive instances are associated with multiple types (as in *his father was a physicist* and *his father was a sportsman*). To compensate for this, we removed entity instances appearing in more than 2 % of the 358 topics, an upper limit we deemed "reasonable" on the number of types an instance can belong to. There are 237 instances per type on average (min: 1, max: 4953). Table 3 shows a random sample of the instances identified in the `wiki` corpus for the types *auteur* (*author*) and *période* (*period of time*). A manual evaluation of half a dozen instance lists reveals a 90 % precision. Errors vary from slight inaccuracies in classification (e.g. while *Jean-Florian Collin* has written a few books, he is primarily known as an architect and politician, not as an *author*) to flagrant extraction issues (e.g. *le contexte des noms de domaine* (*domain name context*) is not a *period of time*).

Table 3. Instances for two types: *auteur* (*author*) and *période* (*period of time*).

Instances for *author*	Instances for *period of time*
Farid Boudjellal	l'estive (*grazing period*)
Jean-Florian Collin	1890 (*1890*)
Richard Matheson	le contexte des noms de domaine (*domain name*
Bernard Yaméogo	*context*)
Hanns Heinz Ewers	la première restauration (*First Restoration*)
Thomas Norton	le chalcolithique (*Copper Age*)
… 416 instances overall	… 152 instances overall

From the relatively precise list of instances for each type *t*, we are able to inspect triples containing these instances at the *arg1* position and gather all corresponding relations in order to build a relation profile P_t.

For the entity type *auteur* (*author*), the 10 most frequent relations gathered are *be, do, have, write, become, give, emeritus, take, put, run* and *say*.[6] A systematic error during part-of-speech tagging gives rise to the erroneous *emeritus*, where the French word *émérite* is mislabeled as a verb rather than as an adjective. For *période* (*period of time*), the most frequent relations are: *be, mark, see, follow, have, do, become, put, av* and *know*. Here, *av* is also due to an unfortunate tagging error.

We distinctly face a situation where uninformative relations (*be, do, have*, etc.) appear in all profiles. At the same time, more type-specific relations emerge (*write, say*, for an *author*; *mark* and *follow* for a *period of time*).

To get a better sense of the quality of these relation profiles[7], we add a tf–idf score to each relation in a given profile P_t. We computed tf–idf by considering each profile as a document, populated with words that correspond to the relations. In other words, for each relation in a given P_t, tf is the relation count and idf is proportional to the inverse of the number of profiles containing the relation. Table 4 shows an excerpt of the profile for the entity type *auteur* (*author*).

5.3 Extracting Entity Instances from `erudit` to Create Dev and Test Sets

Our goal is to label entity instances with the correct entity type. While these types were extracted from the corpus `wiki`, we select instances from the corpus `erudit`, in an effort to assess how well the entity types from one corpus generalize to another one, with a different writing style (see Sect. 4.1).

We performed triple extraction with the adapted REVERB previously described on `erudit`. We then selected all *arg1*s whose frequency is greater than 50. We extracted their relation profiles as explained in Sect. 5.2.

---

[6] We translate the French relations for the sake of clarity.

[7] We also compute these scores for classification purposes (see Sect. 5.3).

Table 4. Relation profile excerpt for the entity type *auteur* (*author*). For each relation, a translation is provided for convenience. Relations are listed in decreasing order of tf–idf score. The relation *gothique* (*gothic*) is due to a POS-tagging error.

author		period of time	
Relation	**Frequency**	**Relation**	**Frequency**
run	293	*gothic*	364
diverge	38	*happen*	220
report	197	*mark*	1709
speak	239	*start*	218
author	95	*follow*	1405
write	635	*change*	95
969 relations	**$\Sigma = 33671$**	**1029 relations**	**$\Sigma = 41954$**

To create a data set, we manually labeled the 120 instances featuring the most relations in their relation profile. We discarded instances that were ambiguous (e.g. the last name *Tremblay* is not enough to identify the entity), were the result of extraction errors (e.g. an adjective mislabeled as a noun) or simply did not belong to any types (e.g. *snow*, *orientation*). Each instance received a label consisting of the entity types extracted in Sect. 5.2 to which it belongs. For example, *Michel Foucault* received the labels *philosopher*, *author*, *writer* (2.85 labels per instance on average).

The classification algorithms described in Sect. 5.4 were tuned on 20 of these instances while the other 100 constituted the test set. We did not need a large number of development instances, since our classifiers do not require the production of a model. The test set was used to assess the impact of a few parameters and to fine-tune some of them.

5.4 Classifiers

Our goal is to classify entity instances (e.g. *Michel Foucault, article*) extracted from erudit into entity types (e.g. *author, scientific study*) characterized by the relation profile observed in the corpus wiki. We tried 4 different approaches to find the closest relation profile for a given instance. Let $P_t = \{r_{t1}, r_{t2}, \ldots, r_{tm}\}$ be the set of all m relations for a given type t. Let $P_i = \{r_{i1}, r_{i2}, \ldots, r_{in}\}$ be the set of all n relations for a given instance i. Let sim be a similarity function between two profiles. We attribute a type t^* to a given instance i by finding $t^* = \operatorname*{argmax}_t \ \text{sim}(P_t, P_i)$. The set of all possible types t includes all the types manually identified during labeling (34 types) to which we added 16 other types randomly selected from the set T of all 358 types identified in Sect. 5.2, for a total of 50 possible types available to each classifier.

For each similarity metric, we added a parameter λ that constrains the number of relations considered in each profile when computing a similarity. The value λ specifies that only the λ-most frequent relations in P_t and P_i should be used. The others are ignored. This allows the algorithm to focus on the most represented relations in each profile. The optimal λ for a given similarity metric is found by exhaustive search on the development dataset. We also investigated results without any such filter.

jaccard is our baseline and consists in computing the generalized Jaccard similarity coefficient between the two relation profiles. We first derive a frequency vector for each profile. For example, for P_t, we obtain $\mathbf{v}_t = \langle freq(r_{t1}), freq(r_{t2}), \ldots \rangle$, where $freq(r_{ti})$ is the count of the relation r_{ti} for the profile t. The similarity can then be computed using the following formula:

$$jaccard(\mathbf{v}_t, \mathbf{v}_i) = \frac{\sum_j \min(\mathbf{v}_t[j], \mathbf{v}_i[j])}{\sum_j \max(\mathbf{v}_t[j], \mathbf{v}_i[j])}$$

cos is the cosine similarity between two profiles, and is computed using the frequency vectors $\mathbf{v}_t$ and $\mathbf{v}_i$ discussed above. We also tried a variant **cos-bin** where, instead of the frequencies of the relations, the vectors are encoded with 1 (the relation is present in profile) or 0 (the relation is absent).

tfidf makes use of the tf–idf scores we introduced in Sect. 5.2. The similarity function here is comparable to the information retrieval scenario where the relations in P_i constitute the query and the different P_t are each a document in a collection. The "most relevant" P_t is therefore the most similar. The similarity score for a given P_i is the sum of the tf–idf scores for each relation found in P_t.

Finally, **kl** is a comparison of the relation distributions in P_t and P_i using the Kullback–Leibler divergence $D_{KL}(Q_t \parallel Q_i)$, where Q_t and Q_i are the probability distributions (relative frequencies) of relations in P_t and P_i respectively. KL divergence does not handle 0 s in these distributions, so we smooth by replacing them by a small value ε whose value was tuned on the development set. We also experimented by throwing out the dimensions with 0 values, with disappointing results (not presented here).

5.5 Results and Error Analysis

The classification errors for all similarity metrics are presented in Table 5. An instance is considered misclassified if the closest type t^* returned by a given similarity metric is not part of the labels attributed manually. The results are unsatisfactory for all metrics, except the KL divergence, with a classification error rate of 34 % on the test set, lower than random (92 %) and the Jaccard baseline (55 %). The metrics generalize quite well from the development set to the test set.

The two authors of this study manually inspected the results for **kl** to identify the kinds of errors the similarity function led to. We identify two kinds of errors: "hard" and "soft" errors (for lack of better terms). Hard errors occur when an entity is unambiguously mislabeled, e.g. *Socrates* is a *municipality*. Soft errors arise when the predicted type is not entirely incompatible with the instance to label, e.g. *United States* is a *group*. This is admittedly a subjective exercise, however the reader can judge for himself by examining a sample of this partition in Table 6. Overall, out of 34 errors, we found that 20 were hard and 14 were soft. Had we tolerated the latter, the classification

Table 5. Classification results for 5 similarity metrics on the development set and the test set. The **kl** algorithm (Kullback–Leibler divergence) yields the best results (in gray). The **random** classifier picks a solution at random while **most frequent** always picks the most frequent label (*city*). The latter two are provided for comparison.

Similarity metric	Classification error (%)	
	dev ($n = 20$)	**test** ($n = 100$)
random	90 %	92 %
most frequent (*city*)	90 %	91 %
jaccard ($\lambda = 50$)	55 %	55 %
cos ($\lambda = 50$)	60 %	64 %
cos-bin ($\lambda = 50$)	55 %	58 %
tfidf ($\lambda = 500$)	50 %	53 %
kl ($\lambda = 100$)	30 %	34 %

Table 6. A sample of a few instances, their manual labels and the predicted type by the kl algorithm. "Hard" classification errors are noted with a double dagger (‡), "soft" ones with a †.

Instance	Type labels	Predicted type (kl)
1960	*year, period of time*	*period of time*
Aragon	*artist, author, writer*	*author*
article	*scientific study, book, compilation*	*scientific study*
Belgium	*country, place, places, toponym*	*city*‡
French Canadians	*people, group*	*characters*†
Germany	*country, place, places, toponym*	*city*‡
Health Canada	*organisation, association, organism, group*	*physician*†
Michel Foucault	*philosopher, writer, author*	*author*
prime minister	*minister, president, master*	*minister*
Socrates	*philosopher, writer, author*	*city*‡
text	*scientific study, book, compilation*	*book*

error for **kl** on the test set would have fallen to 20 %. These 20 hard errors contain 10 misclassifications where a country (*Canada, Mexico*) is classified as a *city*.

The parameter λ significantly affects the performance of the kl similarity metric. There seems to be an optimal value in the interval [50, 1000] at around 40 % classification error, outside of which the performance is poor. Both the development and test set exhibit the same behavior.

6 Discussion

One of the goals of this paper was to attempt open information extraction (OIE) in French and assess the difficulties encountered while doing so. The adaptation of REVERB went smoothly, partly because there are drop-in French replacements for the POS and chunking statistical models the software uses, and partly because its API is expertly written. We also opted not to adapt all of REVERB's filters to French, because we favoured recall over precision in our architecture. However, we feel that implementing the rest would be straightforward should we need it in the future.

Like most OIE approaches, the problem of uninformative and ambiguous triples is significant. We lose a third of extracted triples to pronominal anaphora alone, which amounts to 10M triples for the corpus `wiki`. This highlights the need for a robust anaphora resolver. For French, a recent study on a commercial-grade grammar checker [20] shows that 70 % of these anaphora could be resolved successfully, a possible addition of 7M triples of information in our case. Naturally, the figure of 10M triples lost is a minimum, since it does not take into account other types of anaphora (e.g. *his father*). However, our system behaved reasonably well in the face of these latter problems, thanks to simple frequency thresholds akin to idf (inverse document frequency), reasoning that ubiquitous instances are bound to be non-specific and uninformative.

The second goal of this paper was to explore whether it was possible to extract information from a generic corpus (`wiki`) and use it to infer new knowledge in a different, domain-specific corpus (`erudit`) through the analysis of OIE's resulting triples. We showed that it is indeed possible to identify and characterize entity types by the relations their respective instances are associated with. It then becomes possible to put these profiles to good use and classify instances extracted from the other corpus, for two thirds of these instances. To our knowledge, in this context, this approach is original. It does suffer however from the fact that the instances to classify must be relatively frequent (in order to gather enough information on them). The system described here would be hard-pressed to associate a hapax instance to an entity type, for instance. Moreover, establishing "relation profiles" proves sensitive to systematic extraction errors, notably those committed during part-of-speech tagging. A tagging error that mislabels an adjective for a verb in a specific context (like *gothic* preceded by *author*) is bound to create significant artefacts in relation profiles, since the latter are designed to gather just such systematic specificities, whether they are linguistically motivated or the result of an extraction problem.

There is room for improvement when considering the figure of 34 % of classification error reported here. We identify some possible solutions above. However, the same statistic also shows that there is definite potential in the idea of exploiting the knowledge derived by OIE from a generic corpus and then applying it to a stylistically and thematically different collection of texts.

References

1. Sowa, J.F.: The challenge of knowledge soup. In: epiSTEME-1, pp. 55–90 (2004)
2. Ferrucci, D.A.: IBM's Watson/DeepQA. In: Proceedings of the 38th Annual International Symposium on Computer Architecture. ACM (2011)
3. Poon, H., Quirk, C., DeZiel, C., Heckerman, D.: Literome: PubMed-scale genomic knowledge base in the cloud. Bioinformatics **30**, 2840–2842 (2014)

4. Bollacker, K., Evans, C., Paritosh, P., Sturge, T., Taylor, J.: Freebase: a collaboratively created graph database for structuring human knowledge. In: Proceedings of the 2008 ACM SIGMOD International Conference on Management of Data, pp. 1247–1250 (2008)
5. Suchanek, F.M., Kasneci, G., Weikum, G.: Yago: A core of semantic knowledge. In: 16th International World Wide Web Conference (WWW 2007) (2007)
6. Lehmann, J., Isele, R., Jakob, M., Jentzsch, A., Kontokostas, D., Mendes, P.N., Hellmann, S., Morsey, M., van Kleef, P., Auer, S., Bizer, C.: DBpedia - a large-scale, multilingual knowledge base extracted from Wikipedia. Semantic Web Journal. **6**, 167–195 (2015)
7. Bizer, C., Heath, T., Berners-Lee, T.: Linked data - the story so far. Int. J. Semant. Web Inf. Syst. **5**, 1–22 (2009)
8. Carlson, A., Betteridge, J., Kisiel, B., Settles, B., Hruschka Jr., E.R., Mitchell, T.M.: Toward an architecture for never-ending language learning. In: Proceedings of the Twenty-Fourth Conference on Artificial Intelligence (AAAI 2010) (2010)
9. Navigli, R., Ponzetto, S.P.: BabelNet: the automatic construction, evaluation and application of a wide-coverage multilingual semantic network. Artif. Intell. **193**, 217–250 (2012)
10. Banko, M., Cafarella, M.J., Soderland, S., Broadhead, M., Etzioni, O.: Open information extraction from the web. In: Proceedings of the 20th International Joint Conference on Artificial Intelligence, Hyderabad, India, pp. 2670–2676 (2007)
11. Fader, A., Soderland, S., Etzioni, O.: Identifying relations for open information extraction. In: Empirical Methods in Natural Language Processing, pp. 1535–1545 (2011)
12. Wu, F., Weld, D.S.: Open information extraction using Wikipedia. In: 48th Annual Meeting of the Association for Computational Linguistics, pp. 118–127 (2010)
13. Mausam, M.S., Bart, R., Soderland, S., Etzioni, O.: Open language learning for information extraction. In: Joint Conference on Empirical Methods in Natural Language Processing and Computational Natural Language Learning, pp. 523–534 (2012)
14. Dong, X., Gabrilovich, E., Heitz, G., Horn, W., Lao, N., Murphy, K., Strohmann, T., Sun, S., Zhang, W.: Knowledge vault: a web-scale approach to probabilistic knowledge fusion. In: Proceedings of the 20th ACM SIGKDD International Conference on Knowledge Discovery and Data Mining, pp. 601–610. ACM (2014)
15. Nakashole, N., Weikum, G., Suchanek, F.: PATTY: a taxonomy of relational patterns with semantic types. In: Joint Conference on Empirical Methods in Natural Language Processing and Computational Natural Language Learning, pp. 1135–1145 (2012)
16. Min, B., Shi, S., Grishman, R., Lin, C.: Ensemble semantics for large-scale unsupervised relation extraction. In: EMNLP and CoNLL, pp. 1027–1037 (2012)
17. Akbik, A., Visengeriyeva, L., Kirschnick, J., Löser, A.: Effective selectional restrictions for unsupervised relation extraction. In: Proceedings of the Sixth International Joint Conference on Natural Language Processing, pp. 1312–1320 (2013)
18. Koch, M., Gilmer, J., Soderland, S., Weld, S.D.: Type-aware distantly supervised relation extraction with linked arguments. In: Empirical Methods in Natural Language Processing (EMNLP), pp. 1891–1901 (2014)
19. Hernandez, N., Boudin, F.: Construction automatique d'un large corpus libre annoté morpho-syntaxiquement en français. In: Traitement Automatique des Langues Naturelles (TALN), Sables d'Olonne, France (2013)
20. Pironneau, M., Brunelle, É., Charest, S.: Pronoun anaphora resolution for automatic correction of grammatical errors (Correction automatique par résolution d'anaphores pronominales) [in French]. In: Proceedings of TALN 2014 (Volume 1: Long Papers), vol. 1, pp. 113–124 (2014)

A Novel Genetic Algorithm for the Word Sense Disambiguation Problem

Wojdan Alsaeedan and Mohamed El Bachir Menai[(✉)]

Department of Computer Science, CCIS, King Saud University, P.O. Box 51178,
Riyadh 11543, Saudi Arabia
wojdan@ccis.imamu.edu.sa, menai@ksu.edu.sa

Abstract. Word sense disambiguation (WSD) is a task in natural language processing, which asks to identify the appropriate sense of a word according to a particular context. Several approaches were investigated to tackle the WSD problem, including genetic algorithms. In this paper, we propose a new genetic algorithm, called GAWSD, that benefits from part-of-speech tagging, domain knowledge, and gloss enrichment to find a sense to a target word. The performance of the algorithm was evaluated on fine-grained and coarse-grained standard corpora. The results show that GAWSD outperformed the best known algorithms on the fine-grained corpus. This result sets GAWSD as a competitive algorithm for WSD.

1 Introduction

Word Sense Disambiguation (WSD) is a challenging problem in natural language processing (NLP). WSD refers to the task that automatically assigns a sense, selected from a set of pre-defined word senses, to an instance of a polysemous word in a particular context. WSD is an AI-complete (Artificial Intelligence-complete) problem, which is identical to an NP-complete problem in complexity theory. WSD is an intermediate step in most NLP applications, such as machine translation, information retrieval, and information extraction.

Approaches to solve the WSD have four main components: word sense selection, external use of knowledge sources, context representation, and the method for sense disambiguation. The selection of word senses is concerned with the sense inventory of a given word. The issues that arise when working on sense distinctions are the presentation of a low-level model of detail (fine-grained sense) or a high-level model of detail (coarse-grained sense) and the organization of senses in a dictionary. The external knowledge sources contain a repository of data that associate words with their senses. The two main kinds of external knowledge sources used are structured (e.g. thesauri, machine-readable dictionaries, ontologies) and unstructured resources (e.g. corpora). WordNet (WN) [6] is an example of a lexical database which has been widely used for the WSD. In WN, the words are grouped into sets of synonyms (called synsets). For each synset, there is a gloss which is the textual definition, and lexical and semantic relations that exist between pairs of synsets (e.g. antonymy, hypernymy,

© Springer International Publishing Switzerland 2016
R. Khoury and C. Drummond (Eds.): Canadian AI 2016, LNAI 9673, pp. 162–167, 2016.
DOI: 10.1007/978-3-319-34111-8_21

hyponymy, holonymy relations). Ponzetto and Navigli [15] have automatically generated relations between noun synsets in WN from Wikipedia, called Word-Net++ (WNPP). Magnini and Cavaglia [9] developed WordNet domain (WND) labels for synsets, in which each synset in WN has at least one domain label (e.g. Food, Architecture, Sport). The context representation component consists in converting unstructured input text into a structured format to make it suitable for automatic methods. The key distinction between WSD methods depends on the amount of knowledge and supervision these methods use. They can be classified into knowledge-based methods, machine learning-based methods, and other methods such as domain-driven disambiguation and metaheuristics algorithms.

Various metaheuristic algorithms were investigated for the WSD problem. Gelbukh et al. [7] applied a GA to solve it as an optimization problem. Decadt et al. [5] proposed GAMBL, a word expert approach using a GA to solve WSD. The feature selection and optimization of the parameters of the algorithm are performed jointly using a GA. Zhang et al. [18] proposed an unsupervised GWSD algorithm and used a GA to maximize the semantic similarity of words. Menai [10] proposed a GA for the WSD and applied it to an Arabic corpus. Alsaeedan and Menai [1] proposed a self-adaptive GA for the WSD problem with an automated tuning of its crossover and mutation operators. Nguyen and Ock [13] introduced a method for WSD using an ant colony optimization algorithm (TSP-ACO).

In this paper we present a new genetic algorithm (GA) for solving the WSD problem, called GAWSD. This algorithm makes use of part-of-speech tagging, domain knowledge, and enrichment of synset glosses. We present experimental results of its performance in comparison to others and show its effectiveness. The rest of the paper is organized as follows. The next section describes the proposed GAWSD algorithm. Section 3 reports and discusses the preliminary experimental results. Finally, Sect. 4 concludes the paper.

2   Proposed GA for the WSD

We introduce a new GA for the WSD, called GAWSD, which improves the performance of a recently introduced self-adaptive GA to solve this problem, called $SAGA_{Elitist}$ [1]. This algorithm encompasses an automated tuning of the probabilities of its crossover and mutation operators. Uniform crossover and mutation operators are used with high probabilities to promote exploration, while single-point crossover and mutation operators are used with low probabilities to promote exploitation. The probabilities of crossover and mutation operators are dynamically adapted depending on the largest fitness of the mating parents f', and the largest fitness f_{max} and the mean fitness $\bar{f}$ in the current population (see [1] for more details). $SAGA_{Elitist}$ uses an elitist survivor selection scheme, and a stochastic universal sampling (SUS) as a parent selection method. Algorithm 1 outlines the main steps of the proposed algorithm GAWSD.

To reduce the number of synsets of words in $Text$ examined by the algorithm, part-of-speech tagging (POS tagging or POST) is performed to only select the

---

Algorithm 1. GAWSD

---

input : $Text, TargetWord, Window_{size}, Population_{size}$
output: $Synset_{best}$

$Text \leftarrow$ **POST**$(Text)$ $Text \leftarrow$ **RemoveStopWords**$(Text)$;
$BagWords \leftarrow$ **SelectWords**$(Text, Window_{size}, TargetWord)$;

forall the $W_j \in BagWords$ **do**
$\quad$ | $\quad Synsets(W_j) \leftarrow$ **WN**(W_j);
$\quad$ | $\quad Synsets(W_j) \leftarrow$ **EnrichGlossSynset**$(Synsets(W_j))$
end

$S_{best} \leftarrow$ **SAGA**$_{Elitist}(Population_{size}, Synsets(W_i)_{i=0...Window_{size}})$;
$Synset_{best} \leftarrow$ **Decode**(S_{best});
Return$(Synset_{best})$

---

word synsets that are related to its POS, instead of processing all of them. For instance, according to WN, the word "drinking" has seven synsets. Five of them refer to the synsets of the verb "drinking", and the two remaining ones refer to the synsets of the noun "drinking". Stop words are then removed, and the synsets of neighbor words $Synsets(W_j)$ (W_j in $BagWords$) of $TargetWord$ within a $Window_{size}$ are retrieved from WN. For fine-grained glosses (or short glosses), Nguyen and Ock [13] showed that a relatedness measure might return a zero overlap between synset glosses of related words. To overcome this issue, we propose to enrich the short glosses of synsets with the glosses of related synsets retrieved from WNPP, and the glosses of synsets in the same cluster available in the synset-clustered inventories of WN. WNPP provides relations, extracted from Wikipedia, between noun synsets in WN. The set of all these synsets is then used by the algorithm SAGA$_{Elitist}$ [1] to find the best individual S_{best}, and hence the best sense $Synset_{best}$ for $TargetWord$.

The fitness function is measured by the following semantic relatedness measure. Given two words w_1 and w_2 and their respective synsets $Synsets(w_1)$ and $Synsets(w_2)$, for each pair of synsets $S_1 \in Synsets(w_1)$ and $S_2 \in Synsets(w_2)$, $gloss(S_i)$ represents the bag of words corresponding to the definitions of the synset S_i. The semantic relatedness measure is given by: $Score_{BPDomain}(S_1, S_2) = Relate_{B\&P}(S_1, S_2) + Relate_{Domains}(S_1, S_2)$, where $Relate_{B\&P}(S_1, S_2)$ is the semantic relatedness measure proposed by Banerjee and Pedersen [2], called extended Lesk measure, and $Relate_{Domains}(S_1, S_2)$ represents the related domains. $Relate_{B\&P}(S_1, S_2)$ is given by: $Relate_{B\&P}(S_1, S_2) = \sum_{S':S_1 \xrightarrow{rel} S'} \sum_{S'':S_2 \xrightarrow{rel} S''} |gloss(S') \cap gloss(S'')|$, where the synset S' is either S_1 itself or has a relation rel with S_1, and the synset S'' is either S_2 itself or has a relation rel with S_2. $Relate_{Domains}(S_1, S_2)$ is given by: $Relate_{Domains}(S_1, S_2) = |domains(S_1) \cap domains(S_2)|$, where $domains(S_i)$ is set of domains corresponding to synset S_i. In this work, the domain information is extracted from WND.

3   Experimental Results

The algorithm GAWSD was implemented by using Java programming language, and extJWNL, a Java API to WN. POST was implemented by the Stanford Log-Linear POS tagger[1]. The default English stop words[2] were removed. Glosses and associative relations were extracted from WN and WNPP. Cluster relations were extracted from the coarse-grained sense inventory used in the coarse-grained English all-words task of SemEval-2007. The corpora considered include SemEval2007: task#17_Subtask#3; fine-grained English all words (S07FGAW), and SemEval2007: Task#7; coarse-grained English all words (S07CGAW). The algorithm's parameters were set as follows: $Population_{size} = 50$, $Window_{size} = 4$, and a number of 100 generations as the exit criterion. All the results were averaged over 10 runs. They are reported in terms of average precision $\overline{P}(\%)$, average recall $\overline{R}(\%)$, and average F-measure $\overline{F}(\%)$.

Table 1. Impact of POST, domain information (Domain), and gloss enrichment (GlossEn) on the performance of GAWSD. The best values are stressed in bold.

Corpus	GAWSD variant	$\overline{P}(\%)$	$\overline{R}(\%)$	$\overline{F}(\%)$
S07FGAW	Domain	34.84	38.81	36.72
	POST	39.27	42.62	40.88
	POST & Domain	39.64	43.09	41.29
	POST & Domain & GlossEn	**64.31**	**66.04**	**65.16**
S07CGAW	Domain	75.35	78.20	76.75
	POST	75.39	78.26	76.79
	POST & Domain	75.40	78.23	76.79
	POST & Domain & GlossEn	**77.59**	**79.58**	**78.57**

We evaluated the impact of POST, domain information (Domain), and gloss enrichment (GlossEn) on the performance of GAWSD. Table 1 shows that the best performance of GAWSD on the two corpora was reached when POST, domain information, and gloss enrichment altogether were performed. The most significant improvement was achieved on the fine-grained corpus (S07FGAW). This result consolidates the idea of improving the quality of WSD solutions by enriching short glosses of synsets with other related glosses. The effect of gloss enrichment is less significant in case of the coarse-grained corpus (S07CGAW). We compared GAWSD against the best performing algorithms, which participated in SemEval-2007 competition. The comparison results are shown in Table 2. GAWSD outperformed the rival algorithms on the fine-grained English all words corpus S07FGAW in terms of both average precision and recall. However, The performance of GAWSD on the coarse-grained English all words corpus S07FGAW, in terms of average F-measure, is 4.64 % below the performance of the algorithm of Navigli and Velardi [12].

---

[1] http://nlp.stanford.edu/software/tagger.shtml.
[2] http://www.ranks.nl/stopwords.

Table 2. Comparison results. The best values are stressed in bold.

Corpus	Rival algorithm	$\overline{P}(\%)$	$\overline{R}(\%)$	$\overline{F}(\%)$
S07FGAW	Cha07 [4]	58.70	58.70	58.70
	Cai07 [3]	57.60	57.60	57.60
	Tra07 [16]	59.10	59.10	59.10
	Mih07 [11]	58.30	58.30	58.30
	GAWSD	**64.31**	**66.04**	**65.16**
S07CGAW	Nav05 [12]	**83.21**	**83.21**	**83.21**
	Cha07 [4]	82.50	82.50	82.50
	Cai07 [3]	81.58	81.58	81.58
	Nov07 [14]	81.45	81.45	81.45
	Izq07 [8]	79.55	79.55	79.55
	$SAGA_{Elitist}$ [1]	75.51	78.77	77.11
	GAWSD	77.59	79.58	78.57

The outperforming algorithms on coarse-grained corpus are either machine learning-based or graph-based algorithms. The performance of machine learning-based algorithms relies on the availability of a large number of annotated/ unannotated training corpora. Graph-based algorithms represent semantic relations extracted from knowledge sources. Their performance depends on both the type of graph and knowledge sources used. These algorithms also rely on a large context of the $TargetWord$ ($Window_{size} \geq 20$), which contrasts with GAWSD. Indeed, the proposed algorithm can reach its best performance given a small $Window_{size}$ of 4.

4   Conclusion and Future Work

In this paper, we have presented a new genetic algorithm GAWSD to solve the WSD problem. GAWSD exploits POST, domain information, and enrichment of synset glosses by other related glosses, to search for the best possible sense to a target word. The preliminary experimental evaluation shows that GAWSD outperformed the best state-of-the-art algorithms on a fine-grained corpus. However, it was outperformed by most of them on a coarse-grained corpus. In the future, we plan to examine potential solutions to improve the performance of GAWSD on coarse-grained corpora,and investigate the impact of combining different fitness evaluation functions on its performance.

References

1. Alsaeedan, W., Menai, M.E.B.: A self-adaptive genetic algorithm for the word sense disambiguation problem. In: Ali, M., Kwon, Y.S., Lee, C.-H., Kim, J., Kim, Y. (eds.) IEA/AIE 2015. LNCS, vol. 9101, pp. 581–590. Springer, Heidelberg (2015)
2. Banerjee, S., Pedersen, T.: Extended gloss overlaps as a measure of semantic relatedness. In: Proceedings of the 18th International Joint Conference on Artificial intelligence, IJCAI 2003, San Francisco, CA, USA, pp. 805–810 (2003)

3. Cai, J.F., Lee, W.S., Teh, Y.W.: NUS-ML: improving word sense disambiguation using topic features. In: Proceedings of the 4th International Workshop on Semantic Evaluations, SemEval 2007, ACL, Stroudsburg, PA, USA, pp. 249–252 (2007)

4. Chan, Y.S., Ng, H.T., Zhong, Z.: NUS-PT: exploiting parallel texts for word sense disambiguation in the english all-words tasks. In: Proceedings of the 4th International Workshop on Semantic Evaluations, SemEval 2007, Stroudsburg, PA, USA, pp. 253–256 (2007)

5. Decadt, B., Hoste, V., Daelemans, W., Bosch, A.V.D.: GAMBL, genetic algorithm optimization of memory-based WSD. In: Proceedings of the Senseval-3: 3rd International Workshop on the Evaluation of Systems for the Semantic Analysis of Text, Barcelona, Spain, pp. 108–112 (2004)

6. Fellbaum, C.: WordNet: An Electronic Lexical Database: Language, Speech, and Communication. MIT Press, Cambridge (1998)

7. Gelbukh, A., Sidorov, G., Han, S.Y.: Evolutionary approach to natural language word sense disambiguation through global coherence optimization. Trans. Commun. $1(2)$, 11–19 (2003)

8. Izquierdo, R., Suárez, A., Rigau, G.: GPLSI: word coarse-grained disambiguation aided by basic level concepts. In: Proceedings of the 4th International Workshop on Semantic Evaluations, SemEval 2007, Stroudsburg, PA, USA, pp. 157–160 (2007)

9. Magnini, B., Cavaglia, G.: Integrating subject field codes into WordNet. In: Proceedings of the 2nd International Conference on Language Resources and Evaluation (LREC), Athens, Greece, pp. 1413–1418, June 2000

10. Menai, M.E.B.: Word sense disambiguation using evolutionary algorithms - application to arabic language. Comput. Hum. Behav. 41, 92–103 (2014)

11. Mihalcea, R., Csomai, A., Ciaramita, M.: UNT-Yahoo: supersenselearner: combining senselearner with supersense and other coarse semantic features. In: Proceedings of the 4th Interernational Workshop on Semantic Evaluations, SemEval 2007, Stroudsburg, PA, USA, pp. 406–409 (2007)

12. Navigli, R., Velardi, P.: Structural semantic interconnections: a knowledge-based approach to word sense disambiguation. IEEE Trans. Pattern Anal. Mach. Intell. $27(7)$, 1075–1086 (2005)

13. Nguyen, K.H., Ock, C.Y.: Word sense disambiguation as a traveling salesman problem. Artif. Intell. Rev. $40(4)$, 405–427 (2013)

14. Novischi, A., Srikanth, M., Bennett, A.: LCC-WSD: system description for english coarse grained all words task at semeval 2007. In: Proceedings of the 4th International Workshop on Semantic Evaluations, SemEval 2007, Stroudsburg, PA, USA, pp. 223–226 (2007)

15. Ponzetto, S.P., Navigli, R.: Knowledge-rich word sense disambiguation rivaling supervised systems. In: Proceedings of the 48th Annual Meeting of the ACL 2010, Stroudsburg, PA, USA, pp. 1522–1531(2010)

16. Tratz, S., Sanfilippo, A., Gregory, M., Chappell, A., Posse, C., Whitney, P.: PNNL: a supervised maximum entropy approach to word sense disambiguation. In: Proceedings of the 4th International Workshop on Semantic Evaluations, SemEval 2007, Stroudsburg, PA, USA, pp. 264–267 (2007)

17. Wu, Z., Palmer, M.: Verbs semantics and lexical selection. In: Proceedings of the 32nd Annual Meeting of the ACL 1994, Stroudsburg, PA, USA, pp. 133–138 (1994)

18. Zhang, C., Zhou, Y., Martin, T.: Genetic word sense disambiguation algorithm. In: Proceedings of the 2nd International Symposium on Intelligent Information Technology Application, IITA 2008, vol. 1, pp. 123–127 (2008)

Mining Biomedical Literature: An Open Source and Modular Approach

Hayda Almeida[1], Ludovic Jean-Louis[2], and Marie-Jean Meurs[1,3(✉)]

[1] Centre for Structural and Functional Genomics,
Concordia University, Montreal, QC, Canada
[2] NetMail, Montreal, QC, Canada
[3] Université du Québec à Montréal, Montreal, QC, Canada
meurs.marie-jean@uqam.ca

Abstract. This paper presents the ongoing development of a full-text natural language search engine for biomedical literature. The system aims to provide search on the full-text content of documents belonging to a database composed of scientific articles, while allowing users to submit their search queries using natural language. Beyond the text content of articles, the system engine also utilizes article metadata, empowering the search by considering extra information from picture and table captions. User queries can be submitted to the system in natural language, releasing the user from the burden of translating their search needs into a query language.

Keywords: Natural language processing · Information retrieval · Full-text search · Document index · Natural language query · Search engine · Biomedical literature

1 Introduction

Scientific researchers and health care practitioners heavily rely on the retrieval of biomedical documents maintained in scientific databases to support their activities. Much effort has been put into improving the retrieval of bioliterature [14,15,30]. However, bioliterature search being essentially an information retrieval task, still imposes great challenges. PubMed (http://www.ncbi.nlm.nih.gov/pubmed) is one of the most popular scientific databases, and a substantial resource for biomedical professionals. It contains over 24 million records as of January 2016. In PubMed Central (PMC) (http://www.ncbi.nlm.nih.gov/pmc/) researchers have access to 3,7 million free full-article texts (Jan. 2016), which represent a fraction of all records maintained in PubMed.

The retrieval of documents in open literature databases is a critical step for biomedical research. The retrieved results can be used as input for a variety of tasks, such as data integration [17,18], literature curation [13], and literature triage [1,12]. However, the retrieval of relevant articles is challenging for researchers using these databases. With the goal of improving the search for

© Springer International Publishing Switzerland 2016
R. Khoury and C. Drummond (Eds.): Canadian AI 2016, LNAI 9673, pp. 168–179, 2016.
DOI: 10.1007/978-3-319-34111-8_22

open scientific literature, this work is an attempt to address two issues that scientific researchers can encounter when gathering relevant literature.

First, the search process in PubMed and PMC presents limitations. PubMed makes available a large amount of records, but its search engine retrieves articles by considering only the abstract content. The PMC search engine retrieves articles considering their full text content, but it only holds a portion of all PubMed records. Second, search requests utilized to retrieve information from these databases have to be translated into query language. As query language differs from natural language, not all users are comfortable enough to translate their search needs efficiently, which makes the task of retrieving relevant data even more difficult.

2   Related Work

We present here studies conducted towards improving and supporting document retrieval in open-access scientific literature databases. Also, we present a review of approaches developed to handle complex user queries, that aim to facilitate information search, better address search needs, and improve the retrieval results. Enhancing the document retrieval process will allow to provide more useful results to various research tasks relying on the input of scientific literature search.

2.1   Scientific Database Search

A variety of methods has been studied in an effort to improve document retrieval relevance of scientific databases. The approaches described here have evaluated the use of full-text articles, image captions, and annotations to enrich the search results, as well as techniques to re-rank retrieved documents. Many studies [4,8,11,22,26] reported that performing search in the full content of bio-literature documents, or in the article metadata, can improve the quality of the search results. The use of full-text articles to improve scientific literature retrieval was described in [8]. The authors aimed to extend the Medical Text Indexer (MTI) (http://ii.nlm.nih.gov/MTI), a tool that provides MeSH terms recommendations for experts working on the indexation of biomedical documents at the U.S. National Library of Medicine. In an evaluation conducted with a dataset composed of 500 articles from 17 journal issues, the authors claimed that the use of full-text articles yields improvement on recall of search results.

In other studies [4,11], the image captions content was used instead of the articles full-text to support the document retrieval in scientific databases. The methods described by the authors were implemented in such a way that the query search in image captions is performed separately from the query search on the article text. In [15], Lu elaborated an overview of 28 free web-based systems for retrieval of general biomedical literature. All systems analyzed in this overview utilized PubMed or similar databases as data source. Among all approaches listed in [15], the most common ones used to improve the relevance of document retrieval were document clustering, and result re-ranking. In [25], the authors presented a bioliterature search engine for data from PubMed and PMC. The approach also includes an annotation step, in which relevant entities are extracted from the article content, and used to support users on the search task.

Several studies have compared PubMed search results with Google Scholar search results [20,22]. In [22] the authors emphasize that PubMed searches only target article abstracts, while Google Scholar searches target the full-text content of documents. The reported results show that Google Scholar retrieves twice as many relevant documents than PubMed among the first result rank positions for clinical questions queries. In [20] the authors analyze search results provided by PubMed and Google Scholar on four clinical questions. The authors in [20] also show that Google Scholar results have better relevance than PubMed ones: the top 20 articles from Google Scholar articles tend to have a higher number of citations when compared with PubMed articles. In [26], the authors made use of full-text articles in an annotation task. The task aimed at curating GO terms from article content, and the authors addressed the importance of taking the full-text of articles into account when performing gene function curation. The study observed that article abstracts would contain 30 % of all GO terms found in a document, while the other article sections would contain 70 % of all GO terms annotated in a document.

2.2   Complex Query Processing

Handling complex queries in information retrieval tasks has been studied as a way of facilitating the search process for users. It is hence of critical interest to develop systems that allow users to submit natural language queries to search engines. Research has been conducted toward this goal [7,9,10,16] using query pre-processing, term suggestion, term expansion, and entity annotation. Facilitating literature search in scientific databases is a meaningful concern. Several studies [5,22,30] have demonstrated that searching PubMed can be a difficult and time-consuming task. PubMed users with highly specific search needs end up reformulating queries frequently [5]. Also, it has been noted that only a small number of clinical practitioners uses advanced options to generate PubMed queries [22]. Moreover, the majority of PubMed queries are submitted by inexperienced users [30], who have trouble expressing concepts with MeSH terms, and finally end up performing their search using natural language terms.

In [25], the authors used entity annotation to support handling complex user queries. The annotations are also used to label user search needs, and the different labels determine which document fields should receive a boost to improve result ranking. In [31], three different query expansion methods were applied in a search task handling patient clinical notes. In this work, query expansion was shown to increase recall, but decrease precision. The authors explained this as a possible result of noise introduced by Unified Medical Language System (UMLS) [3] annotations. In [9], the authors described an approach to handle natural language queries. Users submit search questions that are reformulated into a query composed of PubMed search terms and controlled vocabulary terms. After receiving the PubMed results for this search, GO terms are extracted from the retrieved abstracts. All the complex query handling approaches described here presented some limitations. Even though [7,9,16] managed to process natural language queries, the search was restricted to article abstracts. The system described in [25], at the time our study was conducted, and to the best of our knowledge, has not been publicly released, and was last updated in 2011.

3   Methodology

We describe here the strategy implemented in bioMine to improve the retrieval of biomedical literature, and address the issues described in Sects. 1 and 2. bioMine combines two approaches. First, the full-text of journal articles is indexed, along with relevant article metadata. This task is handled by the *document indexation module*. Second, the queries are processed from their natural language format, as submitted by users, and enriched with biomedical terms provided by the UMLS Metathesaurus. This task is handled by the *complex query module*. The corpus of journal articles utilized in the development of bioMine is further described in Sect. 3.1. The *document indexation module* and the *complex query module* are described in more details in Sects. 3.2 and 3.3. bioMine and its modules are implemented in Java.

3.1   Corpus Description

The search engine described here was developed using the open-access scientific articles provided by PubMed and PMC. The articles are part of the PubMed Baseline Database (BD) files, and the PMC Open Access (OA) Subset repository (http://www.ncbi.nlm.nih.gov/pmc/tools/openftlist/). PubMed and PMC are open access databases managed by the U.S. National Institutes of Health's National Library of Medicine (NIH/NLM). PubMed holds life science journal articles, citations, and books. PMC contains the digital archive of biomedical and life sciences journal literature. The PMC OA Subset holds part of the complete PMC collection, in which all documents are available under the Creative Commons (https://creativecommons.org/about/license/) license. As of January 2016, PubMed BD files contain over 24,350,000 entries, with publication years since 1809. The PMC OA Subset contains over 1,200,000 journal articles, with publication years since 1973. PubMed BD documents are available in XML format, while PMC OA documents are available in NXML format. These file formats are standardized according to two different Document Type Definitions (DTD) managed by the Journal Article Tag Suite (JATS) (http://jats.nlm.nih.gov/). In total, 25,403,053 documents from PubMed BD and PMC OA were used to generate the bioMine search index. To generate the index, a set of specific XML tags was extracted from all files, and used to represent each document entry. The list of tags utilized in our experiments is detailed in Sect. 3.2.

3.2   Document Indexation Module

The bioMine indexation module is built based on the open-source search platform Solr (http://lucene.apache.org/solr/). When indexing documents with Solr, a

Table 1. Indexed fields in bioMine, and their availability in PubMed BD and PMC OA

Field	PubMed BD	PMC OA
1 Article title	✓	✓
2 Journal title	✓	✓
3 Abstract	✓	✓
4 Body section titles	✗	✓
5 Body full content	✗	✓
6 Author names	✓	✓
7 Reference authors	✓	✓
8 Reference title	✗	✓
9 Reference IDs	✗	✓
10 Object captions	✗	✓
11 PMCID	✓	✓
12 PMID	✓	✓
13 Article keywords	✗	✓
14 Publication year	✓	✓

document is considered as a set of key/value pairs where the keys are index *fields* and the values represent the indexed content. Since the files provided by PubMed BD and PMC OA datasets are XML documents, we built a parser to process each file, and populate the index set of fields. The bioMine parser retrieves tags from many XML document sections, using their content to semantically represent each article in the index. The XML tags used for the bioMine document indexation are considered as bioMine document fields. Table 1 shows the list of chosen fields, and the availability of fields according to document provenance.

The content extracted from tags in XML and NXML documents are kept as is. Granularity is an important factor when indexing documents, since it defines how many index entries will be associated with each XML document. For instance, each section of an article could be assigned an individual index entry. Since bioMine goal is to support the discovery of articles, we choose to have one index entry for each article instead. The bioMine engine indexes the full content of an article body. This content, in addition to the other document fields, provides a semantic representation of a document content in the index. Searches of journal articles in bioMine are handled by the complex query module, which is further explained in Sect. 3.3.

3.3   Complex Query Module

User queries submitted to bioMine are handled by the complex query module. The bioMine query module accepts and processes queries submitted in natural language. In addition, bioMine queries are separated in types according to the perceived user need, and are processed under different strategies to help improve the result relevance. To identify the user needs, and label a query, this module performs a first syntactic analysis in the user query. This step assigns one of the three bioMine *types* to the user query. Each query type utilizes a different search strategy. After being assigned a type, queries can be expanded with UMLS Metathesaurus terms. The query expansion step assigns to queries the UMLS concepts related to the user query terms. We describe hereafter the query type labelling and generation strategies, and the query expansion step.

Query Type. The complex query module labels user queries with one of the three following types: *Keyword Query* (K_Q), *Open Question Query* (O_Q) or *Statement Query* (S_Q). The three query types are defined based on common user search needs, and they are used as an attempt to increase the recall of documents that are relevant to a given query. bioMine uses the query types in order to guide the search engine towards prioritizing documents having query terms in specific document fields that are searched differently according to the query type. The query labelling process analyzes query terms for syntactic cues that can indicate the user search intent. The syntactic cues can be the presence of punctuation, or stop-words (the stop-word list used in bioMine query module is composed by a combination of an English stop-word list, and PubMed stop-word list [19]). A K_Q is a user query formulated without stop-words. K_Q are submitted to the search engine as is. K_Q example: *enzyme structure function*. An O_Q is a user query that contains interrogative cues (question words or question mark). O_Q are normalized before submitted to the search engine, by having removed the interrogative cues, and stop-words (if any present). O_Q example: *what is the relationship between the structure of an enzyme and its function?* A S_Q is a user query that contains stop-words, but no interrogative cues. S_Q are normalized by having the stop-words removed before being submitted to the search engine. S_Q example: *the relationship between enzyme structure and function.*

Query Type Generation Strategies. Each query type is associated to a different strategy to generate a bioMine query. The query strategies aim to better address the user search needs. They determine: the document fields considered in a search; if boost weights are assigned to certain document fields; and the relevance of search terms that are found if search terms are considered separately or sequentially (query term search is concerned with the presence of search terms in specific fields, while phrase term search looks for search terms sequentially in specific fields).

Boost weights are used to prioritize documents in which the search terms are found in the specific (boosted) document fields. The three query types processed by bioMine, and their strategies are as described in the following table. For all strategies, field **14** is only included in the search if a query has a numeric term of 2 or 4 characters length; and fields **11** and **12** are only included if a query has a numeric term of more than 4 characters length.

	Fields	Phrase terms	Boost?	Query terms	Boost?
K_Q strategy	1, 6	✓	✓	✓	✗
	3, 5	✓	✗	✓	✓
	10, 13	✗	✗	✓	✗
	14	✗	✗	✓	✗
	11, 12	✗	✗	✓	✗
O_Q strategy	1, 10	✓	✗	✓	✓
	3, 5	✓	✓	✓	✓
	6, 13	✗	✗	✓	✗
	14	✗	✗	✓	✗
	11, 12	✗	✗	✓	✗
S_Q strategy	1, 5	✓	✓	✓	✓
	3, 10	✓	✗	✓	✓
	6, 13	✗	✗	✓	✗
	14	✗	✗	✓	✗
	11, 12	✗	✗	✓	✗

Query Expansion. After the query terms are pre-processed during the query type labelling step, they are expanded with UMLS concepts. To perform this expansion step, we utilize the open-source tool MetaMap [2] (https://metamap. nlm.nih.gov). MetaMap is an UMLS Metathesaurus annotator system. The tool is capable of processing natural language input, extracting or mapping a given text content to UMLS concepts. The query, already processed in the type labelling step, is sent to a MetaMap instance, which annotates UMLS concepts related to the query, if any are found. The MetaMap annotations found are added to the query terms (if the original terms do not overlap with the annotation terms).

4   Experimental Evaluation and Preliminary Results

The evaluation of IR systems is commonly performed with the use of reference judgments [27]. Reference judgments are a mapping of queries and correct response documents. Some previous studies reviewed in Sect. 2 made use of reference judgment collections of less than a thousand documents [26,31]. These works had either an annotated collection, or experts available to annotate one. Creating a reference judgment collection can be costly, and even unfeasible, in tasks handling large datasets, since usually annotations are done manually by specialists. Since the dataset used in this work is considerably large, generating a reference judgement with manual annotations would be an effortful and expensive task.

4.1   Evaluation Without Reference Judgments

Previous work [23,29] dedicated effort to develop methods to evaluate IR systems without reference judgments. The authors suggested comparing results of similar systems, and the documents retrieved for a given query. The evaluation methods should consider, for example, presence or absence of retrieved documents, and their ranking position in the result list. The tasks related to biomedical document retrieval described in Sect. 2 lacked enough information about their evaluation methods. In [25] the authors described a full search engine similar to bioMine, but did not present which evaluation was adopted. [10,22,31] have not provided detailed information about, for instance, the indexation process or the document relevance computation, preventing the presented work to be reproduced. On top of this, at the time this study was conducted, the systems developed in these works appeared not to be available as open source software. This makes it difficult to carry on a fair comparison between results obtained by these approaches, and results obtained by bioMine.

4.2 Pseudo-judgments Evaluation

Pseudo-judgements are evaluation collections automatic generated by IR systems, with little or none human influence. For pseudo-judgments, the top K results retrieved by a search are considered as the most relevant, and further evaluated. In [6,21] the authors investigated the use of pseudo-relevance judgments to evaluate IR systems, by using a pool of results. The authors in [6] described a method that consists in generating a set of similar queries, retrieving relevant documents for all similar queries, and finally evaluating the ranked results against human relevance judgments. According to the authors, the system and the human's ranking were correlated. In [21], the authors described an effort to generate pseudo-relevance judgments instead of human relevance judgments using a pool of search results retrieved by a variety of IR systems.

4.3 bioMine Evaluation

Considering both [6,21] works, we suggest a comparable evaluation method, that uses pseudo-judgments, and sets of annotated queries. Queries and their corresponding relevant documents were obtained from the mycoCLAP [24] database. Biocurators working on mycoCLAP have searched extensively biomedical literature databases. They have evaluated several thousands of scientific articles to characterize fungal enzymes having specific properties, and finally map an article with a mycoCLAP enzyme entry. Within all articles mapped to mycoCLAP entries, 9 documents belong to the PMC OA. The great majority of the other documents belong to PubMed BD. We utilized the user search queries generated by mycoCLAP biocurators to retrieve: the 9 documents in mycoCLAP that can be found in PMC OA, as well as 10 randomly selected documents in myco-CLAP that can be found in PubMed BD. Our goal is to evaluate the document retrieval performance for article journals containing full-text, as well as abstract only. Our evaluation dataset, as shown in Table 2, is then composed by a set of 19 mappings of queries and correct response documents.

4.4 Evaluation Metric and Preliminary Results

To compute bioMine performance evaluation, we utilize the Mean Reciprocal Rank (MRR) score, that was previously applied in information retrieval

Table 2. List of queries and correct response documents

mycoCLAP ID	PID	$Q_{\#}$	Biocurator query
AMY13A_CRYFL	PMC3068306	Q_1	Alpha-amylase from Cryptococcus flavus activity characterization
BGL3C_ASPFU	PMC3312866	Q_2	Aspergillus fumigatus beta-glucosidase purification and characterization
MAN5A_ASPNG	PMC2780388	Q_3	Characterization of GH5 beta-mannanase enzyme from Aspergillus niger
MLG16B_ASPFU	PMC3092853	Q_4	Characterization of GH16 beta-glucanase from Aspergillus fumigatus
PGX28B_FUSOX	PMC3180650	Q_5	Purification and characterization of an exo- polygalacturonase from Fusarium oxysporum
PMO9D_PHACH	PMC3223205	Q_6	Phanerochaete chrysosporium GH61 purification and characterization
RHA78E_EMENI	PMC3312857	Q_7	Purification and characterization of an alpha- L-rhamnosidase from Aspergillus nidulans
XYN11A_LEUGO	PMC2291056	Q_8	Xylanase characterization from Leucoagaricus gongylophorus
XYN11B_TRIRE	PMC2702311	Q_9	Recombinant expression and characterization of xylanase from Trichoderma reesei
ZAX43C_PENPU	PMID20562284	Q_{10}	Bifunctional alpha-L-arabinofuranosidase/xylobiohydrolase from Penicillium purpurogenum
MSD47S_ASPPH	PMID10215597	Q_{11}	Enzymatic properties alpha-mannosidase Aspergillus saitoi
CBH6A_MAGOR	PMID20709852	Q_{12}	Characterization of Magnaporthe oryzae cellobiohydrolase
ABF51A_ASPAW	PMID9758835	Q_{13}	Substrate specificity of alpha-L- arabinofuranosidase from Aspergillus awamori
CHI18B_CANAL	PMID7708682	Q_{14}	Cloning and characterization Candida albicans chitinase
AGL13B_CANAL	PMID1400249	Q_{15}	Characterization of Candida albicans maltase
GAN53A_HUMIN	PMID12761390	Q_{16}	Beta-1,4-galactanases from Humicola insolens and Myceliophthora thermophila
BGN5A_NEOSP	PMID12427996	Q_{17}	Neotyphodium sp beta-1,6-glucanase expression and characterization
EBG16A_FLAVE	PMID21653698	Q_{18}	Purification of endo-beta-1,3-galactanase from Flammulina velutipes
XYL3A_ASPOR	PMID9872754	Q_{19}	Aspergillus oryzae beta-xylosidase optimum pH and temperature

tasks [28]. The Reciprocal Rank (RR) score is the inverse rank position ($\frac{1}{POS}$) of a correct response document retrieved by a system. It evaluates to 1 in case the correct response document is ranked at first position in a search result list. For an evaluation considering a set of queries and correct response documents, the MRR can be utilized to compute an average among all RR scores. The search queries listed in Table 2 are the search terms as provided by users searching for scientific literature. We submitted these search queries as is to bioMine. The user search queries are internally processed by the complex query module, and finally submitted to bioMine search index. For the sake of comparison, we submitted the same user search queries to PubMed BD and PMC OA search engines, with default configurations. All queries mapped to a PMID were searched in bioMine and PubMed BD, while the queries mapped to a PMCID were searched in PMC OA and bioMine. The first 20 ranked results were taken into account from each ranked result list provided by bioMine, and PubMed BD or PMC OA. A RR score was computed for each ranked result list, considering the correct response document provided by the biocurators. Finally, we computed the MRR for all

Table 3. Queries submitted to bioMine, PubMed BD, PMC OA, and correct response document ranking

$Q_\#$	PID rank	bioMine rank	bioMine RR score	$Q_\#$	PID rank	bioMine rank	bioMine RR score
Q_1	3	2	0.500	Q_{10}	2	1	1.000
Q_2	1	20	0.050	Q_{11}	N/A	7	0.143
Q_3	1	2	0.500	Q_{12}	1	1	1.000
Q_4	2	8	0.125	Q_{13}	2	1	1.000
Q_5	2	13	0.077	Q_{14}	1	1	1.000
Q_6	9	1	1.000	Q_{15}	2	1	1.000
Q_7	2	5	0.200	Q_{16}	1	N/A	0.000
Q_8	1	17	0.059	Q_{17}	N/A	1	1.000
Q_9	1	10	0.100	Q_{18}	1	N/A	0.000
				Q_{19}	1	1	1.000
Total # of queries = 19				**MRR = 0.513**			

queries submitted to bioMine. According to our results, queries mapped to a PMCID always retrieved the correct response document, either using bioMine or PMC OA search. The correct response document was ranked higher in bio-Mine search compared to the PMC OA search for Q_1 and Q_6, while Q_3 and Q_7 presented the correct document in similar ranking between bioMine and PMC OA search results, with only few positions of difference in the result list. When looking into results of queries mapped to a PMID, we observe that in some cases the first 20 rank did not have the expected document. For PubMed, this issue occurred in search results for queries Q_{11} and Q_{17}. While PubMed did not retrieve the correct response document among the top 20 results, bioMine was able to retrieve it in a fairly high ranked position. For bioMine, this issue occurred in search results of queries Q_{16} and Q_{18}, however for all other documents, we observe that bioMine retrieved the correct response document in a higher ranked position than PubMed. The MRR score for all 19 queries submitted to bioMine is 0.513, which demonstrates that the system is capable of retrieving the correct response document ranked at first position approximately half of the time (Table 3).

5 Conclusion and Ongoing Work

Literature search on open access scientific databases is a task that supports many activities in the life sciences and biomedical domain, but it can still be challenging. In this paper, we described the ongoing development of bioMine, a bioliterature search engine that aims to facilitate the retrieval of scientific literature. bioMine is an attempt to address two main issues. First, while the indexed content from PubMed BD, one of the most popular scientific databases, records only the abstract content of documents, bioMine offers the possibility of searching for literature using also the full-text of scientific articles, obtained from the PMC OA. Second, while open access databases require users to perform searches using a query language, bioMine provides the possibility of using natural language queries thanks to its complex query module. Despite the large size of the indexed corpus, bioMine is still in its infancy, and much work is needed to enhance its performance. For instance, further analysis is currently being carried on to improve the retrieval of full text documents.

Reproducibility. To ensure full reproducibility and comparisons between systems, bioMine is publicly released as an open source software in the following repository: https://github.com/BigMiners/bioMine.

References

1. Almeida, H., Meurs, M.-J., Kosseim, L., Butler, G., Tsang, A.: Machine learning for biomedical literature triage. PLOS ONE **9**(12), 12 (2014)
2. Aronson, A.R., Lang, F.-M.: An overview of MetaMap: historical perspective and recent advances. J. Am. Med. Inform. Assoc. **17**(3), 229–236 (2010)
3. Bodenreider, O.: The Unified Medical Language System (UMLS): integrating biomedical terminology. Nucleic Acids Res. **32**(suppl 1), D267–D270 (2004)
4. Divoli, A., Wooldridge, M.A., Hearst, M.A.: Full text and figure display improves bioscience literature search. PLOS ONE **5**(4), e9619 (2010)
5. Dogan, R.I., Murray, G.C., Névéol, A., Lu, Z.: Behaviour, Understanding PubMed User Search Behaviour through Log Analysis. Database, 2009:bap018 (2009)
6. Efron, M., Winget, M.: Query polyrepresentation for ranking retrieval systems without relevance judgments. J. Am. Soc. Inf. Sci. Technol. **61**(6), 1081–1091 (2010)
7. Fontelo, P., Liu, F., Ackerman, M.: askMEDLINE: a free-text, natural language query tool for MEDLINE/PubMed. BMC Med. Inform. Decis. Mak. **5**(1), 5 (2005)
8. Gay, C.W., Kayaalp, M., Aronson, A.R.: Semi-automatic Indexing of Full Text Biomedical Articles. In: AMIA Annual Symposium Proceedings, vol. 2005, p. 271. American Medical Informatics Association (2005)
9. Gobeill, J., Gaudinat, A., Pasche, E., Vishnyakova, D., Gaudet, P., Bairoch, A., Ruch, P.: Deep Question Answering for Protein Annotation. Database, 2015:bav081 (2015)
10. Griffon, N., Chebil, W., Rollin, L., Kerdelhue, G., Thirion, B., Gehanno, J.-F., Darmoni, S.J.: Performance evaluation of Unified Medical Language System synonyms expansion to query PubMed. BMC Med. Inform. Decis. Mak. **12**(1), 12 (2012)
11. Hearst, M.A., Divoli, A., Guturu, H., Ksikes, A., Nakov, P., Wooldridge, M.A., Ye, J.: BioText search engine: beyond abstract search. Bioinformatics **23**(16), 2196–2197 (2007)
12. Hirschman, L., Burns, G.A.P.C., Krallinger, M., Arighi, C., Cohen, K.B., Valencia, A., Wu, C.H., Chatr-Aryamontri, A., Dowell, K.G., Huala, E., et al.: Text Mining for the Biocuration Workow. Database, 2012:bas020 (2012)
13. Howe, D., Costanzo, M., Fey, P., Gojobori, T., Hannick, L., Hide, W., Hill, D.P., Kania, R., Schaeffer, M., St Pierre, S., et al.: Big data: the future of Biocuration. Nature **455**(7209), 47–50 (2008)
14. Hunter, L., Cohen, K.B.: Biomedical language processing perspective: what is beyond PubMed? Mol. Cell **21**(5), 589 (2006)

15. Lu, Z.: PubMed and Beyond: A Survey of Web Tools for Searching Biomedical Literature. Database, 2011:baq036 (2011)
16. Lu, Z., Wilbur, W.J., McEntyre, J.R., Iskhakov, A., Szilagyi, L.: Finding query suggestions for PubMed. In: AMIA Annual Symposium Proceedings, vol. 2009, p. 396. American Medical Informatics Association (2009)
17. Morris, B.D., White, E.P.: The EcoData retriever: improving access to existing ecological data. PLOS ONE **8**(6), e65848 (2013)
18. Mudunuri, U.S., Khouja, M., Repetski, S., Venkataraman, G., Che, A., Luke, B.T., Girard, F.P., Stephens, R.M.: Knowledge and theme discovery across very large biological data sets using distributed queries: a prototype combining unstructured and structured data. PLOS ONE **8**(12), e80503 (2013)
19. National Center for Biotechnology Information. PubMed [Table, Stopwords] (2005)
20. Nourbakhsh, E., Nugent, R., Wang, H., Cevik, C., Nugent, K.: Medical literature searches: a comparison of PubMed and Google Scholar. Health Inf. Libr. J. **29**(3), 214–222 (2012)
21. Ravana, S.D., Rajagopal, P., Balakrishnan, V.: Ranking retrieval systems using pseudo relevance judgments. Aslib J. Inf. Manage. **67**(6), 700–714 (2015)
22. Shariff, S.Z., Bejaimal, S.A.D., Sontrop, J.M., Iansavichus, A.V., Haynes, R.B., Weir, M.A., Garg, A.X.: Retrieving clinical evidence: a comparison of PubMed and google scholar for quick clinical searches. J. Med. Internet Res. **15**(8), e164 (2013)
23. Spoerri, A.: Using the structure of overlap between search results to rank retrieval systems without relevance judgments. Inf. Process. Manage. **43**(4), 1059–1070 (2007)
24. Strasser, K., McDonnell, E., Nyaga, C., Wu, M., Wu, S., Almeida, H., Meurs, M.-J., Kosseim, L., Powlowski, J., Butler, G., et al.: mycoCLAP, the Database for Characterized Lignocellulose-active Proteins of Fungal Origin: Resource and Text Mining Curation Support. Database, 2015:bav008 (2015)
25. Thomas, P., Starlinger, J., Vowinkel, A., Arzt, S., Leser, U.: GeneView: a comprehensive semantic search engine for PubMed. Nucleic Acids Res. **40**(W1), W585–W591 (2012)
26. Van Auken, K., Schaeffer, M.L., McQuilton, P., Laulederkind, S.J.F., Li, D., Wang, S.-J., Hayman, G.T., Tweedie, S., Arighi, C.N., Done, J., Mller, H.-M., Sternberg, P.W., Mao, Y., Wei, C.-H., Lu, Z.: BC4GO: A Full-text Corpus for the BioCreative IV GO Task. Database, 2014:bau074 (2014)
27. Voorhees, E.M.: Variations in relevance judgments and the measurement of retrieval effectiveness. Inf. Process. Manage. **36**(5), 697–716 (2000)
28. Voorhees, E.M., et al.: The TREC-8 question answering track report. In: TREC, vol. 99, pp. 77–82 (1999)
29. Wu, S., Crestani, F.: Methods for ranking information retrieval systems without relevance judgments. In: Proceedings of the 2003 ACM Symposium on Applied Computing, pp. 811–816. ACM (2003)
30. Yoo, I., Mosa, A.S.M.: Analysis of PubMed user sessions using a full-day PubMed query log: a comparison of experienced and nonexperienced PubMed users. JMIR Med. Inform. **3**(3), e25 (2015)
31. Zeng, Q.T., Redd, D., Rindflesch, T., Nebeker, J.: Synonym, topic model and predicate-based query expansion for retrieving clinical documents. In: AMIA Annual Symposium Proceedings, vol. 2012, p. 1050. American Medical Informatics Association (2012)

Word Normalization Using Phonetic Signatures

Vincent Jahjah[1], Richard Khoury[2]([✉]), and Luc Lamontagne[1]

[1] Department of Computer Science and Software Engineering,
Laval University, Quebec City, QC, Canada
`vincent.jahjah.1@ulaval.ca, luc.lamontagne@ift.ulaval.ca`
[2] Department of Software Engineering, Lakehead University, Thunder Bay, ON, Canada
`richard.khoury@lakeheadu.ca`

Abstract. Text normalization is the challenge of discovering the English words corresponding to the unusually-spelled words used in social-media messages and posts. In this paper, we detail a new word-searching strategy based on the idea of sounding out the consonants of the word. We describe our algorithm to extract the base consonant information from both miswritten and real words using a spelling and a phonetic approach. We then explain how this information is used to match similar words together. This strategy is shown to be time efficient as well as capable of correctly handling many types of normalization problems.

Keywords: Social media · Normalization · Wiktionary · TheFreeDictionary

1   Introduction

Social network messages have become omnipresent in today's world, and their highly relaxed spelling rules and tolerance to extreme irregularities in spelling poses problem when trying to apply traditional Natural Language Processing (NLP) tools and techniques, which were developed for properly-written English text. A sampling of Twitter messages studied in [1, 2] found over 4 million out-of-vocabulary (OOV) words, and, new spelling variations are created constantly, both voluntarily and accidentally.

The challenge of developing algorithms to automatically correct the OOV words found on social media and replace them with the correct in-vocabulary (IV) words is known as normalization. In [3], it was suggested that OOV words are often pronounced similarly to the IV words they represent; for example, "2niite" sounds almost exactly like "tonight". In this paper, we further add the intuition that this similarity is due to the consonants used, rather than the vowels. We propose to make an algorithm that matches OOV words to IV words using the pronunciation of their consonants.

2   Background

Today, billions of written messages are sent online. A significant proportion of them are written in an informal manner with little to no concern for proper spelling. This has given rise to the challenge of normalization, or of developing algorithms to determine which real IV words are intended by the miswritten OOV words found on social

R. Khoury and C. Drummond (Eds.): Canadian AI 2016, LNAI 9673, pp. 180–185, 2016.
DOI: 10.1007/978-3-319-34111-8_23

networks. While the variety of spellings of OOV words is seemingly endless, it was noted in [2] that their creation follows a small set of simple rules. The rules proposed in [2] are *abbreviation* (deleting letters from the word, e.g. "together" → "tgthr"), *phonetic substitution* (substituting letters for other symbols that sound the same, e.g. "2gether"), *graphemic substitution* (substituting a letter for a symbol that looks the same, e.g. "t0gether"), *stylistic variation* (misspelling the word to make it look like one's personal pronunciation, e.g. "togeda", "togethor"), and *letter repetition* (e.g. "togetherrr"). Approaches to normalize the OOV words are varied, and can include various degrees of automatic data gathering [2–4], OOV pre-processing [5], hand crafted data [6, 7], and considerations for the OOV word's context [4, 7]. Some works focus on specific types of normalizations ([8] for letter repetitions, [9–11] for letters deletion).

The system proposed in [7] used algorithms to generate a list of 660,000 OOV versions of 47,000 words, and performed a lexical and syntactic analysis of the messages in order to recognize observed words. In [2], the authors developed an algorithm to discover the most common letter substitutions between pairs of IV and OOV words and compute their probabilities. They then use their probabilistic model to generate the most probable IV words of new OOV words. In [9], the authors used a rule-based algorithm to map between IV and OOV words by removing double letters, unnecessary vowels, and prefixes and suffixes. In [10], they improved on their original idea by training a probabilistic machine translation model instead of using rules.

3   Methodology

The fundamental assumption of our work is that consonants contain more of the phonetic information of the words than vowels do. As a result, we believe OOV words can be more easily recognized if one ignores vowel differences but gets the consonants right. Compare for example the ease of recognizing "byutfl" or "btfl" as variants of "beautiful", as opposed to "eauiu". The core aspect of the methodology we propose is the extraction of the phonetic information of each word, which we call the signature of the word. We further distinguish between two types of signatures, namely the visual signature (VS) which is what the sequence of consonants should look like when written down, and the phonetic signature (PS), which is what the sequence of consonants should sound like when read out loud. For example, the VS of the word "accent" would be "cnt" and its PS would be "Xnt". These signatures are the information that our methodology will then use to match OOV words to IV English words. For example, the OOV words "accnt" and "aksnt" will both have the same signatures as 'accent', and can thus be matched to that IV word.

3.1   Phonetic Data and Signature Extraction

The first stage of our methodology consists in extracting the VS and PS of real English words, to serve as a reference. The algorithm also extracts simplified reconstructions of the word with vowels reinserted at the correct places. For instance, the word "accent" has a visual reconstruction (VR) of "acent" and the phonetic reconstruction (PR) "aXent". Reconstructions will be useful in the word matching stage.

Our algorithm starts with a training data set of English words paired with their IPA spellings, which represent their pronunciations. These pairs can be obtained from online dictionaries, such as Wiktionary and TheFreeDictionary, from which we obtained 78,327 pronunciations for 48,864 words. Steps 1 and 2 of the algorithm extract the VS and VR of each word. The VS is the sequence of consonants in the word as it is spelled, with all vowels and duplicates removed ("accent" → "cnt"). The VR is the VS with vowels included ("accent" → "acent"). They can thus be created together. Step 3 simplifies the IPA version of the word by merging common sets of single sounds into a single symbol. This is done by finding and replacing the following six pairs of IPA symbols: tʃ → C, ks → X, kw → Q, hw → w, tz → Z, and gz → G. In step 4, this simplified IPA is used to create the PS in the same process as in steps 1 and 2, namely by removing duplicate IPA symbols and vowel IPA symbols in our data set (the ambiguous IPA symbols "j" and "w" are treated as consonant sounds for now). Step 5 and 6 create the temporary VS and PS by replacing vowels and vowel IPAs with * in the VR and the simplified IPA respectively (e.g. "one" → "*n*").

Step 7 requires matching consonant letters in the word to sounds in the IPA string. One consonant can be matched to an identical IPA symbol (e.g. "too" → "tu") or to a different IPA symbol ("act" → "akt"). But an IPA symbol might not have a matching consonant (the "j" symbol in "dupe" → "djupe"), or a consonant in the word might not be matched to an IPA symbol (the letter "b" in "subtle" → "sutl"). We use the temporary signatures from steps 5 and 6 to clarify matters (for example "one" → "won" has the temporary signatures "*n*" → "w*n", indicating that a consonant sound occurs before the first vowel without an associated letter). Our algorithm uses a set of about 100 rules and 200 exceptions to handle our 78,000 word-IPA pairs.

3.2 Signature Generation of OOV Words

The second major stage of our methodology consists in generating the signatures of a new OOV word, for which IPA information is unknown. We first account for three of the five main types of word transformations. For the repetition type, we reduce all repeated letters to doubles and singles ("suuuupppp" → "suupp", "sup"). While the IV signatures actually discard duplicates, the matching algorithm in the next stage will prioritize exact matches, so it is important to generate both. For graphemic and phonetic substitutions, we substitute the numbers for letters. When multiple substitutions are possible (1 could stand for L, I, or "one"), the algorithm lists all possible cases. The output will be a list of variations of the OOV word without digits and with limited repetitions ("1337" → "oneet", "leet", "ieet", "onet", "let", "iet").

Next, we consider four ways of generating the signatures of a word. One of these will result in the correct signature for the IV word. First, the VS and VR of each OOV word in the list of variations is generated using steps 1 and 2 of the extraction algorithm. Next, the PS and PR are generated, using an approximate IPA pronunciation generated from the set of rules from step 7 of the extraction algorithm. Third, to deal with the possibility that there are permutations of consonants in the word ("answer" → "asnwer"), our algorithm generates VS and VR pairs of permutations of adjacent consonants ("asnwers" → "answers", "asnwesr"). Finally, we deal with potential suffixes in the OOV word. These suffixes are often altered in standard ways, which means most of them can be easily corrected using nine simple rules: in → ing, a → er, a → or, z → s, n → ing, as → ers, as → ors, d → ed, and ah → er. We remove the suffix ("dansa" → "dans") and make a VR and VS for the root.

3.3 Similar Word Matching

The third and final stage of our methodology consists in determining which IV English word the OOV word sounds like. First we find a set of IV words with identical signatures, along with their occurrence probability (taken from standard word frequency tables such as ProjectGutenberg). Then our algorithm applies a set of heuristics to score the IV as a match to the OOV word. The first heuristic considers whether the signature is visual, phonetic, permuted, or suffixed, corresponding to the four ways of generating the signature in part 3.2. The matches are prioritized in that order. The second heuristic considers on how the first letter of the matching word compares to that of the OOV word. The highest value is given if the first letter of both words is the same, then a lesser value is given, in order: if the IV word begins by a silent vowel truncated in the OOV word ("nough" → "enough"), if an initial letter H was deleted, if the initial vowel was phonetically transformed ("ate" → "eight"), and finally if a non-silent initial vowel was deleted. In any other case, the word is eliminated from consideration. The final heuristic scores the IV word compared to the OOV word's reconstruction. Maximum score is given if the IV word is identical to the reconstruction, then if there is only a suffix difference. Failing that, the algorithm computes the Levenshtein distance between the IV word and reconstruction, using an implementation of the distance that ignores duplicate letters. Two lesser scores are given if the distance is 0 or 1, respectively. If the distance is greater than 1, then no score is given.

4 Experimental Results

In order to test our normalization methodology, we used the test set from [2, 4]. This corpus lists 3525 OOV words that were observed in real tweets, along with their corresponding IV English words, sorted into seven basic types [2], namely abbreviation, phonetic substitution, graphemic substitution, stylistic variation, letter repetition, typographical errors (a default catch-all category), and multiple types at once.

Table 1 gives the accuracy of our algorithm, both overall and per type of normalization problem. A top-n match is defined as the presence of the correct IV word in the n most probable IV words found by our algorithm. The results include only the 3525 OOV words mapping to known IV words in our system.

One thing to note is that the overall top-2 accuracy increases by over 10 % compared to the top-1. This may be due to the ambiguity inherent to OOV spellings. As mentioned earlier, the OOV word "accnt" could map to either "accent" or "account" (incidentally the top two IV words returned for that word), and it is impossible to tell which one the user intended without the sentence context, which is not taken into account in our work.

Table 1. Corpus composition and test results by type.

Type	Words	Top-1	Top-2	Top-5	Top-10
Overall Average	3525	56.6 %	66.1 %	73.6 %	77.2 %
Abbreviation	658	30.8 %	41.4 %	53.1 %	57.2 %
Phonetic substitution	60	66.6 %	76.6 %	86.3 %	93.3 %
Graphemic substitution	63	80.9 %	80.9 %	80.9 %	80.9 %
Stylistic variation	1646	55.7 %	67.4 %	76.7 %	81.3 %
Letter repetition	756	92.1 %	96.8 %	98.5 %	99.2 %
Typo	141	17.0 %	21.9 %	23.4 %	25.5 %
Multiple types	201	32.8 %	43.7 %	51.2 %	56.7 %

Repetition is the type of normalization problem that our system handles best, reaching a top-1 accuracy of 92.1 %. Since our system is designed to sound out consonants, we likewise find that abbreviation-type OOV words that crop vowels, silent letters, and consonants in multi-letter sounds cause no problems. Nearly all of the abbreviation-type OOV words that failed to be recognized were missing voiced consonants. Phonetic substitutions are also handled very well by our system. The main failure case occurs when the substitution affects the first letter of the word. These cases are blocked by the second heuristic of our matching algorithm.

Graphemic substitutions have the unusual behaviour of being only matched in top-1 or not at all. This can be explained by the fact that all substitution cases we accounted for will result in exact signatures and perfect IV word matches, while most substitutions that were not included in our system (such as the letter Q replacing a G) will lead to wrong OOV word signatures that cannot be recognized.

The final normalization type is typos. These include problems such as letters permutations or accidentally repeated syllables ("novemember" → "November"), which our system can handle relatively well, or deliberately using foreign words or slang ("oogie" → "disgusting"), which our system fails completely on. This type of normalization problem violates our fundamental assumption, that OOV words are phonetically-similar to the intended IV ones.

5   Conclusion

In this paper, we presented a novel normalization algorithm for the OOV words commonly found in social media messages. The two fundamental intuitions behind our algorithm are that these OOV words are phonetic approximations of the IV words they represent, and that consonants are more phonetically important than are vowels. Consequently, our algorithm is designed to extract the consonant signature of the OOV word, and to match it with IV words that have identical signatures, ranking the matches using a set of linguistic transformation heuristics. Our results show that this method can recognize 80.0 % of OOV words that appeared in a testing corpus of Twitter messages, with 56.6 % of them ranking the correct IV word in first place. Future work may look at handling these special cases, for example by integrating spell-checking routines to deal with common typos or lists of common abbreviations such as in [7].

References

1. Petrovic, S., Osborne, M., Lavrenko, V.: The Edinburgh Twitter corpus. In: Proceedings of the Naacl Workshop on Computational Linguistics in a World of Social Media, Los Angeles, USA, pp. 25–26 (2010)
2. Liu, F., Weng, F., Wang, B., Liu, Y.: Insertion, deletion, or substitution?: Normalizing text messages without pre-categorization nor supervision. In: Proceedings of the 49th Annual Meeting of the Association for Computational Linguistics: Human Language Technologies, vol. 2, Stroudsburg, USA, pp. 71–76 (2011)
3. Khoury, R.: Phonetic normalization of microtext. In: Proceedings of the 2015 IEEE/ACM International Conference on Advances in Social Networks Analysis and Mining, 25–28 August 2015, Paris, France, pp. 1600–1601
4. Liu, F., Weng, F., Jiang, X.: A broad-coverage normalization system for social media language. In: Proceedings of the 50th Annual Meeting of the Association for Computational Linguistics, Jeju, Korea, pp. 1035–1044 (2012)
5. Clark, E., Araki, K.: Text normalization in social media: progress, problems and applications for a pre-processing system of casual english. In: PACLING 2011. Procedia - Social and Behavioral Sciences, vol. 27, pp. 2–11 (2011)
6. Han, B., Cook, P., Baldwin, T.: Lexical normalization for social media text. ACM Trans. Intell. Syst. Technol. (TIST) 4(1), article no. 5. Digital Publication (2013). http://dl.acm.org/citation.cfm?id=2414425&picked=prox&CFID=768981160&CFTOKEN=83762437
7. Jose, G., Raj, N.S.: Lexico-Syntactic Normalization Model for noisy SMS Text. Dept. of Comput Schi., SCMS Sch. of Eng. & Technol., Ernakulam, India, November 2014
8. Hirankan, P., Suchato, A., Punyabukkana, P.: Detection of wordplay generated by reproduction of letters in social media text. In: 10th International Joint Conference of JCSSE, pp. 6–10, May 2013
9. Pennell, D.L., Liu, Y.: Normalization of text messages for text-to-speech. In: Proceedings of the 35th International Conference on Acoustics, Speech and Signal Processing, Dallas, USA, pp. 4842–4845 (2010)
10. Pennell, D.L., Liu, Y.: Normalization of informal text. Comput. Speech Lang. 28(1), 256–277 (2014)
11. Maitama, J.Z., et al.: Text normalization algorithm for facebook chats in hausa language. In: 5th International Conference of ICT4M, pp. 1–4, November 2014

Forecasting Canadian Elections Using Twitter

Kenton White[1,2]([✉])

[1] Advanced Symbolics, 41 York Street, 4th Floor, Ottawa, ON, Canada
kenton.white@advancedsymbolics.com
[2] School of Electrical Engineering and Computer Science,
University of Ottawa, 800 King Edward Dr., Ottawa, ON, Canada
kenton.white@uottawa.ca

Abstract. Experiments forecasting Canadian elections with Twitter are presented. A methodology for creating a representative Twitter sample is described and validated against census data. This sample and election polls are input into a VARX forecast model. The model covariance error monitors the forecast accuracy, measuring forecast confidence before the election occurs. The model is tested on several Canadian elections.

1 Introduction

Social media provides an unprecedented window into the political discourse – the tantalizing allure of tapping into a collective consciousness. Could it be possible, using platforms like Facebook and Twitter, to predict the next Prime Minister?

Tumasjan et al. found a correlation between share of Twitter traffic for the 6 main parties in the 2009 German national parliament election and the final percentages [13]. Based on this correlation, they speculated that Twitter could be a leading indicator of political opinion. O'Connor et al. [9], saw a similar leading indicator using topic sentiment. They compared the daily sentiment score for consumer confidence, presidential approval, and the 2008 presidential election race with Gallup polls. For consumer confidence and presidential approval, they found that the daily sentiment score was a leading indicator of consumer polls, but not for election polls. Bermingham and Smeaton [2] performed a similar analysis for the 2011 General Irish Election, comparing several features, like mention volume and sentiment, against the election results. They found that share of volume of mentions, followed closely by the share of positive volume, were the best predictors. Sang and Bos [12] studied the 2011 Dutch Senate elections, again comparing several features. Contrary to Bermingham and Smeaton, they found that sentiment scores were the best predictor. In a novel twist, the researchers introduced poll-dependent weights to correct for demographic differences between Twitter users and the Dutch electorate.

Contrasting these successful demonstrations, Gayo-Avello [4] and Metaxas [8] together analyzed the 2010 US Congressional election. After accounting for incumbency effects, they reported neither tweet volume nor sentiment had predictive value. In their analysis, Gayo-Avello and Metaxas made several critiques:

© Springer International Publishing Switzerland 2016
R. Khoury and C. Drummond (Eds.): Canadian AI 2016, LNAI 9673, pp. 186–191, 2016.
DOI: 10.1007/978-3-319-34111-8_24

- *A Priori Model.* Previous studies chose the best method and feature *a posteriori.* Method and features must be defined before the election starts.
- *Testable Models.* Previous studies could not measure forecast accuracy until after the election. A model must monitor the forecast accuracy.
- *Representative Samples.* Polls use a representative sample of the population for their analysis. Twitter forecasts must use a similar sample for analysis.

This paper directly addresses these critiques.

2   Sampling

Twitter studies collect tweets from the "Garden Hose", a down-sampled stream of real time tweets. These streams are also rate limited, being temporarily disabled when the number of streamed tweets exceeds a fixed amount in a set time. Down sampling and rate limiting produces a sample where the systematic errors are unknown, preventing the researcher from correcting for any biases.

Election studies are restricted to tweets within the election geography. These experiments use the self reported location in the Twitter profile. The location field text is queried against a reverse address lookup service[1] to determine if the user falls within the desired geography.

Following the polling professional methodology, analysis should be performed on a random sample of Twitter users. Samples can be created using the social network graph structure. Each user has public connections with other users. These connections are crawled, extracting a sample. Simple random walks on the graph generate biased samples, undersampling users with a small number of connections while oversampling those with a large number of connections.

Markov Chain Monte Carlo (MCMC) methods remove this bias [6], but with an important caveat: the algorithms converge to the stationary distribution by sampling the same user multiple times. Perfect samplers remove the statistical degeneracy from MCMC samplers. Coupling From The Past (CFTP) [10], a well known perfect sampler, can be modified to work on social networks using Conditional Independence Coupling (CIC) [14,15]. CIC retains the unbiased graph sampling from MCMC methods while avoiding degenerate samples. CIC is the sampling method used in these experiments.

3   Demographics

The representativeness of the sample is verified by directly measuring the demographics of 5,000 sampled Twitter users from Toronto. The 5,000 users were hand classified using Mechanical Turk, collecting 3 classifications for each account. Accounts where 2 out of 3 people agreed on all 3 of the characteristics were kept. The final data set had 3,032 accounts. Random error biases were removed using post stratification weighting [7].

---

[1] Microsoft's Bing Maps, https://msdn.microsoft.com/en-us/library/ff701713.aspx.

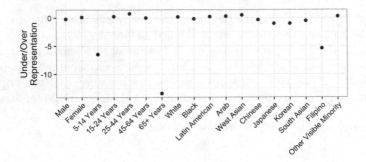

Fig. 1. Under/Over representation in weighted sample data compared against the 2011 Toronto census.

Figure 1 shows the deviation of the weighted Twitter data from the 2011 Toronto census data. The demographics of the online users follow the census demographics [3] with some important exceptions. Users under the age of 14 years were underrepresented, since Twitter requires that all users must be over the age of 14. Users over the age of 65 were also under represented, supporting anecdotal observations that Twitter is primarily used by younger people. Finally, Filipinos were underrepresented due to difficulty classifiers had differentiating Filipino ethnic traits.

4   Forecasting

O'Connor [9] had noted that Twitter data was a leading indicator of some public opinion polls. Achrekar et al. formalized this observation with a Vector Autoregression with Exogeneous Variables (VARX) model forecasting CDC influenza data [1]. VARX models require an output variable (the variable to be forecast) and an input variable (the variable from which the forecast will be derived).

Here a similar VARX technique is used for forecasting elections where election polls are the output variable and Twitter data is the input variable. The VARX model is implemented using the R package DSE [5]. The output variable is a time series of the aggregated daily polls for each major party. The input variable is an appropriate Twitter feature.

Which Twitter feature to use is determined by experiment on the 2014 Ontario Election. A sample of 32,339 Ontario Twitter users with 2,500,821 tweets between May 1, 2014 and 8 pm on June 12, 2014 (when the polls closed) was collected. For each of the 3 main parties (Liberals, Conservatives, and NDP) mentions of the party and the party leader were tracked.

Six separate features were tested:

- Tweet Volume (mentions of a candidate).
- Tweet SoV (unique people mentioning a candidate).
- Positive Volume (positive mentions of a candidate).
- Positive SoV (unique people positively mentioning a candidate).

Table 1. Comparison of different features for the 2014 Ontario election. Winning party is in bold.

	Without post stratification			With post stratification		
	LPC	CPC	NDP	LPC	CPC	NDP
Tweet volume	30.2 %	**50.7 %**	19.1 %	32.4 %	**46.7 %**	20.9 %
Tweet SoV	32.3 %	**50.4 %**	17.4 %	35.1 %	**50.3 %**	14.6 %
Positive volume	34.2 %	**54.9 %**	10.9 %	35.0 %	**51.9 %**	13.1 %
Positive SoV	38.5 %	**50.3 %**	11.2 %	40.5 %	**49.0 %**	10.5 %
Mean sentiment volume	**39.5 %**	37.4 %	23.1 %	38.8 %	**39.0 %**	22.2 %
Mean sentiment SoV	**41.1 %**	35.7 %	23.2 %	**40.1 %**	33.7 %	26.2 %
Election results[a]	**41.3 %**	33.3 %	25.3 %	**41.3 %**	33.3 %	25.3 %

[a] Election results are normalized to the top 3 parties

– Mean Sentiment Volume (mean sentiment score for each candidate).
– Mean Sentiment SoV (person's mean sentiment score for each candidate).

Mentions were normalized against the total people tweeting in a given day. Sentiment is calculated using word polarity [11]. For each of the 6 features the model was run with and without post stratification weighting. (demographics were determined as described in Sect. 3). Results are summarized in Table 1. Mean sentiment SoV provides the best final election prediction. Including demographic weighting does not improve the results. In the subsequent experiments, mean sentiment SoV without demographic weighting will be used.

The covariance matrix is an estimate on the forecast accuracy and can be used as a measure of the forecast confidence. This measure's effectiveness was tested on 2 recent provincial elections – Alberta 2012 and British Columbia 2013 – where the election polls missed the winning party by large margins. Since the model is trained on the election polls, the Twitter forecast should not be substantially different, but the covariance error should be large enough to cast doubt on the forecast accuracy.

A sample of 938,831 tweets from 24,015 Alberta Twitter users between March 22, 2012 and April 23, 2012 and 1,589,339 tweets from 32,603 British Columbia Twitter users between April 3, 2013 and May 14, 2013 was used. The polls are extrapolated to election day using a standard ARIMA model (without information from Twitter), which allows the election polls the benefit of any momentum changes from the days leading up to the election.

Table 2 summarizes the ARIMA and VARX errors. In both elections the VARX model, using the data from Twitter, has a much larger error than the corresponding ARIMA models that only use the poll data. For comparison, the error using an ARIMA model and a VARX model for the Ontario election is also shown. In the Ontario election, where the Twitter forecast agreed well with the election results, the covariance error is lower.

The model was tested live on the 2015 Federal Canadian election. A sample of 34,732,633 tweets from 130,816 Canadians between January 1, 2015 and

Table 2. Comparison of error for Alberta 2012 (top), British Columbia 2013 (middle) and Ontario 2014 (bottom) election. Winning party is in bold.

	Alberta			
	PC	Wildrose	Lib	NDP
ARIMA	$35.8 \pm 0.9\%$	**$38.1 \pm 1.1\%$**	$11.1 \pm 0.6\%$	$11.4 \pm 0.2\%$
VARX	$35.2 \pm 4.6\%$	**$37.6 \pm 6.5\%$**	$12.1 \pm 3.6\%$	$11.1 \pm 1.5\%$
Actual	**44.0%**	34.3%	9.9%	8.5%
	British columbia			
	Lib	NDP	Green	
ARIMA	$36.4 \pm 1.0\%$	**$44.3 \pm 1.9\%$**	$11.2 \pm 1.5\%$	
VARX	$34.3 \pm 3.1\%$	**$42.4 \pm 9.7\%$**	$11.8 \pm 6.3\%$	
Actual	**44.1%**	39.7%	8.2%	
	Ontario			
	Lib	PC	NDP	
ARIMA	**$36.3 \pm 1.0\%$**	$31.8 \pm 1.4\%$	$23.5 \pm 1.6\%$	
VARX	**$37.0 \pm 1.1\%$**	$32.3 \pm 1.3\%$	$23.0 \pm 1.7\%$	
Actual	**38.7%**	31.3%	23.8%	

Table 3. Summary of the 2015 Federal Canadian election. Winning party is in bold.

	Twitter			Polls			Actual		
	LPC	CPC	NDP	LPC	CPC	NDP	LPC	CPC	NDP
Canada	**38.1 ± 0.8**	31.1 ± 0.7	18.7 ± 0.9	**38.0 ± 2.1**	30.7 ± 1.1	20.5 ± 0.6	**39.5**	31.9	19.7
Atlantic	**54.0 ± 2.9**	18.7 ± 2.1	19.7 ± 2.1	**54.0 ± 4.6**	17.7 ± 2.5	21.1 ± 1.7	**59.1**	18.2	17.9
Quebec	**28.4 ± 1.9**	20.5 ± 1.7	24.3 ± 2.1	**30.1 ± 5.6**	18.1 ± 1.6	25.4 ± 4.1	**35.7**	16.7	25.4
Ontario	**45.6 ± 1.8**	32.7 ± 2.1	15.7 ± 2.1	**45.4 ± 0.7**	33.1 ± 1.2	16.6 ± 1.3	**44.8**	35.0	16.6
Prairies	33.5 ± 2.5	**42.5 ± 3.0**	18.4 ± 2.8	36.0 ± 4.4	**40.2 ± 5.6**	18.1 ± 0.9	34.3	**42.9**	19.5
Alberta	28.0 ± 2.0	**54.1 ± 1.9**	17.2 ± 2.0	25.3 ± 6.6	**52.8 ± 5.6**	18.1 ± 4.6	24.6	**59.5**	11.6
BC	**35.6 ± 1.9**	32.2 ± 2.0	23.2 ± 2.1	32.6 ± 4.5	**32.8 ± 4.6**	25.1 ± 2	**35.2**	30.0	25.9

October 19, 2015 was collected. Daily forecast models were run against national and provincial polls. Final forecast numbers were calculated when the polls closed.

Table 3 summarizes the Twitter forecast results. The mean value of the previous day's polls as well as the final election results are provided for reference. The Twitter forecast model correctly predicted the overall Canadian election as well as the provincial results. In contrast, the polls missed the correct winner in British Columbia. Both the polls and the Twitter forecast underestimated the Liberal's win in Quebec. Covariance errors in all of the Twitter forecasts were low, indicating confidence in the forecast result.

References

1. Achrekar, H., Gandhe, A., Lazarus, R., Yu, S.H., Liu, B.: Twitter improves seasonal influenza prediction. In: HEALTHINF, pp. 61–70 (2012)
2. Bermingham, A., Smeaton, A.F.: On using twitter to monitor political sentiment and predict election results. In: Sentiment Analysis where AI meets Psychology (SAAIP) Workshop at the International Joint Conference for Natural Language Processing (IJCNLP 2011), pp. 2–10 (2011)
3. Canada, S.: Toronto, ontario (code 3520005) and ontario (code 35) (table) (2012). http://www12.statcan.gc.ca/census-recensement/2011/dp-pd/prof/index.cfm?Lang=E
4. Gayo Avello, D., Metaxas, P.T., Mustafaraj, E.: Limits of electoral predictions using twitter. In: Proceedings of the Fifth International AAAI Conference on Weblogs and Social Media. Association for the Advancement of Artificial Intelligence (2011)
5. Gilbert, P.D.: Brief User's Guide: Dynamic Systems Estimation (2006). http://cran.r-project.org/web/packages/dse/vignettes/Guide.pdf
6. Gjoka, M., Kurant, M., Butts, C.T., Markopoulou, A.: Walking in facebook: a case study of unbiased sampling of osns. In: Proceedings of the IEEE INFOCOM, 2010, pp. 1–9. IEEE (2010)
7. Lumley, T.: Analysis of complex survey samples. J. Stat. Softw. **9**(1), 1–19 (2004)
8. Metaxas, P.T., Mustafaraj, E., Gayo-Avello, D.: How (not) to predict elections. In: Privacy, Security, Risk and Trust (PASSAT) and 2011 IEEE Third Inernational Conference on Social Computing (SocialCom), pp. 165–171. IEEE (2011)
9. O'Connor, B., Balasubramanyan, R., Routledge, B.R., Smith, N.A.: From tweets to polls: linking text sentiment to public opinion time series. ICWSM **11**, 122–129 (2010)
10. Propp, J.G., Wilson, D.B.: Exact sampling with coupled markov chains and applications to statistical mechanics. Random Struct. Algorithms **9**(1–2), 223–252 (1996)
11. Rinker, T.W.: qdap: Quantitative Discourse Analysis Package. University at Buffalo/SUNY, Buffalo (2013). http://github.com/trinker/qdap 2.2.4
12. Sang, E.T.K., Bos, J.: Predicting the 2011 dutch senate election results with twitter. In: Proceedings of the Workshop on Semantic Analysis in Social Media, pp. 53–60. Association for Computational Linguistics (2012)
13. Tumasjan, A., Sprenger, T.O., Sandner, P.G., Welpe, I.M.: Predicting elections with twitter: what 140 characters reveal about political sentiment. ICWSM **10**, 178–185 (2010)
14. White, J.K., Li, G.: Method and system for sampling online social networks, April 2015. https://patents.google.com/patent/US20150095415A1/en
15. White, K., Li, G., Japkowicz, N.: Sampling online social networks using coupling from the past. In: 2012 IEEE 12th International Conference on Data Mining Workshops (ICDMW), pp. 266–272. IEEE (2012)

Time-Sensitive Topic-Based Communities on Twitter

Hossein Fani[1,3]([⊠]), Fattane Zarrinkalam[1,2], Ebrahim Bagheri[1],
and Weichang Du[3]

[1] Laboratory for Systems, Software and Semantics (LS[3]),
Ryerson University, Toronto, Canada
hosseinfani@gmail.com
[2] Department of Computer Engineering,
Ferdowsi University of Mashhad, Mashhad, Iran
[3] Faculty of Computer Science, University of New Brunswick,
Fredericton, Canada

Abstract. This paper tackles the problem of detecting temporal content-based user communities from Twitter. Most existing content-based community detection methods consider the users who share similar topical interests to be like-minded and use this as a basis to group the users. However, such approaches overlook the potential temporality of users' interests. In this paper, we propose to identify *time-sensitive topic-based communities* of users who have similar temporal tendency with regards to their topics of interest. The identification of such communities provides the potential for improving the quality of community-level studies, such as personalized recommendations and marketing campaigns that are sensitive to time. To this end, we propose a graph-based framework that utilizes multivariate time series analysis to represent users' temporal behavior towards their topics of interest in order to identify like-minded users. Further, Topic over Time (TOT) topic model that jointly captures keyword co-occurrences and locality of those patterns over time is utilized to discover users' topics of interest. Experimental results on our Twitter dataset demonstrates the effectiveness of our proposed temporal approach in the context of personalized news recommendation and timestamp prediction compared to non-temporal community detection methods.

Keywords: Community detection · Topic modeling · Time series analysis · Graph clustering · Twitter

1 Introduction

Information sharing and communication patterns of users in social network platforms such as Twitter can lead to the formation of communities that consist of like-minded or similarly behaving users. Besides models that utilize network structure for identifying communities, various topic-based approaches have been

© Springer International Publishing Switzerland 2016
R. Khoury and C. Drummond (Eds.): Canadian AI 2016, LNAI 9673, pp. 192–204, 2016.
DOI: 10.1007/978-3-319-34111-8_25

investigated in the literature that employ textual content published by the users or jointly with social connections to detect like-minded users [1,13]. However, these approaches consider the users who share similar topics of interest as like-minded users and do not take into account their temporal behaviour towards these topics. For example, some users who are less interested in political issues such as *'Arresting WikiLeaks Chief in the U.K'* might show their inclination only after a few weeks of delay while others might react immediately to the topic. The community of users who contributes to this topic instantaneously should be viewed as a different community from those users who only pay attention to the topic after several weeks. The ability to model these temporal topic-based communities provides the potential for improving the quality of community-level studies, such as personalized recommendations and marketing campaigns [5,22]. For example, in case of news recommendation, it would be unreasonable to recommend a news article about a topic to those users who showed interest in it a few weeks ago and have since moved on. On the other hand would make reasonable sense to recommend the same article to users who are actively pursuing this topic on Twitter at the current point in time.

We propose to identify communities of user who have similar temporal tendency with regards to the active topics on Twitter based on their published tweets. Hence, we support temporality in our community detection process. Further, since tweets are rather short, noisy and do not provide sufficient contextual information for identifying their semantics, we model semantics of tweets by annotating them with unambiguous concepts described in external knowledge bases such as Wikipedia [21].

Our work in this paper makes the following contributions:

- We propose a model to incorporate user's interests towards active topics on Twitter within a temporal framework. We adopt the Topics over Time (TOT) model [19] that jointly captures word co-occurrences and locality of those patterns in time to simultaneously discover topics and model users' temporal inclination towards these topics. User models are utilized for identifying similar users.
- We propose a graph-based framework over multidimensional user time series for discovering time-sensitive topic-based communities in Twitter.
- We demonstrate the performance of our user community detection approach in the context of personalized news recommendation and time stamp prediction in order to compare our work with the state of the art.

This paper is organized as follows: In the next section, we review the related work. The proposed approach is introduced in Sect. 3. Section 4 is dedicated to our evaluations. Finally, in Sect. 5, we conclude the paper.

2   Related Work

Existing user community detection approaches can be broadly classified into two categories [7]: *Topology-based* and *Topic-based* approaches. The *Topology-based* community detection approaches represent the social network as a graph whose nodes are users and edges indicate explicit user relationships. This approach relies only on the network structure of the social network graph. On the other hand, *Topic-based* approaches mainly utilize textual content of the users' posts in the social network to detect communities. Topology-based view to community detection on social networks may not be able to identify communities of users that share similar conceptual interests because there are many users on a social network that have very similar interests but are not explicitly connected to each other. Further, many of the social connections may not be due to users' interests similarity but can be due to other factors such as friendship and kinship that do not necessarily point to inter-user interest similarity [6]. Since, the goal of our proposed approach is to detect communities formed toward the topics extracted from users' information contents, we review topic-based community detection methods in this section.

Most of these works have proposed a probabilistic model to detect topic-based user communities based on textual content or jointly with social connections [1,16,17,23]. For example, Abdelbary et al. [1] have identified users' topics of interest and extract user communities based on the topics utilizing Gaussian Restricted Boltzmann Machine. Yin et al. [20] have integrated community discovery with topic modeling in a unified generative model to detect communities of users who are coherent in both structural relationships and latent topics. In their framework, a community can be formed around multiple topics and a topic can be shared between multiple communities. Sachan et al. [17] have proposed probabilistic schemes that incorporate users' posts, social connections and interaction types to discover latent user communities in social networks. In their paper, they have considered three types of interactions: conventional tweets, reply tweets and retweets.

Another class of work attempts to transform the topic-based community detection problem into a graph clustering problem. Liu et al. [11] have proposed a clustering algorithm based on topic-distance between users to detect topic-based communities in a social tagging network. In this work, LDA is used to extract hidden topics in tags. Peng et al. [15] have proposed a hierarchical clustering algorithm to detect latent communities from tweets. They have used the predefined categories in SINA Weibo and have calculated the pairwise similarity of users based on their degree of interest in each category. All the above methods do not incorporate temporal aspects of users' interests and undermine the fact that users of like-minded communities would ideally show similar contribution or interest patterns for similar topics throughout time. Hu et al. [10] is one of the few that consider this aspect. These authors propose a unified probabilistic generative model to extract latent communities over temporal topics and analyze topic temporal fluctuation across different communities.

3   Proposed Approach

In this section, we present our approach to identify Twitter user communities, within a specific time period T, who share similar topics of interest at the same time. We refer to these communities as *time-sensitive topic-based communities*. To this end, we first model temporal behavior of users' interests towards existing active topics on Twitter in time period T by building their *user-topic contribution time series* and then detect latent communities by means of a graph-based clustering algorithm. These steps are described in the following.

User-Topic Contribution Time Series. Assuming there are K active topics $z_1, z_2, ..., z_K$ present on Twitter in time period T, we view T as L consecutive time intervals, and calculate user's contribution to each topic z at each time interval. Thus, user u is represented by temporally ordered vectors denoted as $\mathbf{Y}^u = [\mathbf{y}_t^u]^{1 \leq t \leq L}$, where $\mathbf{y}_t^u$ is a vector of size K, representing u's degree of interest at time interval t to each topic z. Collectively, $\mathbf{Y}^u$ is a K-variate time series for user u, named the *user-topic contribution time series*.

Considering $\mathbb{M}$, the set of tweets posted during time period T as a text corpus, it is possible to extract topics $\mathbb{Z}$ using topic modeling methods. LDA [3] as a topic modelling method assumes that a document d is a mixture of topics and implicitly uses co-occurrence patterns of terms to extract sets of correlated terms to represent topics. However, the dynamics of topics on Twitter can temporarily impact the co-occurrence patterns of terms. For instance, *'Arrest'* and *'Julian Assange'* have appeared much more frequently on Twitter during the December 2010 time period because of the arrests made with regards to the WikiLeaks case in U.K.; however, this association between the two terms are temporal and cannot be generalized. The LDA method does not model time which can confound co-occurrence patterns and result in unclear, suboptimal topic discovery [19]. We need a topic model that explicitly models time jointly with keyword co-occurrence patterns. To address this issue, we exploit Topics over Time (TOT) [19] to identify active topics on Twitter, which simultaneously captures word co-occurrences and locality of those patterns over time and is hence able to discover more event-specific topics.

On the other hand, applying topic modeling methods such as TOT and LDA to extract topics from tweets might suffer from the sparsity problem [12,18]. Because, they are designed for regular documents rather than short, noisy and informal texts like tweets. To obtain better topics from Twitter without modifying the standard topic modeling methods, we annotate each tweet m from our collection of tweets $\mathbb{M}$ with concepts defined in Wikipedia using an existing semantic annotator and employ these concepts instead of the words/terms. For instance, for a tweet such as *'Sweden issues Warrant for Wikileaks exec Julian Assange's arrest* http://bit.ly/9Ho0WM', a semantic annotator such as TagMe [8] is able to identify and extract four Wikipedia concepts, namely *'Sweden'*[1], *'Arrest_warrant'*, *'WikiLeaks'* and *'Julian_Assange'* as basic units of this tweet. We believe that using concepts instead of words leads to the reduction of noisy content within the topic detection process, because each concept implicitly represents a collection of terms which are collectively more meaningful than a single term [8].

---

[1] https://en.wikipedia.org/wiki/Sweden.

Formally, let $\mathbb{C} = \{c_1, c_2, ..., c_N\}$ be a set of N Wikipedia concepts extracted from $\mathbb{M}$, we consider a concept $c \in \mathbb{C}$ as the basic unit of tweets and model a tweet posted by user u at timestamp ts, $\mathbf{m}_{ts}^u \in \mathbb{M}$, as a vector of N nonnegative integers, where the ith number shows the occurrence frequency of the ith concept. Similar to previous works in the literature [9,12], we aggregate the tweets of a user which are published in a given time interval into a single document for use as training data. More specifically, let T be a specified time period divided into L consecutive time intervals. A document $\mathbf{d}_{ts}^u$ is an aggregate over all tweets of user u posted during time interval $1 \leq t \leq L$, i.e., $\mathbf{d}_t^u = \sum_{t \leq ts < t+1} \mathbf{m}_{ts}^u$.

By running TOT [19] over these documents, we can discover topics $\mathbb{Z} = \{\mathbf{z}_1, \mathbf{z}_2, ... \mathbf{z}_K\}$, where a topic $\mathbf{z}$ is a vector of N real numbers in $\mathbb{R}^{[0,1]}$. The ith number shows the participation score of the ith concept in forming the topic. Collectively, $\mathbb{Z} = \{\mathbf{z} \in \mathbb{R}^{[0,1]^n} : ||\mathbf{z}||^1 = 1\}$ is the set of all topics indexed from 1 to K. $|| \cdot ||^1$ is the L^1-norm of $\mathbf{z}$. In TOT, the vector is the Dirichlet distribution of concepts in the topic with the parameter β; notationally, $\phi_{\mathbf{z} \in \mathbb{Z}} \sim Dir(\beta)$.

Finally, *user-topic contribution time series* of each user u, $\mathbf{Y}^u$, is the stream of document-topic contribution $\mathbf{y}_t^u$ over L consecutive time intervals, i.e., $\mathbf{Y}^u = [\mathbf{y}_t^u]^{1 \leq t \leq L} = [\mathbf{y}_1^u, \mathbf{y}_2^u, ..., \mathbf{y}_L^u]$ where $\mathbf{y}_t^u$ is a vector of K real numbers in $\mathbb{R}^{[0,1]}$. The ith number shows the participation score of the ith topic in forming the document $\mathbf{d}_t^u$ of user u for time interval $1 \leq t \leq L$. TOT [5] assumes a Dirichlet distribution of topics in the document d_t^u with the parameter α; notationally, $\theta_{\mathbf{d}_t^u} \sim Dir(\alpha)$.

We believe that the behavior of the user-topic contribution time series can be considered to be a good measure for finding the similarity between two users in that it allows us to find like-minded users based on their *temporally*-correlated contributions in similar topics. In the following, we use user-topic contribution time series of users to discover time-sensitive topic-based user communities.

Time-Sensitive Topic-Based Communities. In order to extract user communities that consist of like-minded users, who have contributed to the same topics with the same temporal behavior and contribution degrees, we first calculate pairwise similarity between two users by computing the 2D variation of cross correlation measure on their corresponding user-topic contribution time series. Formally, the $2D$ cross-correlation measure of two matrices, such as $A_{I \times J}$ and $B_{I \times J}$, denoted by $XC_{(2I-1) \times (2J-1)}$, is calculated as follows:

$$XC[l, m](A, B) = \sum_{i=0}^{I-1} \sum_{j=0}^{J-1} A[i, j] \times B^*[i - l, j - m] \tag{1}$$

where B^* denotes the complex conjugate of B. Intuitively, the $2D$ cross-correlation measure slides one matrix over the other and sums up the multiplications of the overlapping elements. Positive row index l corresponds to a downward shift of the rows of A over B and negative column index m indicates a leftward shift of the columns. A maximum correlation occurs at $XC[0, 0]$ if the two matrix are the same without any shift.

Now, given the fact that user-topic contribution time series of each user, i.e., $\mathbf{Y}^u$, towards the K topics over L time intervals can be represented as a $K \times L$

matrix, the similarity distance between two users can be calculated through the $2D$ cross-correlation of their user-topic contribution time series without a shift in time. The normalized similarity distance of two users u_1 and u_2, denoted as $usd(u_1, u_2)$, is defined as follows:

$$usd(u_1, u_2) = \frac{XC[0,0](\mathbf{Y}^{u_1}, \mathbf{Y}^{u_2})}{\sqrt{(\mathbf{Y}^{u_1} \cdot \mathbf{Y}^{u_1}) \times (\mathbf{Y}^{u_2} \cdot \mathbf{Y}^{u_2})}} \tag{2}$$

where $\mathbf{Y}^u$ is the user-topic contribution time series for user u.

We are now able to calculate the correlation distance between all user pairs and build a weighted user graph. A user graph, $UG = <\mathbb{V}, \mathbb{E}>$, is a weighted undirected graph where $\mathbb{V}$ is the set of all users and $\mathbb{E} = \{usd(u_i, u_j) | \forall u_i, u_j \in \mathbb{V}, i \neq j\}$. After constructing the user graph, we apply a graph partitioning algorithm namely the Louvain Method (LM) [4] to extract clusters of users that form latent communities. Louvain is a greedy optimization method that initially finds small communities by locally maximizing the modularity and consequently performs the same procedure on the new graph by considering each community extracted in the previous step as a single vertex. We find the Louvain method to be suitable for topic extraction because of its following characteristics: (i) this algorithm can be applied to weighted graphs; (ii) it does not require a priori knowledge of the number of communities, and (iii) it is computationally very efficient when applied to large and dense graphs. While modularity maximization is NP-hard, the complexity of Louvain's greedy implementation is $O(nlogn)$, where n is the number of vertices in the graph [4,14]. To calculate the degree of interest of each extracted user community UC at each time interval t to each topic $\mathbf{z} \in \mathbb{Z}$, named community-topic contribution time series, we sum over the user-topic contribution time series of all users who are the members of the user community, i.e., $\mathbf{y}_t^{UC} = \sum_{u \in UC} \mathbf{y}_t^u$.

4   Experiments

In this section we show the performance of our proposed approach in the applications of news recommendation and timestamp prediction on a dataset obtained from the Twitter.

4.1   Dataset

In our experiments, we use a Twitter dataset which is publicly available[2] and presented by Abel et al. [2]. It consists of approximately 3M tweets posted by 136 K users between Nov. 1 and Dec. 31, 2010. We annotated the text of each tweet with Wikipedia concepts using the TAGME RESTful API [8] which resulted in 351 K concepts.

---

[2] http://wis.ewi.tudelft.nl/websci11/.

4.2 Qualitative Analysis

To make the overall idea of our time-sensitive community detection approach more clear, in Fig. 1, we utilize heat map to visualize the contribution of three sample users in each day for the two months period of our Twitter dataset between Nov. 1 and Dec. 31 (X-axis) to each extracted topic (Y-axis). The number of topics is set to 50. As illustrated in Fig. 1, the users namely @amfumero, @lazarogonzale and @ebarrera highly contribute to similar topics z_1, z_{12} and z_{34}. However, @ebarrera evidently shows a time shift in his contribution to topic z_{12}. While the first two users' contributions to topic z_{12} span from mid November to early December, @ebarrera starts his posts about the topic from mid December to late December. We do believe that the former two users should be the members of a same community different from the community of the latter.

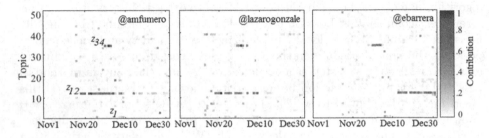

Fig. 1. The heat map of user-topic contribution time series for three Twitter users.

To discover user communities based on the similarity distance between their user-topic contribution time series, we use the implementation of Louvain method available on Pajek[3]. Six of our extracted communities are shown in Fig. 2. Likewise, we use heat map to visualize their community-topic contribution time series in Fig. 3. As seen in this figure, the users of communities UC_1 and UC_2 discuss the same set of topics, z_1, z_{12}, and z_{34}, but in different time intervals with regard to topic z_{12} with a week of delay. Topic-based community detection methods that overlook temporal behavior of users would merge the users of such communities like @amfumero, @lazarogonzale and @ebarrera into a single community. However, our time-sensitive topic-based community detection method incorporates this deviation in users' temporal behavior and therefore group @amfumero and @lazarogonzale into user community UC_1 and @ebarrera to user community UC_2. We believe that this is an important distinguishing feature of our work. To highlight this feature, consider the case of a news recommendation application. It would be unreasonable to recommend a news article on topic z_{12} in December to members of UC_1 who discussed the topic in late November but would make total sense to recommend the same article to members of UC_2 who start to pursue the topic in early December. This is illustrated in Sect. 4.4 by evaluating the performance of a news recommender application. Other communities, UC_3 to UC_6, are formed because there is at least one distinctive topic in their topics of interest. For example, user community UC_3 contributes to topics z_4, z_{34} while members of UC_4 are interested in z_{28}.

---

[3] http://mrvar.fdv.uni-lj.si/pajek/.

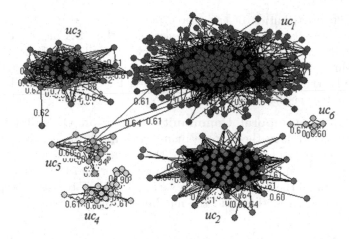

Fig. 2. The output of the Louvain community detection method.

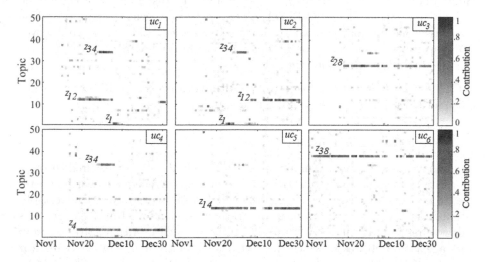

Fig. 3. The heat maps of community-topic contribution time series for respective user communities in Fig. 2.

4.3   Comparison Methods

We investigated the behavior of our temporal topical community detection method by benchmarking it against the following methods:

(CD-LDA) Non-temporal Community Detection Over LDA Topics. To apply LDA, we aggregated all concepts extracted from tweets of a user in each day as a single document and applied the MALLET implementation of LDA[4] on the collection of such documents to discover the latent topics. To compute

---

[4] http://mallet.cs.umass.edu/topics.php.

the probability distribution of each topic per user, we aggregated the tweets published by a user and performed inference using the LDA model. To discover user communities, we then calculated the pairwise similarity of users based on the cosine similarity of their probability distributions and built a weighted user graph. Finally, the Louvain method was applied for extracting the latent user communities. We set the number of topics to 50.

(CD-TOT) Non-temporal Community Detection Over Topics Over Time. The goal of this method is to evaluate the impact of considering temporal issues on extracting topics. We aggregated the concepts which are mentioned in the tweets of a user in each day as a single document and the date of that day as the timestamp of the document. Then we applied TOT to extract topics. The community detection method and the number of topics that are used are the same as CD-LDA.

(TCD-LDA) Temporal Community Detection Over LDA Topics. In this method, we extract topics and the probability distribution of each topic per user per day, similar to CD-LDA. Then, each user u will be represented by user-topic contribution time series toward the detected LDA topics. To discover communities, we used the proposed method introduced in Sect. 3 under time-sensitive topic-based communities.

(TCD-TOT) Temporal Community Detection Over Topics Over Time. This method is our proposed approach in this paper.

4.4   News Recommendation

The performance of the community detection method can be measured through observations made at the application level such as news recommendation. Therefore, to evaluate our work, we deploy the widely used news recommender application and adopt the same evaluation strategy as mentioned in [2,21]. To this end, we first build ground truth by collecting, for each user, news articles from news agencies' websites to which the user has explicitly linked in her tweets (or retweets). Given this gold standard, the objective is to see whether it would be possible to recommend the right news articles to each Twitter user. The right news articles for a given user would be the ones that they had posted in their timeline. Since our main objective is not to propose a news recommender application, we adopt a simple recommender algorithm as proposed in [2] as follows:

Given K topics extracted in time interval T, we represent each user community UC as a K-tuple vector $\mathbf{I}^{UC}$ over the extracted topics which is calculated by aggregating the community-topic contribution time series of each user community UC over the whole time period T of L consecutive time intervals. Formally,

$$\mathbf{I}^{UC} = \sum_{1 \leq t \leq L} \mathbf{y}_t^{UC} \tag{3}$$

where y_t^{UC} is the K-tuple community-topic contribution vector of user community UC in time interval t. We also represent each news article A as a weighted vector $\mathbf{I}^A = (i_1^A, i_2^A, ..., i_K^A)$ over the K extracted topics where i_k^A denotes the degree of A's relatedness to topic $\mathbf{z}_k$ and is calculated based on the frequency of the constituent concepts of $\mathbf{z}_k$ in news article A. Given $\mathbf{I}^A$ and $\mathbf{I}^{UC}$, we recommend article A to the users of user community UC based on the cosine similarity of their corresponding vectors.

We use standard information retrieval metrics: Mean Reciprocal Rank (MRR) which is the inverse of the first position that a correct item occurs within the ranked recommendations and Success at rank K (S@K) that shows the probability that at least one correct item occurs within the top-k ranked recommendations. In the following, the five methods are compared to each other in terms of MRR, S@1 and S@10. Figure 4 summarizes the results.

CD-LDA and CD-TOT use the same community detection method over different kinds of topics. CD-LDA uses LDA which extracts topics by only considering co-occurrence between keywords. However, CD-TOT considers time jointly with keywords co-occurrence patterns by applying TOT as its topic detection method. By comparing these two methods it is possible to evaluate the degree of granularity of topics in topical communities for developing news recommendation applications. As illustrated in Fig. 4, CD-TOT outperforms CD-LDA in terms of three metrics. Our observations show that TOT topics which are better localized in time compared to topics of LDA are more effective in recommending related news based on user's interests. By comparing the results of the TCD-LDA and TCD-TOT, this observation is also confirmed.

The difference between the CD-LDA and TCD-LDA is in their community detection method. TCD-LDA produces communities of users who have contributed to the same topics with the same temporal behavior and contribution degrees. However, communities extracted by CD-LDA are based on the similarity of user's topics of interest. As Fig. 4 shows, TCD-LDA outperforms CD-LDA in terms of all metrics. This means that incorporating temporal aspects of users'

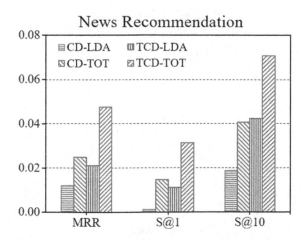

Fig. 4. Comparison between different methods in context of news recommendation.

interests to extract like-minded communities leads to more cohesive communities that consequently results in higher quality news recommendations. This is also confirmed by comparing the results of CD-TOT and TCD-TOT.

Further, when looking at the results in Fig. 4, it can be concluded that our proposed method, i.e. TCD-TOT, that produces temporal topical communities leads to highest quality of recommendations in comparison with the other methods. The reason lies in the fact that our approach incorporate user temporal behavior and contribution degrees of users over TOT topics that takes into account both time and co-occurrence of keywords to discover topics.

4.5   Timestamp Prediction

In this section, adopted from [19], we evaluate our temporal communities in terms of the capability of predicting the timestamp of tweets given their content. To do so, given the extracted user communities and a tweet, we first find the most similar community to the tweet in terms of the degree of their contribution to each topic. Then, within the selected community, we predict the tweet's timestamp by taking the peak time for the tweet's topic within that community. As our test set, we randomly selected 1,000 tweets which were not included when our communities were built (leave-out protocol) and were annotated with at least 5 concepts. The results of comparing TCD-TOT and CD-TOT are illustrated in Fig. 5 in terms of prediction accuracy as a function of the tolerance range in days (e.g. when tolerance range is set to 20 days, we consider a predicted timestamp as accurate if its difference with ground truth timestamp is within 20 days). Based on the results, we can see that applying our proposed temporal community detection method over TOT topics, TCD-TOT, leads to achieving more accurate prediction results compared to the non-temporal community detection method, CD-TOT.

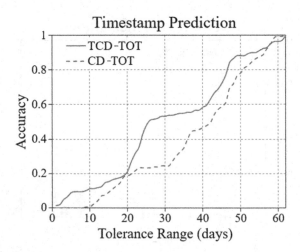

Fig. 5. The results of timestamp prediction in terms of prediction accuracy as a function of the tolerance range in days.

5   Conclusion

In this paper, we have addressed the problem of identifying time-sensitive topic-based user communities on Twitter. We focus on communities which are formed not only based on the different topics of interest to the users, but also based on the temporality of the user contributions. To this end, we have utilized Topic over Time (TOT) that jointly captures term co-occurrences and locality of those patterns over time to discover active topics of Twitter in a given time period based on the tweets published in that time period. Further, we extract user communities through novel multivariate time series modeling of like-minded users. We investigated the performance of the proposed approach in the application of news recommendation and timestamp prediction. The experimental results indicate that the proposed model achieves better performance in comparison with non-temporal community detection methods.

References

1. Abdelbary, H., ElKorany, A., Bahgat, R.: Utilizing deep learning for content-based community detection. In: Science and Information Conference, pp. 777–784 (2014)
2. Abel, F., Gao, Q., Houben, G.-J., Tao, K.: Analyzing user modeling on twitter for personalized news recommendations. In: Konstan, J.A., Conejo, R., Marzo, J.L., Oliver, N. (eds.) UMAP 2011. LNCS, vol. 6787, pp. 1–12. Springer, Heidelberg (2011)
3. Blei, D.M., Ng, A.Y., Jordan, M.I.: Latent dirichlet allocation. In: Advances in Neural Information Processing Systems, NIPS 2001, vol. 14, pp. 601–608 (2001)
4. Blondel, V., Guillaume, J., Lambiotte, R., Lefebvre, E.: Fast unfolding of communities in large networks. J. Stat. Mech. Theor. Exp. **2008**(10), P10008 (2008)
5. Chen, J.M., Tang, Y., Li, J.G., Mao, C.J., Xiao, J.: Community-based scholar recommendation modeling in academic social network sites. In: Huang, Z., Liu, C., He, J., Huang, G. (eds.) WISE Workshops 2013. LNCS, vol. 8182, pp. 325–334. Springer, Heidelberg (2014)
6. Deng, Q., Li, Z., Zhang, X., Xia, J.: Interaction-based social relationship type identification in microblog. In: International Workshop on Behavior and Social Informatics and Computing, pp. 151–164 (2013)
7. Ding, Y.: Community detection: topological vs. topical. J. Infometrics **5**(4), 498–514 (2011)
8. Ferragina, P., Scaiella, U.: Fast and accurate annotation of short texts with wikipedia pages. J. IEEE Softw. **29**(1), 70–75 (2012)
9. Hong, L., Davison, B.: Empirical study of topic modeling in twitter. In: 1st ACM Workshop on Social Media Analytics, pp. 80–88 (2010)
10. Hu, Z., Yao, J., Cui, B.: User group oriented temporal dynamics exploration. In: AAAI 2014, pp. 66–72 (2014)
11. Liu, H., Chen, H., Lin, M., Wu, Y.: Community detection based on topic distance in social tagging networks. TELKOMNIKA Indonesian J. Electr. Eng. **12**(5), 4038–4049 (2014)
12. Mehrotra, R., Sanner, S., Buntine, W., Xie, L.: Improving lda topic models for microblogs via tweet pooling and automatic labeling. In: SIGIR 2013, pp. 889–892. ACM (2013)

13. Natarajan, N., Sen, P., Chaoji, V.: Community detection in content-sharing social networks. In: ASONAM 2013, pp. 82–89 (2013)
14. Noack, A., Rotta, R.: Multilevel local search algorithms for modularity clustering. In: SEA 2009, pp. 257–268 (2009)
15. Peng, D., Lei, X., Huang, T.: Dich: a framework for discovering implicit communities hidden in tweets. World Wide Web **18**(4), 795–818 (2014)
16. Rosen-Zvi, M., Griffiths, T., Steyvers, M., Smyth, P.: The author-topic model for authors and documents. In: UAI 2004, pp. 487–494 (2004)
17. Sachan, M., Contractor, D., Faruquie, T., Subramaniam, L.: Using content and interactions for discovering communities in social networks. In: WWW 2012, pp. 331–340 (2012)
18. Sriram, B., Fuhry, D., Demir, E., Ferhatosmanoglu, H., Demirbas, M.: Short text classification in twitter to improve information filtering. In: SIGIR 10, pp. 841–842 (2010)
19. Wang, X., McCallum, A.: Topics over time: a non-markov continuous-time model of topical trends. In: KDD 2006, pp. 424–433 (2006)
20. Yin, Z., Cao, L., Gu, Q., Han, J.: Latent community topic analysis: integration of community discovery with topic modeling. ACM Trans. Intell. Syst. Technol. **3**(4), 63:1–63:21 (2012)
21. Zarrinkalam, F., Fani, H., Bagheri, E., Kahani, M., Du, W.: Semantics-enabled user interest detection from twitter. In: WI-IAT 2015, pp. 469–476 (2015)
22. Zhao, G., Lee, M., Hsu, W., Chen, W., Hu, H.: Community-based user recommendation in uni-directional social networks. In: CIKM 2015, pp. 189–198 (2013)
23. Zhou, D., Manavoglu, E., Li, J., Giles, C., Zha, H.: Probabilistic models for discovering e-communities. In: WWW 2006, pp. 173–182 (2006)

Reasoning and Learning

On Tree Structures Used by Simple Propagation

Anders L. Madsen[1,2(✉)], Cory J. Butz[3], Jhonatan S. Oliveira[3],
and André E. dos Santos[3]

[1] Hugin Expert A/S, Aalborg, Denmark
anders@hugin.com
[2] Department of Computer Science, Aalborg University, Aalborg, Denmark
[3] Department of Computer Science, University of Regina, Regina, Canada

Abstract. *Simple Propagation* (SP) is a new junction tree-based algorithm for probabilistic inference in discrete Bayesian networks. It is similar to Lazy Propagation, but uses a simpler approach to exploit the factorization during message computation. The message construction is based on a *one-in, one-out*-principle meaning a potential has at least one non-evidence variable in the separator and at least one non-evidence variable not in the separator. This paper considers the use of different tree structures to guide the message passing in SP and reports on an experimental analysis using a set of real-world Bayesian networks.

Keywords: Bayesian networks · Inference · Simple propagation

1 Introduction

The Simple Propagation (SP) algorithm was introduced in [1] as a new algorithm for belief update in Bayesian networks. SP is similar to LP as it proceeds by message passing in a join tree or a junction tree taking advantage of a decomposition of clique and separator potentials, but it takes a simpler approach to the computation of potentials in the message passing phase of belief update. In SP, message construction is based on a *one-in, one-out*-principle meaning a potential has at least one non-evidence variable in the separator and at least one non-evidence variable not in the separator.

In this paper, we empirically analyze the role that the choice of the tree structure plays when SP conducts inference. We consider three kinds of tree structures, namely, optimal (or believed to be close to optimal) junction trees, junction trees produced by the *fill-in-weight* heuristic, and *maximal prime subgraph decomposition* (MPD) trees [4]. When using optimal junction trees or the ones produced by the fill-in-weight heuristic, experimental results shows that in fewer than half the cases, the space cost of SP is almost invariant. Moreover, the time cost of SP in these two categories is approximately the same in 14 out of 28 cases. On the other hand, SP often runs out of memory on MPD trees.

© Springer International Publishing Switzerland 2016
R. Khoury and C. Drummond (Eds.): Canadian AI 2016, LNAI 9673, pp. 207–212, 2016.
DOI: 10.1007/978-3-319-34111-8_26

2 Preliminaries and Notation

A Bayesian network BN $\mathcal{N} = (\mathcal{X}, G, \mathcal{P})$ consists of a set of discrete random variables $\mathcal{X} = \{X_1, \ldots, X_n\}$, where $\mathrm{dom}(X)$ is the state space of X and $||X|| = |\mathrm{dom}(X)|$, an acyclic, directed graph (DAG) $G = (V, E)$, where $V \sim \mathcal{X}$ is the set of vertices and E is the set of edges, and a set $\mathcal{P}$ of conditional probability distributions (CPDs). A BN represents a decomposition of the joint probability distribution $P(\mathcal{X}) = \prod_{X \in \mathcal{X}} P(X \mid \mathrm{pa}(X))$, where $\mathrm{pa}(X)$ denotes the parents of X in G and $\mathrm{fa}(X) = \mathrm{pa}(X) \cup \{X\}$. For example, a BN $\mathcal{N} = (\mathcal{X}, G, \mathcal{P})$ over six variables is shown by G in Fig. 1, where we assume $||X_1|| = ||X_4|| = 3$, $||X_2|| = ||X_5|| = ||X_6|| = 4$, and $||X_3|| = 2$.

Belief update is the task of computing the posterior marginal probability distribution $P(X \mid \epsilon)$ for each non-observed variable $X \in \mathcal{X} \setminus \mathcal{X}_\epsilon$ given evidence ϵ assumed to be instantiations of variables $\mathcal{X}(\epsilon)$. A variable X is a *barren* w.r.t. a set $T \subseteq \mathcal{X}$, evidence ϵ, and DAG G, if $X \notin T$, $X \notin \mathcal{X}_\epsilon$ and X only has barren descendants in G, if any [6]. The notion of barren variables can be extended to graphs with both directed and undirected edges [3].

A junction tree $T = (\mathcal{C}, \mathcal{S})$ of $\mathcal{N}$ is created by moralization and triangulation of G (see, e.g., [2]), where $\mathcal{C}$ denotes the cliques and $\mathcal{S}$ denotes the separators of T. The state space size of clique or separator A is defined as $s(A) = \prod_{X \in A} ||X||$.

3 Simple Propagation

SP [1] can be considered a simplification of LP [5] with respect to how messages are computed. The basic idea is to maintain a decomposition of clique and separator potentials and to exploit independence relations induced by evidence and barren variables during belief update. Each CPD $P \in \mathcal{P}$ is associated with a clique C s.t. $\mathrm{dom}(P) \subseteq C$, where P is reduced to reflect ϵ. Next messages are passed over the computational tree structure relative to a selected root of T.

In SP, the *one-in, one-out*-principle is applied when a clique sends a message to a neighbouring clique over a separator S. The *one-in, one-out*-principle states that a probability potential ϕ has at least one non-evidence variable in S and another variable X not in S and SP can eliminate X. The computation of messages in SP is performed using the Simple Message Computation (SMC) algorithm shown as Algorithm 1. The SUMOUT algorithm corresponds to Variable Elimination summing out X from the product of $\Phi_X = \{\phi \in \mathcal{F} \mid X \in \mathrm{dom}(\phi)\}$ and replacing Φ_X in $\mathcal{F}$ with the result.

	Procedure $SMC(\mathcal{F}, S, \mathcal{X}(\epsilon))$
1	$\mathcal{F} = \mathrm{REMOVEBARREN}(\mathcal{F}, S)$
2	**while** $\exists \phi(\mathcal{Y}) \in \mathcal{F}$ *with* $X \notin (S \setminus \mathcal{X}(\epsilon))$ *and* $X' \in (S \setminus \mathcal{X}(\epsilon))$ **do**
3	$\quad \mid \mathcal{F} = \mathrm{SUMOUT}(X, \mathcal{F})$
	end
4	**return** $\{\phi(\mathcal{Y}) \in \mathcal{F} \mid \mathcal{Y} \subseteq S\}$

Algorithm 1. The Simple Message Computation algorithm.

In addition to passing the potentials created by Algorithm 1, SP also passes any potential ϕ for which $\text{dom}(\phi) \subseteq S$, i.e., the potentials where the domain is a subset of the separator. After a full round of message passing, the posterior marginal $P(X \,|\, \epsilon)$ can be computed from any clique or separator containing X.

4 Computational Tree Structure

The computational tree structure induces a partial order on the set of possible (implicit) elimination orders produced by the SMC algorithm during the message passing step of belief update. As SP maintains a decomposition of clique and separator potentials, it becomes sensitive to the order in which variables are eliminated during message passing. The number of variables to be eliminated when sending a message from A to B is determined by $A \setminus B$ which is influenced by $|A|$ and $|B|$. Thus, the impact of the (implicit) elimination order tends to increase with $|A \setminus B|$. As SP uses the same potentials for message computation as LP, we can rely on the correctness considerations of [4].

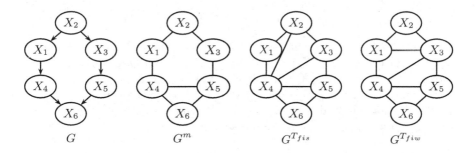

Fig. 1. A DAG G with G^m and triangulations using fill-in-size and fill-in-weight.

To illustrate the impact of using different computational tree structures, consider the moral graph G^m and two triangulations of G^m in Fig. 1. The fill-in-size heuristic may produce the elimination order $\sigma_{fis} = (X_6, X_1, X_5, X_2, X_3, X_4)$, while the order $\sigma_{fiw} = (X_6, X_5, X_2, X_1, X_3, X_4)$ will be produced by the fill-in-weight heuristic (optimal w.r.t. $s(T)$).

The junction trees T_{fis} and T_{fiw} produced are shown in Fig. 2. This figure also shows the MPD tree T_{MPD} for $\mathcal{N}$ constructed from T_{fiw} by merging cliques connected by incomplete separators in G^m. For simplicity, we assume that clique $X_4 X_5 X_6$ is selected as root of each tree.

Due to space limitations, we only consider the first message passed in each of the computational trees shown in Fig. 2. For T_{fis}, the first message is from clique $X_1 X_2 X_4$ to $X_2 X_3 X_4$. The clique potential $\pi_{X_1 X_2 X_4} = (\{P(X_2), P(X_1 \,|\, X_2), P(X_4 \,|\, X_1)\})$ has two CPDs satisfying the *one-in, one-out*-principle: $P(X_1 \,|\, X_2)$ and $P(X_4 \,|\, X_1)$. Notice $P(X_2)$ does not satisfy the principle as all domain variables are in S. Both $P(X_1 \,|\, X_2)$ and $P(X_4 \,|\, X_1)$ satisfy the condition in line 2

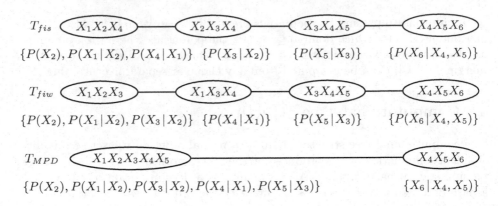

Fig. 2. Three different computational tree structures.

of the SMC algorithm and variable X_1 can be eliminated using the SumOut-algorithm producing $\phi(X_4 \mid X_2)$. The message is then

$$\pi_{X_1 X_2 X_4 \rightarrow X_2 X_3 X_4} = (\{P(X_2), \phi(X_4 \mid X_2)\}). \tag{1}$$

For T_{fiw}, the first message is from clique $X_1 X_2 X_3$ to $X_1 X_3 X_4$. The potential $\pi_{X_1 X_2 X_3} = (\{P(X_2), P(X_1 \mid X_2), P(X_3 \mid X_2)\})$ has two CPDs satisfying the *one-in, one-out*-principle: $P(X_1 \mid X_2)$ and $P(X_3 \mid X_2)$. Notice $P(X_2)$ does not satisfy the principle. Both $P(X_1 \mid X_2)$ and $P(X_3 \mid X_2)$ satisfy the condition in line 2 of the SMC algorithm and variable X_2 can be eliminated using the SumOut-algorithm producing $\phi(X_1, X_3)$. The message is then

$$\pi_{X_1 X_2 X_3 \rightarrow X_1 X_3 X_4} = (\{\phi(X_1, X_3)\}). \tag{2}$$

For T_{MPD}, the first message is from clique $X_1 X_2 X_3 X_4 X_5$ to $X_4 X_5 X_6$. Potential $\pi_{X_1 X_2 X_3 X_4 X_5} = (\{P(X_2), P(X_1 \mid X_2), P(X_3 \mid X_2), P(X_4 \mid X_1), P(X_5 \mid X_3)\})$ has two CPDs satisfying the *one-in, one-out*-principle: $P(X_4 \mid X_1)$ and $P(X_5 \mid X_3)$. Assume variable X_1 selected for elimination first producing the updated set $\mathcal{F} = \{P(X_2), P(X_3 \mid X_2), P(X_5 \mid X_3), \phi(X_4 \mid X_2)\}$. Next, $P(X_5 \mid X_3)$ and $\phi(X_4 \mid X_2)$ satisfy the condition in line 2. Assume variable X_2 is selected for elimination producing the updated set $\mathcal{F} = \{P(X_5 \mid X_3), \phi(X_4, X_3)\}$. Next, $P(X_5 \mid X_3)$ and $\phi(X_3, X_4)$ satisfy the condition in line 2. Variable X_3 is the last variable eliminated producing the updated set $\mathcal{F} = \{\phi(X_4, X_5)\}$. The message is then

$$\pi_{X_1 X_2 X_3 X_4 X_5 \rightarrow X_4 X_5 X_6} = (\{\phi(X_4, X_5)\}). \tag{3}$$

The important point to notice is that different tree structures lead to different messages being constructed by SP, as shown in Eqs. (1), (2), and (3). The next section reports on an empirical evaluation of the performance impact of different tree structures.

5 Experimental Analysis

Table 1 shows statistics on a sample of 10 out of 28 BNs used in the experiments and information on the junction tree $\hat{T}$, the junction tree T_{fiw} and the MPD tree T_{MPD}, where sizes are on a log-scale in base 10. Table 2 shows the average largest factor over 100 sets of random evidence created by SP during belief update (this includes computing marginals using Variable Elimination). The empty entries denote examples where SP ran out of memory on some evidence sets.

In fewer than half the cases, the space cost of SP is almost invariant to the use of $\hat{T}$ or T_{fiw} as the computational structure. These are the cases where $s(\hat{T}) \approx s(T_{fiw})$. In the cases where $s(\hat{T}) << s(T_{fiw})$, there is a large difference in the average size of the largest factor created by SP. SP is not able to perform belief update in 20 out of 28 networks using T_{MPD} as the computational structure.

Table 1. Information on 10 out of the 28 Bayesian networks and the computational tree structures used in the experiments.

$\mathcal{N}$	$|\mathcal{X}|$	$|\hat{C}|$	$|\mathcal{C}_{fiw}|$	$|\mathcal{C}'|$	max $s(\hat{C})$	max $s(C_{fiw})$	max $s(C')$	$s(\hat{T})$	$s(T_{fiw})$	$s(T')$
ADAPT_DX09_T2	671	489	489	284	3.1	3.5	29.6	4	5	30
Amirali_network	681	556	555	461	6.9	7.5	41.5	7	8	42
Heizung.	44	28	28	14	7.6	7.6	29.2	8	8	29
Hepar_II	70	58	58	55	2.6	2.6	2.9	3	3	4
Mildew	35	29	28	15	6.1	6.6	20.6	7	7	21
Munin1	189	162	160	70	7.6	7.9	69.2	8	8	69
andes	223	180	175	79	4.8	5.4	40.0	5	6	40
cc245	245	235	235	232	5.4	5.4	6.0	6	6	6
sacso	2371	1229	1175	980	5.2	6.4	107.5	6	7	107
ship	50	35	35	10	6.6	8.1	35.6	7	8	36

Table 2. Average space cost of belief update using SP.

$\mathcal{N}$	$\mu(\hat{T})$	$\sigma(\hat{T})$	$\mu(T_{fiw})$	$\sigma(T_{fiw})$	$\mu(T_{MPD})$	$\sigma(T_{MPD})$
ADAPT_DX09_T2	520.09	315372	727.13	845205		
Amirali_network	70716.8	3.6E+10	1093871.04	1.3E+13		
Heizung.	4091131	1.3E+14	3542217.5	1.2E+14		
Hepar_II	106.5	15384.4	106.5	15384.4	186.58	71361
Mildew	219317	1.6E+11	580396.08	1.7E+12		
Munin1	2499265	6.2E+13	4158649.69	1.3E+14		
andes	7958.68	3.1E+08	13922.2	1.4E+09		
cc245	21764.6	3.3E+09	21764.6	3.3E+09	37384.3	1.4E+10
sacso	12887.4	1.1E+09	67052.22	4.1E+10		
ship	732996	1.8E+12	8645006.83	8E+14		

The eight networks for which SP can perform belief update have an average size of the largest factor of less than $100,000$ probabilities.

Table 3 shows the average computation cost (wall-clock time) of belief update by SP. In 14 out of 28 cases the time cost of SP using T_{fiw} is approximately the same as the time cost SP using $\hat{T}$. The difference is less than five per cent. Since SP using T_{MPD} is only able to solve the most simple networks, the time cost of SP in this case is low and in many cases comparable to the time cost of $\hat{T}$ and T_{fiw} (at least in absolute terms).

Table 3. Average time cost of belief update using SP (mean and standard deviation).

$\mathcal{N}$	$\mu(\hat{T})$	$\sigma(\hat{T})$	$\mu(T_{fiw})$	$\sigma(T_{fiw})$	$\mu(T_{MPD})$	$\sigma(T_{MPD})$
ADAPT_DX09_T2	0.19	0.011	0.19	0.011		
Amirali_network	0.38	0.043	0.61	0.835		
Heizung.	0.24	0.326	0.23	0.31		
Hepar_II	0.03	0	0.02	0	0.03	0
Mildew	0.04	0.001	0.05	0.006		
Munin1	0.74	3.26	0.98	7.408		
andes	0.13	0.005	0.11	0.004		
cc145	0.07	0.001	0.07	0.001	0.07	0.001
sacso	0.83	1.051	0.74	0.899		
ship	0.16	0.084	1.26	25.073		

6   Conclusion

This paper has investigated the impact of the secondary computational structure used for belief update on time and space performance of SP. The results of a preliminary empirical performance evaluation on a set of real-world Bayesian networks indicate that the tree structure used to control the message passing can have a significant impact on both time and space performance.

References

1. Butz, C.J., Oliveira, J.S., dos Santos, A.E., Madsen, A.L.: Bayesian network inference with simple propagation. In: Proceedings of the Twenty-Ninth International FLAIRS Conference (2016)
2. Jensen, F.V., Jensen, F.: Optimal junction trees. In: Proceedings of UAI, pp. 360–366 (1994)
3. Madsen, A.L.: Variations over the message computation algorithm of lazy propagation. IEEE Trans. Syst. Man Cybern. Part B **36**(3), 636–648 (2006)
4. Madsen, A.L., Butz, C.: On the tree structure used by lazy propagation for inference in Bayesian networks. In: van der Gaag, L.C. (ed.) ECSQARU 2013. LNCS, vol. 7958, pp. 400–411. Springer, Heidelberg (2013)
5. Madsen, A.L., Jensen, F.V.: Lazy propagation: a junction tree inference algorithm based on lazy evaluation. Artif. Intell. **113**(1–2), 203–245 (1999)
6. Shachter, R.D.: Evaluating influence diagrams. Oper. Res. **34**(6), 871–882 (1986)

A Simple Method for Testing Independencies in Bayesian Networks

Cory J. Butz[1]([⊠]), André E. dos Santos[1], Jhonatan S. Oliveira[1],
and Christophe Gonzales[2]

[1] Department of Computer Science, University of Regina, Regina, Canada
{butz,dossantos,oliveira}@cs.uregina.ca
[2] LIP6 - département DESIR, Université Pierre Et Marie Curie, Paris, France
christophe.gonzales@lip6.fr

Abstract. Testing independencies is a fundamental task in reasoning with *Bayesian networks* (BNs). In practice, *d-separation* is often utilized for this task, since it has linear-time complexity. However, many have had difficulties in understanding d-separation in BNs. An equivalent method that is easier to understand, called *m-separation*, transforms the problem from directed separation in BNs into classical separation in undirected graphs. Two main steps of this transformation are pruning the BN and adding undirected edges.

In this paper, we propose *u-separation* as an even simpler method for testing independencies in a BN. Our approach also converts the problem into classical separation in an undirected graph. However, our method is based upon the novel concepts of inaugural variables and rationalization. Thereby, the primary advantage of u-separation over m-separation is that m-separation can prune unnecessarily and add superfluous edges. Hence, u-separation is a simpler method in this respect.

Keywords: Bayesian networks · Testing independencies · d-separation · m-separation

1 Introduction

Pearl [9] states that perhaps the founding of *Bayesian networks* (BNs) [3,4,7] made its greatest impact through the notion of d-separation. *d-Separation* [8] is a graphical method for deciding which conditional independence relations are implied by the *directed acyclic graph* (DAG) of a BN. To test whether two sets X and Z of variables are conditionally independent given a third set Y of variables, denoted $I(X, Y, Z)$, d-separation checks whether every path from X to Z is "blocked" by Y in the DAG. d-Separation considers every variable on each of these paths. Each variable is classified into one of three categories. The same variable can assume different classifications depending on which path is being considered. Depending on the direction of the edges, sometimes blocking works intuitively. On the contrary, sometimes a path is not "blocked" by Y even though it necessarily traverses Y. Unfortunately, the above has led to many

© Springer International Publishing Switzerland 2016
R. Khoury and C. Drummond (Eds.): Canadian AI 2016, LNAI 9673, pp. 213–223, 2016.
DOI: 10.1007/978-3-319-34111-8_27

having difficulties in understanding d-separation [10] even though there exists a linear-time implementation of d-separation [2].

m-Separation [5,13], an equivalent method for testing independencies in a BN, seeks to avoid the confusion associated with the directionality of edges in a DAG by turning the problem into classical separation in an undirected graph. The undirected graph is constructed using a two-step process. First, a sub-DAG is built by pruning the DAG of the BN. Second, the constructed sub-DAG is moralized. The *moralization* [6] of a DAG involves adding an undirected edge between every pair of variables with a common child and then dropping directionality. The independence $I(X, Y, Z)$ holds in the BN if and only if X and Z are separated by Y in the undirected graph.

The main contribution of this paper is an even simpler method for testing independencies in BNs, called *undirected separation* (*u-separation*). Like m-separation, we seek to apply classical separation in an undirected graph built from the given BN. We prove that only *inaugural* [1] variables need be pruned from the BN. Next, we suggest the notion of *rationalization* as a refinement of moralization. Unlike moralization, rationalization only adds an undirected edge between parents of a common child when the common child is in Y of $I(X, Y, Z)$. We establish that u-separation is equivalent to m-separation. Two advantages of u-separation are that it can prune fewer variables than m-separation and add fewer undirected edges than m-separation. Given the prominence of d-separation [9], the novel method of u-separation provides a clearer understanding of BNs.

This paper is organized as follows. In Sect. 2, m-separation is reviewed. Section 3 presents u-separation. The equivalence between m-separation and u-separation is established in Sect. 4. In Sect. 5, advantages of u-separation are provided. Conclusions are given in Sect. 6.

2   Background

Let $U = \{v_1, v_2, \ldots, v_n\}$ be a finite set of variables. Let $\mathcal{B}$ be a *directed acyclic graph* (DAG) on U. A *directed path* from v_1 to v_k is a sequence $v_1, v_2, \ldots, v_k$ with directed edges (v_i, v_{i+1}) in $\mathcal{B}$, $i = 1, 2, \ldots, k - 1$. For each $v_i \in U$, the *ancestors* of v_i, denoted $An(v_i)$, are those variables having a directed path to v_i, while the *descendants* of v_i, denoted $De(v_i)$, are those variables to which v_i has a directed path. For a set $X \subseteq U$, we define $An(X)$ and $De(X)$ in the obvious way. The *children* $Ch(v_i)$ and *parents* $Pa(v_i)$ of v_i are those v_j such that $(v_i, v_j) \in \mathcal{B}$ and $(v_j, v_i) \in \mathcal{B}$, respectively. An *undirected path* in a DAG is a path ignoring directions. A *path* in an undirected graph is defined similarly. The *moralization* [6] of a DAG is the undirected graph formed by adding an undirected edge between each pair of parents of a common child and then dropping directionality. A singleton set $\{v\}$ may be written as v, $\{v_1, v_2, \ldots, v_n\}$ as $v_1 v_2 \cdots v_n$, and $X \cup Y$ as XY.

A *Bayesian network* (BN) [3,4,7] is a DAG $\mathcal{B}$ on U together with *conditional probability tables* (CPTs) $P(v_1 | Pa(v_1))$, $P(v_2 | Pa(v_2))$, ..., $P(v_n | Pa(v_n))$. For example, Fig. 1 shows a BN, where CPTs $P(a)$, $P(b)$, ..., $P(j|i)$ are not provided.

We call $\mathcal{B}$ a BN, if no confusion arises. The product of the CPTs for $\mathcal{B}$ on U is a *joint probability distribution* $P(U)$ [7]. The *conditional independence* [12] of X and Z given Y holding in $P(U)$ is denoted $I_P(X, Y, Z)$. It is known that if $I(X, Y, Z)$ holds in $\mathcal{B}$, then $I_P(X, Y, Z)$ holds in $P(U)$.

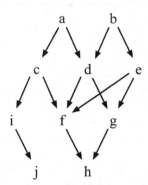

Fig. 1. A BN $\mathcal{B}$ modified from [3].

m-Separation [5,13] is an equivalent method to *d-separation* [7] for testing independencies in BNs. Let X, Y, and Z be pairwise disjoint sets of variables in a BN $\mathcal{B}$. m-Separation tests $I(X, Y, Z)$ with four steps: (i) prune $\mathcal{B}$ onto $XYZ \cup An(XYZ)$; (ii) construct the moralization of the sub-DAG from (i); (iii) delete Y and its incident edges from the undirected graph built in (ii); and (iv) if there exists a path from (any variable in) X to (any variable in) Z in (iii), then $I(X, Y, Z)$ does not hold; otherwise, $I(X, Y, Z)$ holds.

Example 1. Consider testing $I(a, de, g)$ using m-separation in the BN $\mathcal{B}$ of Fig. 1. In step (i), the sub-DAG constructed on $\{a, d, e, g\} \cup An(\{a, d, e, g\}) = \{a, b, d, e, g\}$ is illustrated in Fig. 2(i). In step (ii), the moralization of the sub-DAG is depicted in Fig. 2(ii), where undirected edges (a, b) and (d, e) were added and then directionality was dropped. In step (iii), variables d and e and their

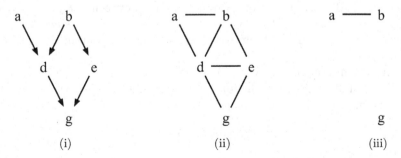

Fig. 2. When testing $I(a, de, g)$ in the BN $\mathcal{B}$ in Fig. 1, m-separation builds the sub-DAG in (i), determines the moralization in (ii), and deletes $Y = \{d, e\}$ and the incident edges, giving (iii).

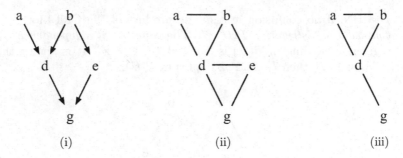

Fig. 3. When testing $I(a, e, g)$ in the BN $\mathcal{B}$ in Fig. 1, m-separation builds the sub-DAG in (i), determines the moralization in (ii), and deletes $Y = \{e\}$ and the incident edges, yielding (iii).

incident edges (a, d), (b, d), (b, e), (d, e), (d, g), and (e, g) are deleted, yielding the undirected graph in Fig. 2(iii). Since there does not exist a path from a to g in Fig. 2(iii), $I(a, de, g)$ holds in $\mathcal{B}$ by m-separation.

Example 2. Consider testing $I(a, e, g)$ using m-separation in the BN $\mathcal{B}$ of Fig. 1. In step (i), the sub-DAG constructed is illustrated in Fig. 3(i). In step (ii), the moralization of the sub-DAG is depicted in Fig. 3(ii), where undirected edges (a, b) and (d, e) were added and then directionality was dropped. In step (iii), variable e and the incident edges (b, e), (d, e), and (e, g) are deleted, giving Fig. 3(iii). Since there exists a path from a to g in Fig. 3(iii), $I(a, e, g)$ does not hold in $\mathcal{B}$ by m-separation.

3   u-Separation

Our simplification of m-separation, called *undirected separation (u-separation)*, requires the notion of inaugural variables and a refinement of moralization.

A variable v_k is called a *v-structure* [11] in a BN $\mathcal{B}$, if $\mathcal{B}$ contains directed edges (v_i, v_k) and (v_j, v_k), but not a directed edge between variables v_i and v_j in $\mathcal{B}$. For example, variable f is a v-structure in BN $\mathcal{B}$ of Fig. 1, since $\mathcal{B}$ contains directed edges (d, f) and (e, f), for instance, and does not contain either directed edges (d, e) or (e, d).

Given an independence $I(X, Y, Z)$ to be tested in a BN $\mathcal{B}$, a variable v is *inaugural* [1], if at least one of the following two conditions is satisfied: (i) v is a v-structure and

$$(\{v\} \cup De(v)) \cap XYZ = \emptyset; \tag{1}$$

or (ii) v is a descendant of a variable satisfying (i). With respect to a given independence $I(X, Y, Z)$ being tested against a BN $\mathcal{B}$, we denote by V the set of all inaugural variables.

Example 3. Consider testing $I(a, de, g)$ in the BN $\mathcal{B}$ of Fig. 1. Variable f is inaugural, since it is a v-structure and, by (1),

$$(\{f\} \cup \{h\}) \cap \{a, d, e, g\} = \emptyset.$$

Consequently, by condition (ii), h is also inaugural, since h is a descendant of f. On the contrary, variable d is a v-structure, but is not inaugural, since

$$(\{d\} \cup \{f, g, h\}) \cap \{a, d, e, g\} \neq \emptyset.$$

Therefore, the set of all inaugural variables in $\mathcal{B}$ for $I(a, de, g)$ is $V = \{f, h\}$.

Definition 1. The *rationalization* of a DAG with respect to an independence $I(X, Y, Z)$ is the undirected graph constructed by adding an undirected edge between parents of a common child v, if $v \in Y$, and then dropping directionality.

Example 4. Given the DAG in Fig. 4(i) and the independence $I(a, de, g)$, the rationalization is in Fig. 4(ii). Here, an undirected edge (a, b) was added between variables a and b, since common child $d \in Y = \{d, e\}$. Directionality was then dropped.

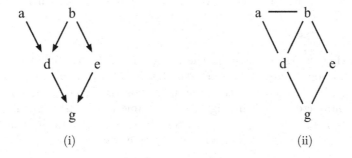

Fig. 4. When testing $I(a, de, g)$ in the DAG in (i), rationalization gives (ii).

Note the key difference between moralization and rationalization. Moralization of Fig. 2(i) adds undirected edges (a, b) and (d, e) as shown in Fig. 2(ii). In contrast, rationalization of Fig. 4(i) only adds undirected edge (a, b) in Fig. 4(ii).

We now formally introduce u-separation.

u-Separation tests independencies in BNs. Let X, Y, and Z be pairwise disjoint sets of variables in a BN $\mathcal{B}$. Then, u-separation tests $I(X, Y, Z)$ with four steps: (i) prune all inaugural variables from $\mathcal{B}$; (ii) construct the rationalization of the sub-DAG in (i); (iii) delete Y and its incident edges from the undirected graph built in (ii); and (iv) if there exists a path from (any variable in) X to (any variable in) Z in (iii), then $I(X, Y, Z)$ does not hold; otherwise, $I(X, Y, Z)$ holds.

Example 5. Let us test $I(a, de, g)$ in the BN $\mathcal{B}$ of Fig. 1 using u-separation. By Example 3, f and h are the only inaugural variables. Thus, step (i) of u-separation prunes inaugural variables f and h from $\mathcal{B}$, yielding the sub-DAG in Fig. 5(i). In step (ii), the rationalization of the constructed sub-DAG gives the undirected graph in Fig. 5(ii). Variables d and e and their incident edges are deleted in step (iii), yielding Fig. 5(iii). In step (iv), there does not exist a path from a to g. Thus, $I(a, de, g)$ holds in $\mathcal{B}$ by u-separation.

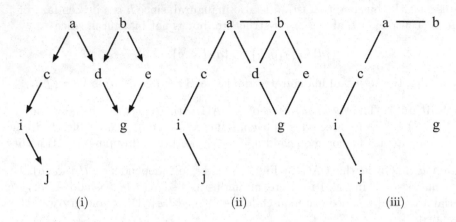

(i)   (ii)   (iii)

Fig. 5. When testing $I(a, de, g)$ in the BN $\mathcal{B}$ in Fig. 1, u-separation builds the sub-DAG in (i) by pruning all inaugural variables, determines the rationalization in (ii), and deletes $Y = \{d, e\}$ and the incident edges, yielding (iii).

Example 6. Let us test $I(a, e, g)$ in the BN $\mathcal{B}$ of Fig. 1 using u-separation. Step (i) of u-separation prunes inaugural variables f and h from $\mathcal{B}$, yielding the sub-DAG in Fig. 6(i). In step (ii), the rationalization of the constructed sub-DAG yields the undirected graph in Fig. 6(ii). Variable e and its incident edges are deleted in step (iii), yielding Fig. 6(iii). Since there exists a path from a to g in Fig. 6(iii), $I(a, e, g)$ does not hold in $\mathcal{B}$ by u-separation.

4   Equivalence of u-Separation and m-Separation

We introduce pertinent notation to establish the equivalence of u-separation and m-separation. Given independence $I(X, Y, Z)$ to be tested in a BN $\mathcal{B}$ on U, we define the following set:

$$W = U - (XYZ \cup An(XYZ)). \tag{2}$$

In other words, W is the set of all variables pruned from $\mathcal{B}$ when m-separation builds the sub-DAG in step (i).

Lemma 1. *Given $I(X, Y, Z)$ to be tested in a BN $\mathcal{B}$. Let V be the set of all inaugural variables. Let W be the set of all variables pruned by m-separation in step (i). Then, $V \subseteq W$.*

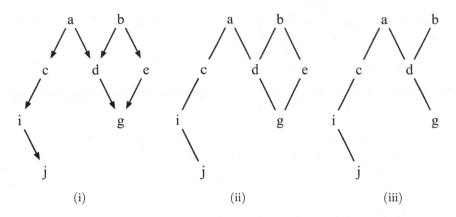

Fig. 6. When testing $I(a, e, g)$ in the BN $\mathcal{B}$ in Fig. 1, u-separation builds the sub-DAG in (i) by pruning all inaugural variables, determines the rationalization in (ii), and deletes $Y = \{e\}$ and the incident edges, yielding (iii).

Proof. Let $v \in V$. By (1), $(v \cup De(v)) \cap XYZ = \emptyset$. Thus, $v \cap XYZ = \emptyset$ and $De(v) \cap XYZ = \emptyset$. Now, $De(v) \cap XYZ = \emptyset$ implies that $v \notin An(XYZ)$. Thus, as $v \notin XYZ$ and $v \notin An(XYZ)$, $v \in W$.

Lemma 1 implies that every variable pruned by u-separation is also pruned by m-separation.

Example 7. Given $I(a, de, g)$ to be tested in the BN $\mathcal{B}$ of Fig. 1, step (i) of u-separation constructs the sub-DAG in Fig. 5(i) by pruning $V = \{f, h\}$. On the other hand, step (i) of m-separation builds the sub-DAG in Fig. 2(i) by pruning $W = \{c, f, i, j, h\}$. By Lemma 1, $V \subseteq W$.

Now, we consider the set of variables pruned by m-separation but not by u-separation. Given $I(X, Y, Z)$ to be tested in a BN $\mathcal{B}$. Let V be the set of all inaugural variables. Let W be the set of all variables pruned by m-separation in step (i). Then the set S of variables pruned by m-separation but not by u-separation is defined as:

$$S = W - V. \tag{3}$$

Lemma 2. *Given $I(X, Y, Z)$ to be tested in a BN $\mathcal{B}$. Then $S \cap (XYZ \cup An(XYZ)) = \emptyset$, where S is defined in (3).*

Proof. By (2), $W \cap (XYZ \cup An(XYZ)) = \emptyset$. Now, by (3), $S \subseteq W$. Hence, the claim follows.

Recall that S denotes the set of variables pruned by m-separation but not by u-separation. Lemma 2 means that none of these variables appear in $I(X, Y, Z)$ nor are ancestors of variables therein.

Example 8. Consider W and V from Example 7 when testing $I(a, de, g)$ in the BN $\mathcal{B}$ of Fig. 1. Then, $V = \{f, h\}$, $W = \{c, f, i, j, h\}$, and $S = \{c, i, j\}$. Moreover, $\{a, d, e, g\} \cup An(\{a, d, e, g\}) = \{a, b, d, e, g\}$. As promised by Lemma 2, $\{c, i, j\} \cap \{a, b, d, e, g\} = \emptyset$.

Lemma 3. *Given $I(X, Y, Z)$ to be tested in a BN $\mathcal{B}$, let S be defined as in (3). Then no variable in S is a v-structure in $\mathcal{B}$.*

Proof. By contradiction, suppose $v \in S$ is a v-structure in $\mathcal{B}$. By (3), $v \notin V$. Then, by (1), $(v \cup De(v)) \cap XYZ \neq \emptyset$. Thus, either

$$v \cap XYZ \neq \emptyset \tag{4}$$

or

$$De(v) \cap XYZ \neq \emptyset, \tag{5}$$

or both, hold. Suppose (4) holds. Then $v \in XYZ$. By Lemma 2, $S \cap (XYZ \cup An(XYZ)) = \emptyset$. A contradiction, since by assumption $v \in S$. Now suppose (5) holds. Then there exists a $v' \in De(v)$ such that $v' \in XYZ$. Since $v \in An(v')$, we know $v \in An(XYZ)$. A contradiction to Lemma 2, since by assumption $v \in S$.

Lemma 3 implies that there are no v-structures among the variables pruned by m-separation but not by u-separation.

Example 9. Consider $S = \{c, i, j\}$ from Example 8 when testing $I(a, de, g)$ in the BN $\mathcal{B}$ of Fig. 1. No variable in S is a v-structure in $\mathcal{B}$.

We now show the equivalence of u-separation and m-separation.

Theorem 1. *Testing independence $I(X, Y, Z)$ in a BN $\mathcal{B}$ with m-separation is equivalent to testing independence $I(X, Y, Z)$ in $\mathcal{B}$ with u-separation.*

Proof. We establish the equivalence between m-separation and u-separation by showing the equivalence of their corresponding steps.

Consider the sub-DAGs built in step (i) of m-separation and u-separation. All paths are the same between the two sub-DAGs, except for those involving variables in S. We now show that these paths do not affect the result of testing $I(X, Y, Z)$. By contradiction, suppose there is an undirected path from X to Z involving S. Then there exists two undirected edges (v_1, s_1) and (v_2, s_2) on this path, where $s_1, s_2 \in S$ and $v_1, v_2 \notin S$. Note that s_1 and s_2 may be the same variable. Now (v_1, s_1) must be a directed edge in $\mathcal{B}$. Otherwise, s_1 would be parent of v_1, and thus would not be pruned by step (i) in m-separation. Similarly, (v_2, s_2) also must be a directed edge in $\mathcal{B}$. This immediately implies that there exists a variable s_3 on the path such that $s_3 \in S$ and s_3 is a v-structure in $\mathcal{B}$. A contradiction to Lemma 3. Therefore, there does not exist an undirected path from X to Z going through S.

Now consider the undirected edges added to the sub-DAGs in step (ii) of m-separation and u-separation. By Lemma 3, S does not contain v-structures in $\mathcal{B}$.

Therefore, the set of v-structures considered is the same for both m-separation and u-separation. For any v-structure v with $v \in Y$ of $I(X, Y, Z)$, both rationalization and moralization will add an edge connecting its parents. The only difference is the case when v-structure v is not in Y. This means variable v is not deleted in step (iii). Let (v_1, v) and (v_2, v) be directed edges in $\mathcal{B}$, namely, v_1 and v_2 are parents of v in $\mathcal{B}$, with no directed edge between v_1 and v_2. There are two cases to consider. Suppose that neither v_1 nor v_2 are in Y. This means that v_1 and v_2 both are not deleted in step (iii). Hence, adding edge (v_1, v_2) is redundant, since there already is an undirected path (v_1, v), (v, v_2) from v_1 to v_2. Now suppose that at least one of v_1 and v_2 are in Y. In this case, either v_1 or v_2, or both, are deleted in step (iii) along with their incident edges. Thus, adding edge (v_1, v_2) is wasteful in step (ii), since it will be deleted in step (iii). Therefore, in either case, adding (v_1, v_2) is immaterial.

Steps (iii) and (iv) are the same in both methods.

Theorem 1 establishes that testing independence with u-separation is equivalent to testing independence with m-separation.

Example 10. Consider the BN $\mathcal{B}$ in Fig. 1. $I(a, de, g)$ holds by m-separation in Example 1 and holds by u-separation in Example 5. Similarly, $I(a, e, g)$ does not hold by m-separation in Example 2 and does not hold by u-separation in Example 6.

5   Advantages

Salient features of u-separation compared to m-separation are described.

In its attempt to transform directed separation in DAGs into classical separation in undirected graphs, m-separation can be heavy-handed in two aspects.

First, in step (i), m-separation can prune variables from the BN unnecessarily. Variables not affecting the outcome of the separation test do not need be pruned. Doing so is wasteful.

Example 11. Consider testing $I(a, de, g)$ in the BN $\mathcal{B}$ of Fig. 1 using m-separation. Here, variables $\{c, f, h, i, j\}$ are pruned, yielding the sub-DAG in Fig. 2(i). Closer inspection reveals that variables c, i, and j would not affect the separation test between variables a and g. Thus, it was wasteful of m-separation to prune these three variables.

u-Separation corrects this shortcoming by pruning only inaugural variables from the BN. In Example 11, u-separation only prunes variables $\{f, h\}$ instead of $\{c, f, h, i, j\}$.

The second undesirable characteristic of m-separation is that moralization can be excessive. More specifically, moralization may add edges to the undirected graph that have no bearing on the outcome of the separation test.

Example 12. Recall how m-separation tests $I(a, de, g)$ in Example 1. Step (ii) of m-separation includes adding undirected edge (d, e), while step (iii) involves deleting edge (d, e).

Example 13. Recall how m-separation tests $I(a, e, g)$ in Example 2. Step (ii) of m-separation involves adding undirected edge (a, b). However, edge (a, b) is immaterial in Fig. 3(ii), since edges (a, d) and (b, d) are also present in step (iv).

Closer inspection of Example 2 highlights the excessiveness of m-separation using moralization to test independencies. Moralization adds two edges, (a, b) and (d, e) in Fig. 3(ii). Edge (d, e) is subsequently deleted by m-separation, while edge (a, b) is superfluous.

u-Separation remedies this failing by introducing the notion of rationalization. Rationalization is a refinement of moralization, since the edges added by rationalization are a subset of those added by moralization. When testing $I(X, Y, Z)$, whereas m-separation adds an undirected edge between variables with a common child, u-separation does so only when the child is in Y. This means that the v-structure variable will be subsequently deleted and the added edge between the parents is required to encode the dependence between them. In Example 5, for instance, u-separation only adds edge (a, b) rather than edges (a, b) and (d, e).

It is perhaps worth mentioning here that moralization is ideally suited when applied for building a *join tree* [6] of a given DAG. For this purpose, moralization ensures that a clique containing all variables in each CPT of the BN is formed in the undirected graph. On the contrary, moralization can be overkill when applied for testing independencies in BN.

6   Conclusion

We have suggested *u-separation* as a new method for testing independencies in BNs. Our emphasis here, like that of m-separation, is on simplicity rather than speed. We have demonstrated how u-separation is simpler than m-separation. By utilizing the notion of *inaugural* variables, Lemma 1 ensures that u-separation will never prune more variables than m-separation. This means that m-separation can prune variables unnecessarily to apply separation in undirected graphs, as Example 7 shows.

Moreover, our analysis shows how m-separation's use of moralization is wasteful. An edge added in step (ii) can be deleted in step (iii), as illustrated in Example 12. Furthermore, Example 13 highlights how m-separation can add an edge that is extraneous. To overcome this excessiveness, u-separation suggests the use of *rationalization* rather then moralization. Rationalization is better suited for testing independencies in BNs, whereas moralization is perfectly suited for building a join tree in BN inference.

References

1. Butz, C.J., dos Santos, A.E., Oliveira, J.S., Gonzales, C.: Testing independencies in Bayesian networks with i-Separation. In: Proceedings of the Twenty-Ninth International FLAIRS Conference (2016)

2. Geiger, D., Verma, T.S., Pearl, J.: d-separation: from theorems to algorithms. In: Fifth Conference on Uncertainty in Artificial Intelligence, pp. 139–148 (1989)
3. Kjærulff, U.B., Madsen, A.L.: Bayesian Networks and Influence Diagrams: A Guide to Construction and Analysis, 2nd edn. Springer, New York (2013)
4. Koller, D., Friedman, N.: Probabilistic Graphical Models: Principles and Techniques. MIT Press, Cambridge (2009)
5. Lauritzen, S.L., Dawid, A.P., Larsen, B.N., Leimer, H.G.: Independence properties of directed Markov fields. Networks **20**, 491–505 (1990)
6. Lauritzen, S.L., Spiegelhalter, D.J.: Local computation with probabilities on graphical structures and their application to expert systems. J. Roy. Stat. Soc. **50**, 157–244 (1988)
7. Pearl, J.: Probabilistic Reasoning in Intelligent Systems: Networks of Plausible Inference. Morgan Kaufmann, Burlington (1988)
8. Pearl, J.: Fusion, propagation and structuring in belief networks. Artif. Intell. **29**, 241–288 (1986)
9. Pearl, J.: Belief networks revisited. Artif. Intell. **59**, 49–56 (1993)
10. Pearl, J.: Causality. Cambridge University Press, Cambridge (2009)
11. Verma, T., Pearl, J.: Equivalence and synthesis of causal models. In: Sixth Conference on Uncertainty in Artificial Intelligence, pp. 220–227. GE Corporate Research and Development (1990)
12. Wong, S.K.M., Butz, C.J., Wu, D.: On the implication problem for probabilistic conditional independency. IEEE Trans. Syst. Man Cybern. Part A: Syst. Humans **30**(6), 785–805 (2000)
13. Zhang, N.L., Poole, D.: A simple approach to Bayesian network computations. In: Proceedings of the Tenth Canadian Artificial Intelligence Conference, pp. 171–178 (1994)

Flexible Approximators for Approximating Fixpoint Theory

Fangfang Liu[1]([⊠]), Yi Bi[2], Md. Solimul Chowdhury[3], Jia-Huai You[3],
and Zhiyong Feng[2]

[1] School of Computer Engineering and Science, Shanghai University,
Shanghai, China
ffliu@shu.edu.cn
[2] School of Computer Science and Technology, Tianjin University, Tianjin, China
[3] Department of Computing Science, University of Alberta, Edmonton, Canada

Abstract. Approximation fixpoint theory (AFT) is an algebraic framework for the study of fixpoints of operators on bilattices, which has been applied to the study of the semantics for a number of nonmonotonic formalisms. A central notion of AFT is that of *stable revision* based on an underlying approximating operator (called approximator), where the negative information used in fixpoint computation is by default. This raises a problem in systems that combine different formalisms, where both default negation and established negation may be present in reasoning. In this paper we extend AFT to allow more flexible approximators. The main idea is to formulate and propose ternary approximators, of which traditional binary approximators are a special case. The extra parameter allows separation of two kinds of negative information, by entailment and by default, respectively. The new approach is motivated by the need to integrate different knowledge representation and reasoning (KRR) systems, in particular to support combined reasoning by nonmonotonic rules with ontologies. However, this small change by allowing flexible approximators raises a mathematical question - whether the resulting AFT is a sound fixpoint theory. The main result of this paper is a proof that answers this question positively.

1 Introduction

AFT, also known as the theory of consistent approximations, is a powerful framework for the study of fixpoints for nonmonotonic logics, which has applied to study semantics for various types of logic programs [1–4,6,14], argumentation systems [7,17], and default and autoepistemic logic [5]. Under this theory, the semantics of a logic program is defined by respective fixpoints closely related to an *approximator* on a bilattice. The approach is highly general as it only depends on mild conditions on approximators. The *well-founded fixpoint* of an approximator defines a well-founded semantics (WFS) and *exact stable fixpoints* define an answer set (or stable) semantics. As different approximators may represent different intuitions, AFT provides a way to treat semantics uniformly and allows to explore alternatives by different approximators. The properties of a semantics hold even without a concrete approximator. For example, the least fixpoint approximates all other fixpoints and, mathematically this property holds for any approximator. Practically, since the WFS is generally easier to compute, it can be employed as an approximation or as a mechanism of constraint propagation in computing exact stable fixpoints.

In this paper we address a theoretical question that arises in appyling the current AFT to integrating different KRR formalisms, in particular to the so-called FOL-programs, which are combined knowledge bases $KB = (L, \Pi)$, where L is a theory of a decidable fragment of first-order logic and Π a set of rules possibly containing some arbitrary first-order formulas. This is one of the recent approaches to combining answer set programming (ASP) with description logics (DLs) [8,12,13,15,20,21]. As an illustration, assume L contains a formula that says all students are entitled to educational discount, $\forall x \ St(x) \supset EdDiscount(x)$. Suppose in an application anyone who is not employed but registered for a class is given the benefit of a student. We can write a rule: $St(X) \leftarrow TakeClass(X, Y), not \ HasJob(X)$. Thus, that such a person enjoys educational discount can be inferred directly from knowledge base L. Note that unemployment is typically verified by default, e.g., by lack of tax record for income.

In AFT, to define a semantics for a nonmonotonic logic is to define an approximator, which is required to be monotone on the underlying bilattice. Then the well-known Knaster-Tarski fixpoint theory [18] can be generalized to chain-complete posets. The least fixpoint of an approximator $\mathcal{A}$ is called the *Kripke-Kleene fixpoint of $\mathcal{A}$*. In logic programming, this fixpoint corresponds to what is typically called Kripke-Kleene semantics due to the fact that it does not compute *unfounded atoms*. E.g., given a logic program $P = \{a \leftarrow a\}$, its Kripke-Kleene semantics is that a is *undefined* or *unknown* in 3-valued logic, while the more desirable well-founded and stable semantics both assign a to *false*, since making it false does not invalidate the rule.

To be able to capture unfounded atoms is a key feature of ASP semantics. In the context of ASP, a pair (u, v) on the bilattice built from the power set of atoms under the subset relation is viewed as a 3-valued interpretation (also called a *partial interpretation*): atoms in u are true, those in v are *possibly true*, thus those not in v are false and the atoms that are not true but possibly true take the truth value *undefined* (or called *unknown*). Roughly speaking, to capture unfounded atoms is to compute those atoms that can be assigned to false without invalidating any rule. In AFT, this is accomplished by what is called a *stable revision operator*, which maps a given pair (u, v) to a new pair by two least fixpoints, denoted $(lfp(\mathcal{A}^1(\cdot, v)), lfp(\mathcal{A}^2(u, \cdot)))$, where $\mathcal{A}^1(\cdot, v)$ and $\mathcal{A}^2(u, \cdot)$ are the projection operators of $\mathcal{A}$ on its first and second components, respectively. Intuitively, given (u, v), $lfp(\mathcal{A}^1(\cdot, v))$ computes the atoms that must be true and $lfp(\mathcal{A}^2(u, \cdot))$ computes those that are possibly true. Then, any atoms that are not possibly true are false.

However, a drawback in the construction of $lfp(\mathcal{A}^2(u, \cdot))$ is that the operator $\mathcal{A}^2(u, \cdot)$ has no access to v. As a result, in applying AFT to logic programs, two kinds of false atoms are mixed together and used in the computation of $lfp(\mathcal{A}^2(u, \cdot))$. E.g., to compute $\mathcal{A}^2(u, u)$, the first step in the iterative construction of $lfp(\mathcal{A}^2(u, \cdot))$[1], two kinds of false atoms in (u, u) are used in the computation: for an atom a, if $a \not\in v$ then it is false by the given partial interpretation (u, v); if $a \not\in u$ and $a \in v$ (a is not true but possibly true) then a being false in (u, u) (since $a \not\in u$) is made by default.

Though this notion of stable revision works well for many nonmonotonic logics, problem arises in combining ASP with classic logic, e.g., for FOL-programs, where one faces two undesirable possibilities in computing $\mathcal{A}^2(u, v)$ - *under estimate* or *over estimate* of negative knowledge.

To illustrate the point of "over estimate", assume an FOL-program $KB = (L, \Pi)$ where $L = \{P(a)\}$ and $\Pi = \{P(a) \leftarrow P(a);\ P(b) \leftarrow F\}$, where F is some arbitrary formula. Stable revision starts from the least element on the underlying bilattice, which is $(\bot, \top)$ (in this context, $\bot$ denotes the empty set and $\top$ the set of all atoms). Given $(\bot, \top)$, we need to compute $(lfp(\mathcal{A}^1(\cdot, \top)), lfp(\mathcal{A}^2(\bot, \cdot)))$, where the first step in computing the latter is to compute $\mathcal{A}^2(\bot, \bot)$.[2] If approximator $\mathcal{A}$ is defined in a way that $\mathcal{A}^2(\bot, \bot)$ uses L and partial interpretation $(\bot, \bot)$ as premises, we will have inconsistent premises, as $P(a)$ is false in $(\bot, \bot)$ which contradicts L. If we allow absurdity to lead to the derivation of $P(b)$, we may then compute an unintended conclusion: $P(b)$ is possibly true. We will have more to say about under estimate later, in Sect. 3.

This paper addresses the above issue by proposing the notion of *ternary* approximators, where the extra parameter holds the information on (already computed) negative atoms. Though this increases representation flexibility, a theoretical question is whether the resulting AFT is a sound fixpoint theory, in that a monotone approximator is guaranteed to possess fixpoints and a least fixpoint. The main result of this paper is a mathematical development that shows the resulting AFT is indeed sound.

The next section introduces background on bilattices, followed by Sect. 3 on extending AFT. As an example, Sect. 4 applies the extended AFT to FOL-programs. This is followed by a discussion of related work and future directions.

2   Background

We assume familiarity with Knaster-Tarski fixpoint theory [18]. Briefly, a *lattice* $\langle \mathcal{L}, \leq \rangle$ is a *poset* in which every two elements have a least upper bound (lub) and a greatest lower bound (glb). A *chain* in a poset is a linearly ordered subset of $\mathcal{L}$. A poset $\langle \mathcal{L}, \leq \rangle$ is *chain-complete* if it contains a least element $\bot$ and every chain $C \subseteq \mathcal{L}$ has a lub in $\mathcal{L}$. A lattice $\langle \mathcal{L}, \leq \rangle$ is *complete* if every subset $S \subseteq \mathcal{L}$ has a lub and a glb.

---

[1] By the Knaster-Tarski fixpoint theory, the least fixpoint can be computed iteratively from the least element of the underlying lattice - in this case, u is the least element in the lattice domain represented by the interval $[u, \top]$.

[2] Again, because $\bot$ is the least element of the domain $[\bot, \top]$.

Let $\langle \mathcal{L}, \leq \rangle$ be a complete lattice. The structure $\langle \mathcal{L}^2, \leq, \leq_p \rangle$ denotes the induced (product) bilattice, where $\leq_p$ is called the *precision order* and defined as: for all $x, y, x', y' \in \mathcal{L}$, $(x, y) \leq_p (x', y')$ if $x \leq x'$ and $y \leq y'$. The $\leq_p$ ordering is a complete lattice ordering on $\mathcal{L}^2$. In this paper, we refer to a lattice $\langle \mathcal{L}, \leq \rangle$ simply by $\mathcal{L}$, which is always assumed to be complete, and denote the induced bilattice by $\mathcal{L}^2$.

We say that a pair $(x, y) \in \mathcal{L}^2$ is *consistent* if $x \leq y$, *inconsistent* otherwise, and *exact* if $x = y$. We denote the set of all consistent pairs by $\mathcal{L}^c$. Note that the restriction to $\mathcal{L}^c$ does not form a sublattice. It is a chain-complete poset, with maximal elements being exact pairs. A consistent pair $(x, y) \in \mathcal{L}^c$ defines an *interval*, denoted by $[x, y]$, which defines the set $\{z \mid x \leq z \leq y\}$. A consistent pair (x, y) in $\mathcal{L}^c$ can be seen as an *approximation* of every $z \in \mathcal{L}$ such that $z \in [x, y]$. In this sense, the precision order $\leq_p$ corresponds to the precision of approximation, while an exact pair approximates the only element in it.

An operator $\mathcal{O}$ on a complete lattice or a chain-complete poset $\mathcal{L}$ is monotone if for all $x, y \in \mathcal{L}$, $x \leq y$ implies $\mathcal{O}(x) \leq \mathcal{O}(y)$. Such a monotone operator possesses fixpoints and a least fixpoint. We denote the least fixpoint of $\mathcal{O}$ by $lfp(\mathcal{O})$. An element $x \in \mathcal{L}$ is a pre-fixpoint of $\mathcal{O}$ if $\mathcal{O}(x) \leq x$; it is a post-fixpoint of $\mathcal{O}$ if $x \leq \mathcal{O}(x)$.

3   An Extended Theory of Approximation

The original AFT is built on $\mathcal{L}^c$. We generalize AFT to $\mathcal{L}^2$ by addressing two issues. The first is on the notion of approximator, and the second is on enriching algebraic manipulation by a stable revision operator.

An *approximator* $\mathcal{A}$ is a $\leq_p$-monotone operator on $\mathcal{L}^2$ that approximates an operator $\mathcal{O}$ on $\mathcal{L}$. In the original theory, it is required that $\mathcal{A}(z, z) = (\mathcal{O}(z), \mathcal{O}(z))$, for all $z \in \mathcal{L}$; i.e., $\mathcal{A}$ extends $\mathcal{O}$ on all exact pairs. This is desired if we only deal with consistent pairs. However, if $\mathcal{A}(z, z)$ is inconsistent, it is possible that $\mathcal{O}(z)$ lies outside of $[x, y]$, as an exact pair in general may not be a maximal element in $\mathcal{L}^2$. This leads to the first attempt to define an approximator on $\mathcal{L}^2$ in Definition 1 below as presented in [2].

We shall remark that the authors of [2] formulated an extension that treats all pairs in $\mathcal{L}$. Since a primary application area of our work is on integrating different formalisms where inconsistency frequently arises, it is natural to build our work on top of [2].

Definition 1. *Let $\mathcal{O}$ be an operator on $\mathcal{L}$. We say that $\mathcal{A} : \mathcal{L}^2 \to \mathcal{L}^2$ is an approximator of $\mathcal{O}$ iff the following conditions are satisfied:*

- *For all $x \in \mathcal{L}$, if $\mathcal{A}(x, x)$ is consistent then $\mathcal{A}(x, x) = (\mathcal{O}(x), \mathcal{O}(x))$.*
- *$\mathcal{A}$ is $\leq_p$-monotone.*

Example 1 (Borrowed from [2]). To see why the consistency condition "$\mathcal{A}(x, x)$ is consistent" in the definition is critical, consider a complete lattice where $\mathcal{L} = \{\bot, \top\}$ and $\leq$ is defined as usual. Let $\mathcal{O}$ be the identify function on $\mathcal{L}$. Then we have two fixpoints, $\mathcal{O}(\bot) = \bot$ and $\mathcal{O}(\top) = \top$. Let $\mathcal{A}$ be an identity function on $\mathcal{L}^2$ everywhere except $\mathcal{A}(\top, \top) = (\top, \bot)$. Thus, $\mathcal{A}(\top, \top)$ is inconsistent. It is easy to check that $\mathcal{A}$ is $\leq_p$-monotone. Since $\mathcal{A}(\bot, \bot) = (\mathcal{O}(\bot), \mathcal{O}(\bot))$, and $(\bot, \bot)$ is the only exact pair such that $\mathcal{A}(\bot, \bot)$ is consistent, $\mathcal{A}$ is an approximator of $\mathcal{O}$, according to the above definition. But $\mathcal{A}(\top, \top) \neq (\mathcal{O}(\top), \mathcal{O}(\top))$, even though $\mathcal{O}(\top) = \top$. If the consistency condition is not imposed on the definition, mappings like the operator $\mathcal{A}$ above would be ruled out as approximators, which means we fail to accommodate inconsistencies as we set out to do.[3]

A central idea of AFT is the notion of *stable revision operator*, denote by $St_\mathcal{A}$, for an approximator $\mathcal{A}$. The goal is to determine *persistently reachable* elements as well as *non-reachable* ones in a chain-complete poset, so that a $\leq_p$-monotone operator $St_\mathcal{A}$ has fixpoints and a least fixpoint. The latter is called the *well-founded fixpoint* of $\mathcal{A}$ and the fixpoints of $St_\mathcal{A}$ are called the *stable fixpoints* of $\mathcal{A}$. Let us denote by $\mathcal{A}^1$ and $\mathcal{A}^2$ the projection of an operator $\mathcal{A}$ on $\mathcal{L}^2$ on its first and second components, respectively, i.e., $\mathcal{A}^1(\cdot, v)$ is $\mathcal{A}$ with v fixed, and $\mathcal{A}^2(u, \cdot)$ is $\mathcal{A}$ with u fixed.

A consistent pair (u, v) can be viewed as an approximation to any exact pair in the interval $[u, v]$, where u is a *lower estimate* and v an *upper estimate*. The operator $St_\mathcal{A}(u, v)$ aims at generating a new pair of lower and upper estimates by the respective fixpoint constructions, as expressed by

$$St_\mathcal{A}(u, v) = (lfp(\mathcal{A}^1(\cdot, v)), lfp(\mathcal{A}^2(u, \cdot)). \tag{1}$$

Since approximator $\mathcal{A}$ is $\leq_p$-monotone, so is operator $St_\mathcal{A}$, whose least fixpoint can be computed from the least element $(\bot, \top)$ in the given bilattice.

Now let us extend the approach to $\mathcal{L}^2$. It is clear that the operator $\mathcal{A}^1(\cdot, v)$ is defined on $\mathcal{L}$, and if $\mathcal{A}$ is $\leq_p$-monotone on $\mathcal{L}^2$ then $\mathcal{A}^1(\cdot, v)$ is monotone (i.e., $\leq$-monotone) on $\mathcal{L}$. As $\mathcal{L}$ is a complete lattice, according to the Knaster-Tarski fixpoint theorem, $lfp(\mathcal{A}^1(\cdot, v))$ is well-defined. However, there are two problems with $lfp(\mathcal{A}^2(u, \cdot))$. When $St_\mathcal{A}(u, v)$ is applied:

1. The operator $\mathcal{A}^2(u, \cdot)$ has no access to v, which restricts how an approximator may be defined; and
2. It is possible that $\mathcal{A}^2(u, \cdot) \notin [u, \top]$ and if so, $\mathcal{A}^2(u, \cdot)$ is not an operator on $[u, \top]$ and $lfp(\mathcal{A}^2(u, \cdot))$ is ill-defined.

As remarked in Introduction, without access to v, we may either *under estimate*, or *over estimate*, negative knowledge. There, we illustrated the problem with *over estimate*. Here, let us illustrate the problem with *under estimate*. In the next example, we sketch the inferences only intuitively (as we are not in a position to provide all relevant definitions); we will come back to this example in Sect. 4 (cf. Example 3) once the underlying approximator is defined.

---

[3] This example specifies a system in which states are represented by a pair of factors - high and low. Here, all states are *stable* except the one in which both factors are high. This state may be transmitted to an "inconsistent state" with the first factor high and the second low. This state is the only inconsistent one, and it itself is stable.

Example 2. Consider an FOL-program $KB = (L, \Pi)$ where

$$L = \{\forall x \, C(x) \supset (A(x) \vee D(x))\} \text{ and } \Pi \text{ has the following rules:}$$

$$
\begin{array}{lll}
A(a) \leftarrow A(a). & B(a) \leftarrow not \ A(a). & D(a) \leftarrow not \ B(a). \\
C(a) \leftarrow not \ C'(a). & C'(a) \leftarrow not \ C(a). &
\end{array}
$$

From L, if $A(a)$ and $D(a)$ are false, then $C(a)$ must be false thus it should not be possibly true. Thus, a condition on deriving a possibly true atom is that its negation is not entailed by L along with already computed (in this case, negative) information, which is computed as follows: $A(a)$ is false by closed world reasoning. Then, we derive $B(a)$ which leads to the inference that $D(a)$ is false. In terms of stable revision, we have the following sequence starting with the least element $(\perp, \top)$ (for brevity, we write a for $A(a)$, b for $B(a)$, and so on):

$$(\perp, \top) \Rightarrow (\perp, \{c, c', b, d\}) \Rightarrow (\{b\}, \{c, c', b, d\}) \Rightarrow (\{b\}, \{c, c', b\}) \Rightarrow (\{b\}, \{c', b\}) \Rightarrow \ \ldots$$

where, e.g., atoms $C(a), C'(a), B(a), D(a)$ in the second pair are possibly true, due to possible derivations by rules. In the last pair, $C(a)$ is not possibly true because $\neg C(a)$ is entailed by L and the preceding partial interpretation. Without access to v of (u, v) in computing $lfp(\mathcal{A}^2(u, \cdot))$ of (1), we would not be able to infer $\neg C(a)$ hence block the derivation of $C(a)$ for being possibly true. This would lead to a problematic situation where on the one hand $C(a)$ is possibly true and on the other it is provably false.

To tackle this problem, we generalize stable revision by adding an extra parameter v in the definition of $\mathcal{A}$, i.e., we now define $\mathcal{A}$ as a ternary operator $\mathcal{A} : \mathcal{L}^3 \to \mathcal{L}^2$, which is called by $St_{\mathcal{A}}(u, v)$ with the first parameter fixed to v; i.e., it is the operator $\mathcal{A}(v, \cdot, \cdot)$. Let us alternatively write it as $\mathcal{A}_v$. Then, the expression in (1) becomes

$$St_{\mathcal{A}}(u, v) = (lfp(\mathcal{A}_v^1(\cdot, v)), lfp(\mathcal{A}_v^2(u, \cdot))) \tag{2}$$

where the least fixpoints are constructed by respective sequences

$$x_0 = \perp, x_1 = \mathcal{A}_v^1(x_0, v), ..., x_{\alpha+1} = \mathcal{A}_v^1(x_\alpha, v), ... \tag{3}$$
$$y_0 = u, y_1 = \mathcal{A}_v^2(u, y_0), ..., y_{\alpha+1} = \mathcal{A}_v^2(u, y_\alpha), ... \tag{4}$$

Clearly, the subscript v in $\mathcal{A}_v^1(x_i, v)$ does not add any new information, as v is already a parameter of the operator,[4] but v in $\mathcal{A}_v^2(u, y_i)$ $(i \geq 0)$ does.

We now formalize this new notion of approximator.

Definition 2. *Let $\mathcal{O}$ be an operator on $\mathcal{L}$. We say that $\mathcal{A} : \mathcal{L}^3 \to \mathcal{L}^2$ is an approximator of $\mathcal{O}$ iff the following conditions are satisfied:*

---

[4] Note that we never need a parameter to carry (already computed) true atoms, as in computing $lfp(\mathcal{A}^1(\cdot, v))$ we do not make default assumptions, and the monotonicity of the operator $\mathcal{A}^1$ guarantees that any previously computed true atoms are derived again.

(i) For all $v \in \mathcal{L}$, if $\mathcal{A}_v(v, v)$ is consistent, then $\mathcal{A}_v(v, v) = (\mathcal{O}(v), \mathcal{O}(v))$.

(ii) For all $v \in \mathcal{L}$, $\mathcal{A}_v$ is $\leq_p$-monotone.

(iii) For all $v, v' \in \mathcal{L}$ such that $v \leq v'$, and all $(x, y) \in \mathcal{L}^2$, $\mathcal{A}_{v'}(x, y) \leq_p \mathcal{A}_v(x, y)$.

Conditions (i) and (ii) are similar to those in Definition 1, and (iii) ensures no loss of approximation accuracy with smaller v (which is more informed in what are not in v).[5] It leads to the notion of monotonicity over different, fixed values v in $\mathcal{A}_v$.

Lemma 1. *Let* $\mathcal{A} : \mathcal{L}^3 \to \mathcal{L}^2$ *be an approximator, and* $v, v' \in \mathcal{L}$ *s.t.* $v \leq v'$. *For all* $(x, y), (x', y') \in \mathcal{L}^2$, *if* $(x, y) \leq_p (x', y')$, *then* $\mathcal{A}_{v'}(x, y) \leq_p \mathcal{A}_v(x', y')$.

Proof. We can show $\mathcal{A}_{v'}(x, y) \leq_p \mathcal{A}_v(x, y) \leq_p \mathcal{A}_v(x', y')$. The first inequality is by part (iii) of Definition 2 and the second by $\leq_p$–monotonicity of $\mathcal{A}_v$.   □

We will build our proposal above to an existing solution to the second problem provided in [2]. Essentially, we need to replace the notion of approximators in [2], by that of ternary approximators formulated above. Mathematically, this is not a trivial process. We now give a detailed mathematical development.

In [6], a desirable property, called $\mathcal{A}$-*reliability*, is introduced to ensure that $\mathcal{A}^2(u, \cdot)$ is an operator on $[u, \top]$. Let us generalize this to $\mathcal{L}^2$. Given an approximator $\mathcal{A}$, $(u, v) \in \mathcal{L}^2$ is called $\mathcal{A}$-*reliable* if $(u, v) \leq_p \mathcal{A}_v(u, v)$.

However, this property is not strong enough. For example, consider Example 1 again: let $\mathcal{L} = \{\bot, \top\}$ and $\mathcal{A}_x$, for any $x \in \mathcal{L}$, be an identity mapping everywhere except that $\mathcal{A}_\top(\top, \top) = (\top, \bot)$. It can be seen that all $(u, v) \in \mathcal{L}^2$ are $\mathcal{A}$-reliable, but $\mathcal{A}_\top^2(u, \cdot)$ is not defined on $[u, \top]$; e.g., when (u, v) is $(\top, \top)$, we have $\mathcal{A}_\top^2(\top, \top) = \bot$, which is outside the interval $[\top, \top]$.

To ensure that $\mathcal{A}_v^2(u, \cdot)$ is defined on $[u, \top]$, it is sufficient that $A_v^2(u, u) \geq u$ holds, i.e., the first application of the operator $\mathcal{A}_v^2(u, \cdot)$ in (4) yields an element in $[u, \top]$.

Lemma 2. *Let* $\mathcal{A} : \mathcal{L}^3 \to \mathcal{L}^2$ *be an approximating operator. For any* $(u, v) \in \mathcal{L}^2$, *if* $A_v^2(u, u) \geq u$, *then for every* $z \in [u, \top]$, $\mathcal{A}_v^2(u, z) \in [u, \top]$.

Proof. We have $u \leq A_v^2(u, u) \leq A_v^2(u, z)$. The first inequality is by the condition in the lemma and the second by $(u, z) \leq_p (u, u)$ and $\leq_p$-monotonicity of $\mathcal{A}_v$.   □

Question remains as what if the condition $A_v^2(u, u) \geq u$ does not hold, in which case $\mathcal{A}_v^2(u, \cdot)$ is not an operator on $[u, \top]$. To resolve this issue, let us consider the operator $\mathcal{A}_v^2(u, \cdot)$ on $[\bot, \top]$. $\mathcal{A}_v^2(u, \cdot)$ is clearly defined on $[\bot, \top]$. Furthermore, since $\mathcal{A}_v$ is $\leq_p$-monotone on $\mathcal{L}^2$, $\mathcal{A}_v^2(u, \cdot)$ is $\leq$-monotone on $[\bot, \top]$. To verify, $\forall y \in [\bot, \top]$ and $\forall y,' y' \in [\bot, \top]$ such that $y' \leq y$, from $\leq_p$-monotonicity of $\mathcal{A}_v$, we have $\mathcal{A}_v(x, y) \leq_p \mathcal{A}_v(x, y')$, it follows $\mathcal{A}_v(u, y) \leq_p \mathcal{A}_v(u, y')$ and thus $\mathcal{A}_v^2(u, y') \leq \mathcal{A}_v^2(u, y)$.

We are now in a position to define the notion of stable revision.

---

[5] For example, if v is a set of possibly true atoms, smaller v means more atoms that are false.

Definition 3. *Let* $\mathcal{A} : \mathcal{L}^3 \rightarrow \mathcal{L}^2$ *be an approximator of some operator on* $\mathcal{L}$. *Stable revision operator* $St_{\mathcal{A}} : \mathcal{L}^2 \rightarrow \mathcal{L}^2$ *is defined as:*

$$St_{\mathcal{A}}(u,v) = \begin{cases} (lfp(\mathcal{A}_v^1(\cdot,v)), lfp(\mathcal{A}_v^2(u,\cdot))) \text{ where } \mathcal{A}_v^2(u,\cdot) \text{ is on } [u,\top] & u \leq \mathcal{A}_v^2(u,u) \\ (lfp(\mathcal{A}_v^1(\cdot,v)), lfp(\mathcal{A}_v^2(u,\cdot))) \text{ where } \mathcal{A}_v^2(u,\cdot) \text{ is on } [\bot,\top] & otherwise \end{cases}$$

Before we show the main results of this section, let us introduce and extend another desirable property given in [6]. An element $(u,v) \in \mathcal{L}^2$ is called $\mathcal{A}$-*prudent* if $u \leq lfp(\mathcal{A}_v^1(\cdot,v))$. The reason for this property is that in general there is no guarantee that $lfp(\mathcal{A}_v^1(\cdot,v))$ improves u, which is now ensured under $\mathcal{A}$-prudence.

Let us denote by $\mathcal{L}^{rp}$ the set of all $\mathcal{A}$-reliable and $\mathcal{A}$-prudent pairs in $\mathcal{L}^2$.

The first lemma below establishes the chain property of $St_{\mathcal{A}}$ on $\mathcal{L}^{rp}$, while the next shows its $\leq_p$-monotonicity. For brevity, we may write St for $St_{\mathcal{A}}$.

Lemma 3. *For any pair* $(u,v) \in \mathcal{L}^{rp}$, *let* $St(u,v) = (u',v')$. *Then, we have that* $(u,v) \leq_p (u',v')$, *and* (u',v') *is* $\mathcal{A}$-*reliable and* $\mathcal{A}$-*prudent.*

Proof (Sketch). By $\mathcal{A}$-prudence, we have $u \leq lfp(\mathcal{A}_v^1(\cdot,v))$. To show $v' \leq v$, by $\mathcal{A}_v$-reliability, we have $\mathcal{A}_v^2(u,v) \leq v$, i.e., v is a pre-fixpoint of $\mathcal{A}_v^2(u,\cdot)$. It follows that if $\mathcal{A}_v^2(u,\cdot)$ is an operator on $[u,\top]$ then it must be the case that $v \in [u,\top]$, thus $lfp(\mathcal{A}_v^2(u,\cdot)) \leq v$. If $\mathcal{A}_v^2(u,\cdot)$ is an operator on $[\bot,\top]$, surely $v \in [\bot,\top]$, then we still have $lfp(\mathcal{A}_v^2(u,\cdot)) \leq v$. It then follows $(u,v) \leq_p (u',v')$.

For the second assertion, as $u' = lfp(\mathcal{A}_v^1(\cdot,v)) = \mathcal{A}_v^1(u',v)$, we have $\mathcal{A}_v^1(u',v) \leq \mathcal{A}_v^1(u',v')$ by $v' \leq v$. Similarly, as $v' = lfp(\mathcal{A}_v^2(u,\cdot)) = \mathcal{A}_v^2(u,v')$, where $\mathcal{A}_v^2(u,\cdot)$ is either an operator on $[u,\top]$ or an operator on $[\bot,\top]$, we have $\mathcal{A}_v^2(u',v') \leq \mathcal{A}_v^2(u,v')$ by $u \leq u'$, thus $(u',v') \leq_p \mathcal{A}_v(u',v')$. Next, by condition (iii) of Definition 2, we have $(u',v') \leq_p \mathcal{A}_v(u',v') \leq \mathcal{A}_{v'}(u',v')$.

Finally, for $\mathcal{A}$-prudence, that $u' \leq lfp(\mathcal{A}_{v'}^1(\cdot,v'))$ can be proved by transfinite induction on the sequences $lfp(\mathcal{A}_v^1(\cdot,v))$ and $lfp(\mathcal{A}_{v'}^1(\cdot,v'))$.   □

Lemma 4. *For any pairs* $(u,v),(u',v') \in \mathcal{L}^{rp}$ *such that* $(u,v) \leq_p (u',v')$, *we have* $St(u,v) \leq_p St(u',v')$.

Proof (Sketch). In all cases, we can show $lfp(\mathcal{A}_v^1(\cdot,v)) \leq_p lfp(\mathcal{A}_{v'}^1(\cdot,v'))$ by transfinite induction on the two sequences. To show $lfp(\mathcal{A}_{v'}^2(u',\cdot)) \leq lfp(\mathcal{A}_v^2(u,\cdot))$, we need to consider four cases: (i) $\mathcal{A}_v^2(u,\cdot)$ and $\mathcal{A}_{v'}^2(u',\cdot)$ are defined on $[u,\top]$ and $[u',\top]$ respectively, (ii) both $\mathcal{A}_v^2(u,\cdot)$ and $\mathcal{A}_{v'}^2(u',\cdot)$ are defined on $[\bot,\top]$ (not in the subdomains as in (i)), (iii) $\mathcal{A}_v^2(u,\cdot)$ is defined on $[u,\top]$ while $\mathcal{A}_{v'}^2(u',\cdot)$ is defined on $[\bot,\top]$, and (iv) $\mathcal{A}_v^2(u,\cdot)$ defined on $[\bot,\top]$ while $\mathcal{A}_{v'}^2(u',\cdot)$ defined on $[u',\top]$.

Let $v_f = lfp(\mathcal{A}_v^2(u,\cdot))$ and $v_f' = lfp(\mathcal{A}_{v'}^2(u',\cdot))$. We then have $v_f = \mathcal{A}_v^2(u,v_f) \geq \mathcal{A}_v^2(u',v_f) \geq \mathcal{A}_{v'}^2(u',v_f)$ by $u \leq u'$ and $v' \leq v$; i.e., v_f is a pre-fixpoint of $\mathcal{A}_{v'}^2(u',\cdot)$. On the other hand, we have $v_f' = \mathcal{A}_{v'}^2(u',v_f') \leq \mathcal{A}_v^2(u,v_f')$, i.e., v_f' is a post-fixpoint of $\mathcal{A}_v^2(u,\cdot)$. For case (i), as v_f' is in the domain of $\mathcal{A}_{v'}^2(u',\cdot)$, it is also in the domain of $\mathcal{A}_v^2(u,\cdot)$ by $u \leq u'$; it follows $v_f' \leq lfp(\mathcal{A}_v^2(u,\cdot))$ as it is a post-fixpoint, thus $lfp(\mathcal{A}_{v'}^2(u',\cdot)) \leq lfp(\mathcal{A}_v^2(u,\cdot))$. For case (ii), as $\mathcal{A}_v^2(u,\cdot)$ and $\mathcal{A}_{v'}^2(u',\cdot)$ are both defined on $[\bot,\top]$, that $v_f' \leq v_f$ naturally follows. For case (iii), since v_f

is surely in the domain of $\mathcal{A}_{v'}^2(u', \cdot)$ and v_f is a pre-fixpoint of $\mathcal{A}_{v'}^2(u', \cdot)$, it follows $lfp(\mathcal{A}_{v'}^2(u', \cdot)) \leq v_f$. For case (iv), also as v_f' is in the domain of $\mathcal{A}_v^2(u, \cdot)$ and v_f' a post-fixpoint of $\mathcal{A}_v^2(u, \cdot)$, we have $v_f' \leq lfp(\mathcal{A}_v^2(u, \cdot))$. We therefore conclude $St(u, v) \leq_p St(u', v')$. $\square$

Let C be a chain in $\mathcal{L}^{rp}$. Denote by C^1 and C^2, respectively, the projection of C on its first and second elements. It is clear that $(lub(C^1), glb(C^2)) = lub(C)$. By adapting a proof of [6] (the proof of Proposition 3.10), it can be shown that for any chain C of $\mathcal{L}^{rp}$, $(lub(C^1), glb(C^2)) = (u, v)$ is $\mathcal{A}$-reliable and $\mathcal{A}$-prudent. It follows from Lemma 3 that the operator St is defined on $\mathcal{L}^{rp}$. Then, from Lemma 4 we conclude the following.

Theorem 1. *The structure $\langle \mathcal{L}^{rp}, \leq_p \rangle$ is a chain-complete poset that contains the least element $(\bot, \top)$, and $St_\mathcal{A}$ is a well-defined, increasing, and $\leq_p$-monotone operator in the poset.*

This completes our theoretical work, which shows that the extended AFT is a sound fixpoint theory; we thus can define:

Definition 4. *The least fixpoint of $St_\mathcal{A}$ on $\mathcal{L}^{rp}$ is called the* well-founded fixpoint *of $\mathcal{A}$ (and the corresponding semantics the $\mathcal{A}$-WFS) , and the exact fixpoints of $St_\mathcal{A}$ on $\mathcal{L}^{rp}$ are called the* exact stable fixpoints *of $\mathcal{A}$ (and the corresponding semantics $\mathcal{A}$-answer set semantics).*

4   FOL-Programs

In this section we briefly explore how the extended AFT may be applied to the study of semantics for FOL-programs.

We assume a language of a decidable fragment of first-order logic, denoted $\mathcal{L}_\Sigma$, where $\Sigma = \langle F^n; P^n \rangle$, and F^n and P^n are disjoint countable sets of function and predicate symbols, each of which comes with a fixed arity. Constants are 0-ary functions. *Terms* are variables, constants, or functions in the form $f(t_1, ..., t_n)$, where each t_i is a term and $f \in F^n$. *First-order formulas*, or just *formulas*, are defined as usual, so are the notions of *satisfaction*, *model*, and *entailment*.

Let Φ_P be a finite subset of P^n and Φ_C a nonempty finite set of constants from F^n. An *atom* is of the form $P(t_1, ..., t_n)$ where $P \in \Phi_P$ and each t_i is either a constant from Φ_C or a variable.

An *FOL-program* is a combined knowledge base $KB = (L, \Pi)$, where L is a first-order theory of $\mathcal{L}_\Sigma$ and Π a *rule base*, which is a finite set of rules of the form $H \leftarrow A_1, ..., A_m, not\ B_1, ..., not\ B_n$, where H and B_i are atoms and A_i are formulas. For any rule r, we denote by $hd(r)$ the head of the rule and $body(r)$ its body, and define $pos(r) = \{A_1, ..., A_m\}$ and $neg(r) = \{B_1, ..., B_n\}$.

A *ground instance* of a rule in Π is obtained by replacing a free variable with a constant in Φ_C. The process of replacing a rule by all its ground instances is called *grounding*. From now on, we assume Π is already grounded. When we refer to an atom/literal/formula, by default we mean it is a ground one.

Given an FOL-program $KB = (L, \Pi)$, the *Herbrand base* of Π, denoted HB_Π, is the set of all ground atoms $P(t_1, ..., t_n)$, where $P \in \Phi_P$ and $t_i \in \Phi_C$.

For ease of presentation, we assume that Φ_P only contains predicate symbols that occur in Π, and it contains at least all predicate symbols that occur in Π but not in L. Under this assumption, no predicate symbol that appears in L but not in Π may be in the underlying Herbrand base, and every predicate symbol appearing in Π but not in L is necessarily in the underlying Herbrand base. Recall that answer sets and WFS are only concerned with atoms in the underlying Herbrand base.

Any subset $I \subseteq HB_\Pi$ is called an *interpretation* of Π. If I is a set of (ground) atoms, we define $\bar{I} = HB_\Pi \backslash I$, and $\neg.I = \{\neg A \mid A \in I\}$.

Given lattice $\langle 2^{HB_\Pi}, \subseteq \rangle$, the induced bilattice is $\langle (2^{HB_\Pi})^2, \subseteq_p \rangle$. A consistent pair (I, J) in $(2^{HB_\Pi})^2$ represents a *partial interpretation* $I \cup \neg.\bar{J}$. Let $(I, J) \in (2^{HB_\Pi})^2$ and L a first-order theory. We say that (I, J) is consistent with L if $L \cup I \cup \neg.\bar{J}$ is consistent, and (I', J') is a *consistent extension* of (I, J) if $I \subseteq I' \subseteq J' \subseteq J$ (if (I, J) is inconsistent, such (I', J') does not exist).

Definition 5. *Let $KB = (L, \Pi)$ be an FOL-program, $(I, J) \in (2^{HB_\Pi})^2$, and ϕ a literal. We define two entailment relations, which extend to conjunctions of literals.*

- *$(I, J) \models_L \phi$ iff, if ϕ is an atom A then $A \in I$, if ϕ is a negative literal not A then $A \notin I$,[6] and if ϕ is an FOL-formula then $L \cup I \cup \neg.\bar{J} \models \phi$.*
- *$(I, J) \Vdash_L \phi$ iff for all consistent extensions (I', J') of (I, J), $(I', J') \models_L \phi$.*

Operator to be Approximated: Let $KB = (L, \Pi)$ be an FOL-program. Define an operator $\mathcal{K}_{KB}: 2^{HB_\Pi} \rightarrow 2^{HB_\Pi}$ as follows: for any $I \in 2^{HB_\Pi}$

$$\mathcal{K}_{KB}(I) = \{hd(r) \mid r \in \Pi, (I, I) \models_L body(r)\} \cup \{A \in HB_\Pi \mid (I, I) \models_L A\}$$

This operator is essentially the immediate consequence operator augmented by *direct positive consequences*. Elements in 2^{HB_Π} are (2-valued) interpretations, which are inherently weak in representing inconsistency. In the extended AFT, it is the inconsistent pairs that tie up this loose end, by explicitly representing negative information.

In the definitions below, we assume an FOL-program $KB = (L, \Pi)$, $(I, J) \in (2^{HB_\Pi})^2$, and $v \in 2^{HB_\Pi}$ such that $J \subseteq v$.

Definition 6 (Operator $\Phi_{KB,v}$: Standard Semantics). *For all $H \in HB_\Pi$,*

- *$H \in \Phi^1_{KB,v}(I, J)$ iff one of the following holds*
  (a) $(I, v) \models_L H$.
  (b) $\exists r \in \Pi$ with $hd(r) = H$, s.t. $\forall \phi \in body(r)$, $(I, J) \Vdash_L \phi$.
- *$H \in \Phi^2_{KB,v}(I, J)$ iff $(I, v) \not\models_L \neg H$ and one of the following holds*
  (a) $\exists I', J'(I \subseteq I' \subseteq J' \subseteq J), (I', J' \cup \bar{J}) \models_L H$.

---

[6] Default negation here is evaluated independently of L, which has been called *local closed world reasoning* [11].

(b) $\exists r \in \Pi$ *with* $hd(r) = H$, *s.t.* $\forall \phi \in body(r)$, $\exists I', J'(I \subseteq I' \subseteq J' \subseteq J)$, $(I', J' \cup \bar{J}) \models_L \phi$.

Operator $\Phi^1_{KB,v}$ computes atoms that must be true, either due to *directly entailed* w.r.t. L (part (a)), or *persistently derivable* (part (b)). Operator $\Phi^2_{KB,v}$ on the other hand computes possibly true atoms, under the condition that their complements are not entailed by (I, v), that are either *potentially entailed* (part (a)), or *possibly derivable* (part (b), in which each body literal may be derived from a different (I', J')).

Note that the pair (I, J) in $\Phi^2_{KB,v}(I, J)$ serves as an interval $[I, J]$ in the sense that any atom in it may be assigned arbitrarily, as expressed in $(I', J' \cup \bar{J})$, as a partial interpretation that extends I, in that the atoms in I' are assigned to true and those in J but not in J' are assigned to false.

If (I, v) is inconsistent with L, then the result is $(HB_\Pi, \emptyset)$.

Lemma 5. Φ_{KB} *is an approximator of* $\mathcal{K}_{KB}$.

Example 3. Consider Example 2 again: $KB = (L, \Pi)$ where $L = \{\forall x\, C(x) \supset (A(x) \vee D(x))\}$ and $\Pi = \{A(a) \leftarrow A(a). \ B(a) \leftarrow not\, A(a). \ D(a) \leftarrow not\, B(a). \ C(a) \leftarrow not\, C'(a). \ C'(a) \leftarrow not\, C(a).\}$ The stable revision operator, $St_{\Phi_{KB}}$ in this case, generates the below sequence (recall that we write a for $A(a)$, b for $B(a)$, and so on):

$$(\emptyset, HB_\Pi) \Rightarrow (\emptyset, \{c, c', b, d\}) \Rightarrow (\{b\}, \{c, c', b, d\}) \Rightarrow (\{b\}, \{c, c', b\}) \Rightarrow (\{b\}, \{c', b\})$$
$$\Rightarrow (\{c', b\}, \{c', b\})$$

$A(a)$ is false by the second pair, which leads to the derivation of $B(a)$ and then to the inference that $D(a)$ is false, in the next two steps. Next, as $\neg C(a)$ is entailed by L, $C(a)$ is no longer possible true. Finally, $C'(a)$ is derived to be true.

The above example shows how "under estimate" is corrected under the extended AFT here, which improves the work of [2]. In the next example, we illustrate how inconsistency is handled.

Example 4. Let $KB = (\{\neg A(a)\}, \Pi)$ where $\Pi = \{A(a) \leftarrow not\, B(a); \ B(a) \leftarrow B(a); \ C(a) \leftarrow\}$. Let $\Phi_P = \{A, B, C\}$ and $\Phi_C = \{a\}$. The well-founded fixpoint of Φ_{KB} is computed by the stable revision operator $St_{\Phi_{KB}}$ as follows:

$$(\emptyset, HB_\Pi) \Rightarrow (\{C(a)\}, \{C(a)\}) \Rightarrow (HB_\Pi, \{C(a)\}) \Rightarrow (HB_\Pi, \emptyset)$$

From the second pair, we get $lfp(\Phi^1_{KB,v'}(\cdot, \{C(a)\})) = HB_\Pi$, due to inconsistency from the derivation of $A(a)$. Let us denote the third pair by (u, v). Because $\Phi^2_{KB,v}(u, u) = \emptyset$ (thus $\not\geq u$), the second case in Definition 3 is triggered, and leads to $lfp(\Phi^2_{KB,v}(u, \cdot)) = \emptyset$, which together with $lfp(\Phi^1_{KB,v}(\cdot, v))$ yields the last pair as the fixpoint.

The well-founded and answer set semantics based on the operator Φ_{KB} are called *standard*, because they are generalizations of the WFS and answer set semantics for normal logic programs.

Theorem 2. *Let $KB = (\emptyset, \Pi)$ be a normal program, i.e., $L = \emptyset$ and Π is a set of normal rules. Then, the Φ-WFS of KB coincides with the WFS of Π [19], and Φ-answer set semantics coincides with the standard stable model semantics of Π [10].*

5 Related Work and Future Directions

AFT has applied to DL-programs [8], which can be represented by HEX-programs [1] and aggregate programs [14], where an approximator can be defined so that the well-founded fixpoint defines the WFS [9] and the exact stable fixpoints define the well-supported answer set semantics [16]. In [12], a well-founded semantics for combing rules with DLs is defined. In both approaches above, syntactic restrictions are imposed so that the least fixpoint is always constructed over sets of consistent literals.

Our work here is built on top of the approach in [2]. Because there is no explicit representation of entailed negative information in approximators, the phenomenon of over estimate and under estimate does arise in the approach of [2].

In our extended AFT, inconsistency handling relies on how an approximator is defined. So far, the approximators defined allow trivialization to take place, but practically one may define approximators so that non-trivialization is supported. E.g., in Example 4, one may define the approximator Φ_{KB} in a way that $(\{C(a)\}, \{C(a)\})$ is stable revised to $(\{A(a), C(a)\}, \{C(a)\})$, where $A(a)$ being true but not possible true indicates where inconsistency initially occurred. Further investigation in this direction is needed.

Recently, the theory of *grounded fixpoints* has generated some interest [4]. Again, the work in general does not treat inconsistent pairs on bilattices. It is interesting to investigate how the approach proposed in this paper may be applied. In addition, the notion of *unfounded atoms* by non-derivation of rules with arbitrary formulas in bodies [3,4] is interesting. In general, the relationship between unfoundedness and stable revision for various classes of programs, including FOL-programs, requires a further study.

In this paper, we only provided an initial attempt to show applications of the extended AFT. We will look into other applications where a separation of established negation and default negation is desirable. Tightly coupled multi-context systems is a potential target.

References

1. Antic, C., Eiter, T., Fink, M.: HEX semantics via approximation fixpoint theory. In: Cabalar, P., Son, T.C. (eds.) LPNMR 2013. LNCS, vol. 8148, pp. 102–115. Springer, Heidelberg (2013)
2. Bi, Y., You, J.-H., Feng, Z.: A generalization of approximation fixpoint theory and application. In: Kontchakov, R., Mugnier, M.-L. (eds.) RR 2014. LNCS, vol. 8741, pp. 45–59. Springer, Heidelberg (2014)
3. Bogaerts, B., Vennekens, J., Denecker, M.: Grounded fixpoints. In: Proceedings of AAAI 2015, pp. 1453–1459 (2015)

4. Bogaerts, B., Vennekens, J., Denecker, M.: Grounded fixpoints and their applications in knowledge representation. Artif. Intell. **224**, 51–71 (2015)
5. Denecker, M., Marek, V.W., Truszczynski, M.: Uniform semantic treatment of default and autoepistemic logics. Artif. Intell. **143**(1), 79–122 (2003)
6. Denecker, M., Marek, V.W., Truszczynski, M.: Ultimate approximation and its application in nonmonotonic knowledge representation systems. Inf. Comput. **192**(1), 84–121 (2004)
7. Dung, P.M.: On the acceptability of arguments and its fundamental role in nonmonotonic reasoning, logic programming and n-person games. Artif. Intell. **77**(2), 321–358 (1995)
8. Eiter, T., Ianni, G., Lukasiewicz, T., Schindlauer, R., Tompits, H.: Combining answer set programming with description logics for the semantic web. Artif. Intell. **172**(12–13), 1495–1539 (2008)
9. Eiter, T., Lukasiewicz, T., Ianni, G., Schindlauer, R.: Well-founded semantics for description logic programs in the semantic web. ACM Trans. Comput. Logic **12**(2) (2011). Article 11
10. Gelfond, M., Lifschitz, V.: The stable model semantics for logic programming. In: Proceedings of ICLP 1988, pp. 1070–1080 (1988)
11. Knorr, M., Alferes, J.J., Hitzler, P.: Local closed world reasoning with description logics under the well-founded semantics. Artif. Intell. **175**(9–10), 1528–1554 (2011)
12. Lukasiewicz, T.: A novel combination of answer set programming with description logics for the semantic web. IEEE TKDE **22**(11), 1577–1592 (2010)
13. Motik, B., Rosati, R.: Reconciling description logics and rules. J. ACM **57**(5), 1–62 (2010)
14. Pelov, M.B.N., Denecker, M.: Well-founded and stable semantics of logic programs with aggregates. Theory Pract. Logic Program. **7**, 301–353 (2007)
15. Rosati, R.: DL+log: tight integration of description logics and disjunctive datalog. In: Proceedings of KR 2006, pp. 68–78 (2006)
16. Shen, Y.-D., Wang, K.: Extending logic programs with description logic expressions for the semantic web. In: Aroyo, L., Welty, C., Alani, H., Taylor, J., Bernstein, A., Kagal, L., Noy, N., Blomqvist, E. (eds.) ISWC 2011, Part I. LNCS, vol. 7031, pp. 633–648. Springer, Heidelberg (2011)
17. Strass, H.: Approximating operators and semantics for abstract dialectical frameworks. Artif. Intell. **205**, 39–70 (2013)
18. Tarski, A.: A lattice-theoretical fixpoint theorem and its applications. Pac. J. Math. **5**(2), 285–309 (1955)
19. van Gelder, A., Ross, K., Schlipf, J.: The well-founded semantics for general logic programs. J. ACM **38**(3), 620–650 (1991)
20. Vennekens, J., Denecker, M., Bruynooghe, M.: FO(ID) as an extension of DL with rules. Ann. Math. Artif. Intell. **58**(1–2), 85–115 (2010)
21. Yang, Q., You, J.-H., Feng, Z.: Integrating rules and description logics by circumscription. In: Proceedings of AAAI 2011 (2011)

Learning Statistically Significant Contrast Sets

Mohomed Shazan Mohomed Jabbar[✉] and Osmar R. Zaïane

Department of Computing Science, University of Alberta, Edmonton, Canada
{mohomedj,zaiane}@ualberta.ca

Abstract. Contrast set learning is important to discover control variables that can distinguish different groups in a dataset. Association rule mining has an inherent connection to the contrast set learning problem and has also been used to address it. All of the association rule based contrast set learning techniques use support-confidence based methods and inherit their limitations. In recent years statistically significant rule mining has become a viable alternative to address those limitations. We propose a novel contrast set learning approach based on statistically significant rule mining that eliminates the limitations in using traditional rule mining approaches and identifies statistically significant contrast sets. We evaluated our method by building a classifier using the discovered contrast sets. The performance of our classifier, while our method is not for classification per se, reveals the effectiveness of our approach in distinguishing the groups.

Keywords: Contrast sets · Association rules · Characterizing groups

1 Introduction

Finding meaningful groups in a dataset is one of the main tasks in data mining known as clustering. Characterizing discovered clusters remains a challenge. For instance, when given a clinical dataset with two clusters, healthy and non-healthy, an intriguing question researchers would like to ask is: "what are the key variables that can differentiate between healthy and non-healthy people?". The difference between contrasting groups in such situations can be described using conditional probabilities [5]. As an example consider, $P(Smoking \wedge Exposure - to - carcinogens | Healthy)$ and $P(Smoking \wedge Exposure - to - carcinogens | Non-healthy)$. These conditional probabilities can be interpreted as association rules as follows: $Healthy \implies (Smoking \wedge Exposure - to - carcinogens)$ and $Non-healthy \implies (Smoking \wedge Exposure - to - carcinogens)$. The consequent of these rules could be representing a contrast set [5].

Contrast set mining was introduced as emerging pattern discovery [2]. A well known contrast set learning algorithm is STUCCO [1]. The contrast set learning problem is intrinsic to the association rule mining problem where several of the previous methods directly exploit this connection. Magnum Opus [4] and CIGAR [3] are two such approaches. Some of these previous techniques are based on the *first-kind* of association rules: $Group \implies Contrast - set$ [1,3] whereas the

© Springer International Publishing Switzerland 2016
R. Khoury and C. Drummond (Eds.): Canadian AI 2016, LNAI 9673, pp. 237–242, 2016.
DOI: 10.1007/978-3-319-34111-8_29

others have primarily based on the *second-kind* of rules: $Contrast - set \implies Group$ [4,5]. Recent works in the literature have empirically proved that only the *second-kind* of contrast sets are possible [5]. Most of the existing techniques are based on the traditional Apriori [7] like rule discovery techniques, which depend on two threshold values called support and confidence and inherits the limitations imposed by them.

In recent years, statistically significant association rule mining has become a viable alternative to address such limitations in traditional approaches. Kingfisher [8] is one such proposed pioneering technique. Kingfisher uses Fisher's exact test to measure the statistical significance of a rule against the null hypothesis. Null hypothesis outlines that the antecedents and the consequent of a given rule are independent. When developing a contrast set learning method, statistically significant association rule mining techniques could be of great use not just because it can eliminate the limitations posed by traditional methods, but also the detected contrast sets would be statistically significant.

2   Problem Definition

In association rule analysis and contrast set learning, we deal with a transaction database D such that each sample transaction E in D can be defined as a vector of size m. Let $A = \{A_1, A_2, ..., A_m\}$ be a set of attributes called *items*. Then a transaction E can be defined as a vector consisting of attribute-value pairs $A_1 = V_1, A_2 = V_2, ..., A_m = V_m$ where each $V_j \in \{V_{j1}, V_{j2}, ..., V_{jn}\}$ and n is a finite number. Given these definitions an **association rule** is an implication of the form $X \implies Y$ where $X \subset A$, $Y \subset A$ and $X \cap Y = \emptyset$. Confidence c in $X \implies Y$ is the percentage of data instances in D containing X also contains Y (i.e. $P(Y|X)$). Support s for $X \implies Y$ is the percentage of data instances in D containing $X \cup Y$. Traditional algorithms discover strong association rules by verifying that their s and c exceed some user defined threshold. **Classification Association Rules (CARs)** are an important class of association rules. Given a set of class labels $C = \{c_1, c_2, ..., c_q\}$ where each instant E in D is associated with a class label c_i and q is the number of classes, a CAR can be defined as an association rule of the form $X \implies c_i$.

A **contrast set** is a conjunction of attribute-value pairs defined on mutually exclusive classes from C such that no A_i occurs more than once. Some of the early research works identify contrast sets as sets belonging to rules of the form $Class \implies Contrast - set$. However according to recent literature we have only considered contrast sets that are identified as sets belonging to rules of the form $Contrast - set \implies Class$ [5]. The STUCCO algorithm detects significant contrast sets by imposing two constraints, known as **deviation conditions**, on the candidate patterns. When both the conditions are met it is defined as a deviation and the candidate pattern is identified as a valid contrast set [1]. These conditions are as follows:

$$\exists_{i,j} P(X|c_i) \neq P(X|c_j) \tag{1}$$

$$\max_{i,j} |support(X, c_i) - support(X, c_j)| \geq min\_dev \tag{2}$$

3   Proposed Approach

We propose a technique, which uses statistically significant CARs, to discover valid contrast sets from a given transaction dataset. Our approach consists of two phases: (1) Generate statistically significant CARs from the given dataset; (2) Mine statistically significant contrast sets from those rules.

3.1   Classification Association Rule Generation

In phase 1 we mine statistically significant CARs. The statistical significance of CARs can be determined by testing for its dependency. Hence, the CAR, $X \implies c_i$ is considered statistically significant at level α, if the probability p of observing an equal or stronger dependency in a dataset complying with the null hypothesis is not greater than α. In the null hypothesis it is assumed that X and c_i are independent. The probability p (i.e. p-value) can be calculated from a cumulative hypergeometric distribution using Fisher's exact test [8].

The significance level α is set to be 0.05. Calculated p-values can be used to prune non significant rules using this α. Kingfisher algorithm [8] accomplishes this task by using an enumeration tree with branch and bound search. In our specific case to detect CARs, we define the shape of the rules to be outputted by the Kingfisher algorithm. This constrained version of the Kingfisher we used obtains only rules with the type $X \implies c_i$ [6].

3.2   Discovering Contrast Sets

In phase 2 we discover statistically significant contrast sets using the CARs obtained in phase 1. To achieve this, we propose a new set of deviation conditions, based on the original conditions from STUCCO, as follows:

$$\exists_{i,j} p_F(X \implies c_i) \neq p_F(X \implies c_j) \tag{3}$$

$$\min_{i,j} |p_F(X \implies c_i) - p_F(X \implies c_j)| \leq max\_dev \tag{4}$$

where p_F refers to the p-value of the particular rule. Equation 3 attempts to capture the statistical significance of the contrast set while Eq. 4 attempts to capture whether the statistical significance of the candidate contrast set across different classes is sufficiently large given the maximum deviation threshold. Note that in contrast to the original conditions introduced in STUCCO, we use a maximum deviation threshold because lower the p_F value better the statistical significance is.

Original deviation conditions from STUCCO was unable to capture the contrast sets which only exist in one group. Based on some of the recent works in the literature [5] we addressed this issue by introducing the condition in Eq. 5 which uses the maximum threshold value or level of significance of the Kingfisher algorithm (i.e. $Kingfisher(p_F)$). Whenever a candidate contrast set is absent in a group, this threshold value can provide a lower bound to the p_F value of that candidate set. This new condition is as follows:

$$\min_{i,j} |p_F(X \implies c_i) - Kingfisher(p_F)| \leq max\_dev \qquad (5)$$

For the maximum deviation threshold we experimented with using the $Avg.(p_F)$ value and $Avg.(p_F) + \sigma(p_F)$ of the rules generated in Phase 1. $\sigma(p_F)$ is the standard deviation of the p-value distribution.

4 Experiments

We conducted experiments on 18 public datasets from the UCI ML Repository[1] to evaluate our method in identifying valid contrast sets. As previously explained, in our experiments, we used a constrained version of the Kingfisher algorithm in the first phase of our method with a 5 % level of significance. In the second phase of our method we discovered contrast sets using the three deviation conditions (Eqs. 3, 4 and 5) we introduced earlier. All the experiments we conducted are 10 fold cross validated and results are averaged in Table 1.

We implemented a baseline method which uses support-confidence based rules to learn contrast sets. To be fair we have obtained the minimum confidence and support from the statistically significant rule set and used them as threshold values to obtain association rules from Apriori like algorithms. Then using Eqs. 1 and 2 we identified contrast sets. However, when we compared the contrast sets detected by our method and this baseline approach, from UCI datasets, it is revealed that there is a little to no overlap between them.

To further investigate our approach, we used the contrast sets obtained by our method to build an associative classifier, CS^2 (**C**lassification based on **S**tatistically **S**ignificant **C**ontrast **S**ets), following similar works [10]. Having a better classification accuracy would mean that we have identified contrast sets which can meaningfully differentiate classes. As shown in Algorithm 1 (line 7–12), CS^2 first recognizes the subset of contrast sets which can contribute to classify a given data instance. Then, based on class, it categorizes this subset of rules. Aggregate function *sum* can be applied to each of these categories to obtain a representative measure for each class. Next a class can be assigned to the data instance by using another aggregate function *min* on the representative measures. We have compared the classification accuracy of CS^2 with several other standard classifiers on UCI datasets even though our target is not to build a classifier but to discover accurate contrast sets. Their classification accuracies are reported in Table 1. We chose classifiers, C4.5 [9], CBA [10] and CPAR [11], to compare. Average accuracy in all 18 datasets indicates that CS^2 has a classification accuracy very close to that of other standard classifiers. It even outperforms CBA algorithm, proving that statistically significant rules can provide quality contrast sets. Results with contrast sets used as baseline were very low and not reported in Table 1.

---

[1] http://archive.ics.uci.edu/ml/.

Algorithm 1. CS^2 Algorithm

INPUT: Database D, Object O, Attributes A, Classes C, Level-of-Significance α
1: CAR = Kingfisher(D, A, C, α)
2: CSet = $\emptyset$
3: **for all** rule r in *CAR* **do**
4:     **if** r suffice Eq. 3, 4 or 5
5:         CSet = CSet $\cup$ r
6: **end for**
7: CSet$_{new}$ = $\emptyset$
8: **for all** c-set c in *CSet* **do**
9:     **if** c.antecedent $\subseteq$ O.antecedent
10:         CSet$_{new}$ = CSet$_{new}$ $\cup$ c
11: **end for**
12: Divide CSet$_{new}$ to subsets based on class labels: $S_1, S_2, ... S_n$
13: **for all** S_i in $S_1, S_2, ... S_n$ **do**
14:     sum all the *ln* p$_F$ values in each subset
15: **end for**
16: Assign the class with lowest some of p$_F$ to O
RETURN *O.label*

Table 1. Comparison of classification results: C4.5, CBA, CPAR and CS^2.

Dataset	#cls	#rec	C4.5	CBA	CPAR	CS^2	
						$Avg.(p_F)$	$Avg.(p_F) + \sigma(p_F)$
adult	2	48842	78.8	**84.2**	77.3	83.4	83.7
anneal	6	898	76.7	94.5	**95.1**	84.1	76.8
breast	2	699	91.5	**94.1**	93.0	82.1	79.1
cylBands	2	540	69.1	**76.1**	70.0	63.8	63.8
flare	9	1389	82.1	**84.2**	63.9	73	64.7
glass	7	214	65.9	**68.4**	64.9	61.7	63.5
heart	5	303	**61.5**	57.8	53.8	55.5	57.8
hepatitis	2	155	84.1	42.2	75.5	81.5	**85.3**
horseColic	2	368	70.9	78.8	**81.2**	77.1	76
ionosphere	2	351	84.6	32.5	**88.9**	78.6	74.3
iris	3	150	91.3	93.3	**94.7**	93.3	93.3
led7	10	3200	**73.8**	73.1	71.3	71.6	73.1
mushroom	2	8124	92.8	46.7	**98.5**	94.7	95.8
pageBlocks	5	5473	92.0	90.9	**92.5**	90	89.7
penDigits	10	10992	70.5	**92.3**	80.5	68.4	70.3
pima	2	768	71.7	**74.6**	74.0	67.5	65.6
wine	3	178	75.8	49.6	88.2	91.4	**92**
zoo	7	101	91.0	40.7	**94.1**	81.2	92
Average			79.1	70.7	79.9	77.7	77.6

5   Conclusion and Future Work

We proposed to mine statistically significant contrast sets by using statistically significant association rules. Our results indicated that the contrast sets learned using our method and the contrast sets learned using traditional association rules have little to no similarity. However, we proved that, on average, our CS^2 classifier based on the proposed approach had close or better performance with other standard classifiers, providing evidence to the quality of the contrast sets we learned. Note again that our purpose is not to build a new classifier but classification was used as a means for validation. In our current work, we only explored the possibility of using association rules of the *second-kind* to mine contrast sets. However, since we introduced the use of statistically significant rules, the possibility of finding and using *first-kind* of association rules should be explored. Ideally, a combination of both type of rules may provide a better set of candidate contrast sets.

References

1. Bay, S.D., Pazzani, M.J.: Detecting group differences: mining contrast sets. Data Min. Knowl. Disc. **5**(3), 213–246. Springer (2001)
2. Dong, G., Li, J.: Efficient mining of emerging patterns: discovering trends and differences. In: 5th ACM SIGKDD International Conference on Knowledge Discovery and Data Mining, pp. 43–52. ACM (1999)
3. Hilderman, R.J., Peckham, T.: A statistically sound alternative approach to mining contrast sets. In: 4th Australia Data Mining Conference (AusDM 2005), pp. 157–172 (2005)
4. Webb, G.I., Butler, S., Newlands, D.: On detecting differences between groups. In: 9th ACM SIGKDD International Conference on Knowledge Discovery and Data Mining, pp. 256–265. ACM (2003)
5. Satsangi, A., Zaïane, O.R.: Contrasting the contrast sets: an alternative approach. In: 11th International Database Engineering and Applications Symposium (IDEAS 2007), pp. 114–119. IEEE (2007)
6. Li, J., Zaïane, O.R.: Associative classification with statistically significant positive and negative rules. In: 24th ACM International Conference on Information and Knowledge Management, pp. 633–642. ACM (2015)
7. Agrawal, R., Srikant, R.: Fast Algorithms for mining association rules. In: 20th International Conference in Very Large Data Bases (VLDB), pp. 487–499 (1994)
8. Hämäläinen, W.: Kingfisher: an efficient algorithm for searching for both positive and negative dependency rules with statistical significance measures. Knowl. Inf. Syst. **32**(2), 383–414 (2012)
9. Quinlan, J.R.: C4. 5: Programs for Machine Learning. Morgan Kaufmann Publishers Inc., (1993)
10. Liu, B., Hsu, W., Ma, Y.: Integrating classification and association rule mining. In: 4th International Conference on Knowledge Discovery and Data Mining, pp. 80–86 (1998)
11. Yin, X., Han, J.: CPAR: classification based on predictive association rules. In: 3rd SIAM International Conference on Data Mining, pp. 369–376 (2003)

A Connection Calculus for the Description Logic $\mathcal{ALC}$

Fred Freitas[1]([✉]) and Jens Otten[1,2]

[1] Informatics Center, Federal University of Pernambuco (CIn - UFPE),
Recife, Brazil
fred@cin.ufpe.br, jeotten@cs.uni-potsdam.de
[2] Institut für Informatik, University of Potsdam, Potsdam, Germany

Abstract. This paper presents a connection calculus for the description logic (DL) $\mathcal{ALC}$. It replaces the usage of Skolem terms and unification by additional annotation and introduces blocking, a typical feature of DL provers, by a new rule, to ensure termination in the case of cyclic ontologies. Besides the connection calculus, a simplified clausal form normalization is presented. Furthermore, termination, soundness and completeness of the calculus are proven.

Keywords: Description logic · Connection method · Inference system · Reasoning

1 Introduction

Description logics (DL) are widely used for knowledge representation and modelling ontologies, e.g., they are the standard ontology language for the semantic web [1]. One approach to reason with DL is the embedding of DL into classical first-order logic (FOL) and using existing provers for FOL. On the one hand, when using a FOL prover to answer DL queries, no alterations are required, since DL correspond to decidable fragments of FOL [6, 8]. Moreover, the main FOL reasoners were exhaustively tested, and thus, their usage is reliable. On the other hand, in a preliminary study, Tsarkov et al. reported that, even running one of the fastest FOL provers, the FOL approach displays a performance much slower than a specially crafted DL reasoner [12].

This paper proposes a connection calculus θ-CM for $\mathcal{ALC}$, an important fragment of DL. It relies on annotations instead of Skolem terms, and includes a rule for copying clauses that implements blocking, similarly to existing tableau-based DL reasoners. It is a significantly enhanced version of the connection calculus proposed earlier [3], which essentially translates a DL formula into a FOL matrix, thus including Skolem terms and unification, the latter here replaced by a similar procedure, θ-substitution.

The calculus was implemented in RACCOON (ReAsoner based on the Connection Calculus Over Ontologies) [4]. Therefore, such a calculus and its implementation can serve any Semantic Web application that deals with $\mathcal{ALC}$, an important DL language.

Section 2 presents the DL $\mathcal{ALC}$; its normalization is shown in Sect. 3. Section 4 explains the formal connection calculus for $\mathcal{ALC}$, while its termination, soundness and completeness are proven in Sect. 5. Section 6 concludes with a summary.

R. Khoury and C. Drummond (Eds.): Canadian AI 2016, LNAI 9673, pp. 243–256, 2016.
DOI: 10.1007/978-3-319-34111-8_30

2   The Description Logic $\mathcal{ALC}$

An ontology O in $\mathcal{ALC}$ is a set of axioms over a signature (N_C, N_R, N_O), where N_C is the *set of concept names* (unary predicate symbols), N_R is the *set of role or property names* (binary predicate symbols); N_O is the set of *individual names* (constants) [1]. *Concept expressions* are inductively defined as follows. N_C includes $\top$, the *universal concept* that subsumes all concepts, and $\bot$, the *bottom concept* subsumed by all concepts; all concept names belong to N_C. If $r \in N_R$ is a *role* and $C, D \in N_C$ are *concepts*, then these formulae are also *concepts*: (i) $C \sqcup D$ (ii) $C \sqcap D$, (iii) $\neg C$, (iv) $\forall r.C$; (v) $\exists r.C$.

A knowledge base in DL consists of a set of basic axioms (TBox), and a set of axioms specific to a particular situation (ABox). Two axiom types are allowed in a TBox $\mathcal{T}$: (i) $C \sqsubseteq D$; (ii) $C \equiv D$, standing for $C \sqsubseteq D$ and $D \sqsubseteq C$. An ABox $\mathcal{A}$ w.r.t. a TBox $\mathcal{T}$ is a finite set of assertions of two types: (i) a *concept assertion* is a statement of the form $C(a)$, where $a \in N_O$, $C \in N_C$ and (ii) a *role assertion* $r(a, b)$, where a, $b \in N_O$, $r \in N_R$. An $\mathcal{ALC}$ formula is either an axiom or an assertion; an ontology O is an ordered pair $(\mathcal{T}, \mathcal{A})$.

There are two major ways of defining the semantics of $\mathcal{ALC}$. The first one relies on the definitions of interpretation, model, etc., over a domain Δ [1]. Another way is by mapping $\mathcal{ALC}$ constructs to FOL (see [1], 2.2.1.3) and exploiting the semantics defined for FOL, see, e.g., [2]. This approach defines a translation ϕ that maps a concept C to a unary predicate $\phi_C(x)$ with a free variable x. If C is a concept and r a role, then, e.g., $\exists r.C$ is translated into the FOL formula $\phi_{\exists r.C}(y) = \exists x r(y, x) \wedge \phi_C(x)$ [1].

The work described in this paper uses an adaption of the translation approach, hence, taking advantage of existing concepts of connection calculi for FOL.

As for notation, in this paper, words starting with a capital letter denote concepts; roles start with small letters. Individuals are denoted by the lowercase letters a, b, c, d; variables are denoted by x, y, z, k, u, v. Terms are either variables or individuals.

3   Normal Form and Matrix for $\mathcal{ALC}$

Definition 1 (Query). A *query* α (a TBox or ABox axiom) against an ontology O is an $\mathcal{ALC}$ formula for which the logical consequence $O \models \alpha$ should be proven.

Definition 2 ($\mathcal{ALC}$ Disjunctive Normal Form, Clause, Matrix, Graphical Matrix). *Literals* (of $\mathcal{ALC}$) are atomic concepts or roles, possibly negated. Literals involved in an universal restriction $\forall r.C$ or in a existential restriction $\exists r.C$ are underlined. In case a restriction involves more than one clause, literals are indexed with the same new *column index number* at the top. Literals can participate at most in two universal restrictions in left-hand side (LHS) axiom's sub-formula or in two existential ones in the right-hand side (RHS); therefore, they can have at most two indices, e.g. $L^{i,j}$. An $\mathcal{ALC}$ formula in *disjunctive normal form (DNF)* is a disjunction of conjunctions (like $C_1 \vee \ldots \vee C_n$), where each C_i has the form $L_1 \wedge \ldots \wedge L_m$ and each L_i is a literal. The ($\mathcal{ALC}$) *matrix* of an $\mathcal{ALC}$ formula in DNF is its representation as a set $\{C_1, \ldots, C_n\}$, where each *clause* C_i has the form $\{L_1, \ldots, L_m\}$ with literals L_i. In the *graphical matrix* representation, clauses are represented as columns, and restrictions as lines; restrictions with indexes are horizontal, while those without are vertical (see Example 1).

Remark 1. To deduce $O \models \alpha$ the validity of the formula $C_1 \wedge \ldots \wedge C_n \to \alpha$ $(O \to \alpha)$, i.e. of $\neg O \vee \alpha$, must be proven. The effects for the DNF are: (i) axioms of the form $E \sqsubseteq D$ translate into $E \wedge \neg D$; (ii) ABox assertions are negated; (iii) free variables are existentially quantified; (iv) FOL Skolemization is applied to universal variables; and (v) the query α is not negated.

Example 1 (Query, Clause, $\mathcal{ALC}$ Matrix). The query $\{\exists hasPet.Cat \sqsubseteq CatOwner,$ $OldLady \sqsubseteq \exists hasPet.Animal \sqcap \forall hasPet.Cat\} \models OldLady \sqsubseteq CatOwner$ reads in FOL as:

$$\left.\begin{array}{l} \forall x((\exists y hasPet(x,y) \wedge Cat(y)) \to CatOwner(x)) \\ \forall z(OldLady(z) \to \forall k(hasPet(z,k) \to Cat(k))) \\ \forall z(OldLady(z) \to \exists v(hasPet(z,v) \wedge Animal(v))) \end{array}\right\} \models \forall u(OldLady(u) \to CatOwner(u))$$

and is represented by the FOL matrix (where a is a Skolem terms, f a function symbol):

$$\{\{hasPet(x,y), Cat(y), \neg CatOwner(x)\}, \{OldLady(z), hasPet(z,v), \neg Cat(v)\},$$
$$\{OldLady(z), \neg hasPet(z,f(z))\}, \{OldLady(z), \neg Animal(f(z))\}, \{\neg OldLady(a)\},$$
$$\{CatOwner(a)\}\}$$

and by the following $\mathcal{ALC}$ matrix (the column index marks the two clauses involved in he same restriction; variables are omitted as they are specified implicitly) (Fig. 1):

$$\begin{bmatrix} hasPet| & OldLady & OldLady & OldLady & \neg OldLady(a) & CatOwner(a) \\ Cat & \neg hasPet & \neg Animal & hasPet| & & \\ \neg CatOwner & & & \neg Cat| & & \end{bmatrix}$$

Fig. 1. The query from Example 1 represented as an $\mathcal{ALC}$ matrix

$$\{\{hasPet, Cat, \neg CatOwner\}, \{OldLady, hasPet, \neg Cat\}, \{OldLady, \underline{\neg hasPet}^1\},$$
$$\{OldLady, \underline{\neg Animal}^1\}, \{\neg OldLady(a)\}, \{CatOwner(a)\}\}$$

Definition 3 (Impurity, Pure Conjunction/Disjunction). *Impurity* in an $\mathcal{ALC}$ formula is a disjunction in a conjunction, or a conjunction in a disjunction. A *pure conjunction* (PC) or *disjunction* (PD) does not contain impurities (see [3] for a formal definition).

Example 2 (Impurity, Pure Conjunction/Disjunction). (a) $\exists r.A$ and $\bigwedge_{i=1}^n A_i$ are PCs if A and each A_i is also a PC. (b) $\forall r.(D_0 \sqcup \ldots \sqcup D_n \sqcup (C_0 \sqcap \ldots \sqcap C_m) \sqcup (A_0 \sqcap \ldots \sqcap A_p))$ is not a PD as it contains two impurities: $(C_0 \sqcap \ldots \sqcap C_m)$ and $(A_0 \sqcap \ldots \sqcap A_p)$.

Definition 4 (Two-Lined Disjunctive Normal Form). An $\mathcal{ALC}$ axiom is in *2-lined DNF* iff it is in DNF and in one of the normal forms (NFs): (i) $\hat{E} \sqsubseteq \check{D}$; (ii) $E \sqsubseteq \hat{E}$;

(iii) $\check{D} \sqsubseteq E$, where E is a concept name[1], $\hat{E}$ is a pure conjunction, and $\check{D}$ is a pure disjunction[2].

Example 3 (Two-Lined Disjunctive Normal Form). The axioms (i) $\hat{E} \sqsubseteq \check{D}$ (1NF); (ii) $E \sqsubseteq \exists r.\hat{E}$ (2NF) and (iii) $\forall r.\check{D} \sqsubseteq E$ (3NF), where $\hat{E} = \wedge_{i=1}^{n} C_i$ and $\check{D} = \vee_{j=1}^{m} D_j$ (Fig. 2).

$$
i)\ 1NF: \begin{bmatrix} C_1 \\ \vdots \\ C_n \\ \neg D_1 \\ \vdots \\ \neg D_m \end{bmatrix} \quad
ii)\ 2NF: \begin{bmatrix} E & \cdots & \cdots & E \\ \neg r & \neg C_1 & \cdots & \neg C_n \end{bmatrix} \quad
iii)\ 3NF: \begin{bmatrix} \neg r & D_1 & \cdots & D_m \\ \neg E & \cdots & \cdots & \neg E \end{bmatrix}
$$

Fig. 2. Examples of the three two-lined normal forms' representations in $\mathcal{ALC}$

Example 4 (Two-Lined DNF). Table 1 shows examples of quantification restrictions. Vertical lines represent existential restrict ions $(\exists r.C)$, horizontal lines represent universal restrictions $(\forall r.C)$ on the LHS axiom's sub-formula or the opposite on the RHS. Lines may overlap. Note also that, if written in FOL, Skolem functions should appear in the two last NFs in Table 1 (e.g., $\neg r(x, f(x))$ would replace $\exists y \ldots \neg r(x, y)$).

Remark 2. The motivation for relying on these NFs is a two-fold: it saves memory by avoiding redundancies in the matrix, and it helps proving the system's soundness, completeness and termination, by restricting the problematic cases to 2-lined columns.

Normalized, "purified" TBoxes may add new, introduced concepts; however, they are *conservative extensions* [5] of their originals, since to every model of the former there is a (sometimes distinct) model of the latter, and validity is preserved. Besides, for these NFs, in the worst case, the number of new concepts grows linearly with the number of impurities; in the average case, this is better than other normalizations (e.g., in [10], the number of new axioms grows linearly with the axioms' length).

Definition 5 (Cycle, Cyclic/Acyclic Ontologies and Matrices). If A and B are atomic concepts in an ontology O, A *directly uses* B, if B appears in the right-hand side of a subsumption axiom whose left-hand side is A. Let the relation *uses* be the transitive closure of *directly uses*. A *cyclic ontology* or *matrix* has a cycle when an atomic concept *uses* itself; otherwise it is *acyclic* [1]; e.g., $O = \{A \sqsubseteq \exists r.B, B \sqsubseteq \exists s.A\}$ is a cyclic ontology. Besides, in acyclic ontologies all subsumption axioms have a concept name in its LHS.

---

[1] The symbols E and $\hat{E}$ were chosen here to designate a concept name and a pure conjunction rather than the usual C and $\hat{C}$, to avoid confusion with clauses, that are also denoted by C.

[2] If $\exists r.\top \sqsubseteq \hat{E}$ or $\forall r. \sqsubseteq \check{D} \in \bot\mathcal{T}$, then the matrix must include axioms $A \sqsubseteq \top$, for all $A \in N_C$, too. Conversely, $\exists r.\top \sqsubseteq \hat{E}$ or $\forall r.\top \sqsubseteq \check{D}$ requires axioms $\bot \sqsubseteq A$, for all $A \in N_C$, in the matrix.

Table 1. Examples of quantification restrictions

Axiom	Matrix	Negated FOL mapping
$\exists r.\hat{E} \sqsubseteq \forall s.\check{D}$ with $\hat{E}$ a pure conjunction, $\check{D}$ a pure disjunction	$\begin{bmatrix} r \\ E_1 \\ \vdots \\ E_n \\ s \\ \neg D_1 \\ \vdots \\ \neg D_m \end{bmatrix}$	$\exists x \exists y \exists z$ $(r(x,y) \wedge$ $E_1(y) \wedge...\wedge E_n(y)$ $\wedge$ $(s(x,z) \wedge$ $\neg D_1(z) \wedge...\wedge \neg D_m(z))$
$A \sqsubseteq \exists r.\hat{E}$ A is a concept name, $\hat{E}$ as above	$\begin{bmatrix} A & \cdots & \cdots & A \\ \neg r & \neg E_1 & \cdots & \neg E_n \end{bmatrix}$	$\exists x \forall y ((A(x) \wedge \neg r(x,y))$ $\vee (A(x) \wedge \neg E_1(y))$ $\vee...\vee (A(x) \wedge \neg E_n(y)))$
$\forall r.\check{D} \sqsubseteq A$ $A, \check{D}$ as above	$\begin{bmatrix} \neg r & D_1 & \cdots & D_m \\ \neg A & \cdots & \cdots & \neg A \end{bmatrix}$	$\forall x \exists y$ $(\neg r(x,y) \wedge \neg A(x)) \vee$ $(D_1(y) \wedge \neg A(x)) \vee...\vee$ $(D_m(y) \wedge \neg A(x))$

4   The $\mathcal{ALC}$ θ-Connection Calculus ($\mathcal{ALC}$ θ-CM)

The $\mathcal{ALC}$ θ-Connection Method (henceforth $\mathcal{ALC}$ θ-CM) differs from the FOL Connection Method (CM) by replacing Skolem functions and unification by θ-substitutions, and, just as typical DL systems, employs blocking to assure termination.

Definition 6 (Path, Connection, θ-Substitution, θ-Complementary Connection). A *path through a matrix M* contains exactly one literal from each clause in M. A *connection* is a pair of literals $\{E, \neg E\}$ with the same concept/role name, but different *polarities*. A *θ-substitution* assigns each (possibly omitted) variable an individual or another variable. A *θ-complementary* connection is a pair of $\mathcal{ALC}$ literals $\{E(x), \neg E(y)\}$ or $\{p(x,v), \neg p(y,u)\}$, with $\theta(x) = \theta(y), \theta(v) = \theta(u)$. The complement $\bar{L}$ of a literal L is E if $L = \neg E$, and it is $\neg E$ if $L = E$.

Remark 3 (θ-Substitution). Simple term unification without Skolem functions is used to calculate θ-substitutions. The application of a θ-substitution to a literal is an application to its variables, i.e. $\theta(E) = E(\theta(x))$ and $\theta(r) = r(\theta(x), \theta(y))$, where E is an atomic concept and r is a role. Furthermore, $x^\theta = \theta(x)$.

Definition 7 (Set of Concepts, Skolem Condition). The *set of concepts* $\tau(x)$ of a variable or individual x contains all concepts that were instantiated by x so far, or, more formally, $\tau(x) \stackrel{\text{def}}{=} \{E \in N_C | E(x) \in Path\}$. The *Skolem condition*, ensures that at most one concept is underlined in the graphical matrix form. This condition is formally defined as $\forall a \left| \left\{ \underline{E^i} \in N_C | \underline{E^i(a)} \in Path \right\} \right| \leq 1$, where i is a column index.

Lemma 1 (Equivalence Between θ-Substitution in $\mathcal{ALC}$ θ-CM and Unification in CM for $\mathcal{ALC}$ Formulae). θ-substitution is equivalent to unification for $\mathcal{ALC}$ formula, i.e., both procedures either return the same results or behave the same way w.r.t. their calculi, when given the same inputs.

Proof. The cases occurring in $\mathcal{ALC}$, all covered by θ-substitution, are in Table 2.

Table 2. Equivalence for $\mathcal{ALC}$ between CM unification and θ-substitution in $\mathcal{ALC}$ θ-CM

Unification		θ-substitution	
Input	Output	Input	Output
$L_1 = E(x)$ $L_2 = \neg E(a)$	$\sigma(L_1) = E(a)$	$L_1 = E \text{ or } E(x)$ $L_2 = \neg E(a)$	$\theta(L_1) = E(a)$
$L_1 = E(x)$ $L_2 = \neg E(y)$	$\sigma(L_1) = E(y)$	$L_1 = E \text{ or } E(x)$ $L_2 = \neg E \text{ or } \neg E(y)$	$\theta(L_1) = E(y)$
$L_1 = E(b)$ $L_2 = \neg E(a)$	Not unifiable	$L_1 = E(b)$ $L_2 = \neg E(a)$	No θ-substitution: Not unifiable
$L_1 = E(x)$ $L_2 = \neg E(f(y))$	$\sigma(L_1) = E(f(y))$	$L_1 = E$ $L_2 = \underline{\neg E^i}$	$\theta(L_1) = E(y)$ $\tau(y) = \tau(y) \cup \{\neg E^i\}$
$L_1 = E(g(x))$ $L_2 = \neg E(f(y))$	Not unifiable	$L_1 = E^k$ $L_2 = \underline{\neg E^j}$	No θ-substitution, as Skolem condition does not hold: $\forall a \left\| \left\{ \underline{E^i}\|E^i(a) \in Path \right\} \right\| \leq 1$

Note that in all cases θ-substitution and unification yield the same substitution or no substitution; the only exception resides in the case where $L_1 = E(x)$ and $L_2 = \neg E(f(y))$ ($L_1 = E, L_2 = \underline{\neg E^i}$ in the notation without variables). However, Lemma 2 shows that they are equivalent, in the sense that unification in FOL CM prevents the same connections that $\mathcal{ALC}$ θ-CM and θ-unification prevent. ∎

Definition 8 ($\mathcal{ALC}$ Connection Calculus). Figure 3 shows the formal $\mathcal{ALC}$ connection calculus ($\mathcal{ALC}$ θ-CM), adapted from the FOL CM [9]. The rules of the calculus are applied in an analytic, bottom-up way. The basic structure is the tuple $< C, M, Path >$, where clause C is the open sub-goal, M the matrix corresponding to the query $O \vDash \alpha$ (O is an $\mathcal{ALC}$ ontology) and $Path$ is the *active path*, i.e. the (sub-)path currently checked. The index $\mu \in \mathbb{N}$ of a clause C^μ denotes that C^μ is the μ-th copy of clause C, increased when Cop is applied for that clause (the variable x in C^μ is denoted x_μ) – see example of copied clauses in Fig. 8. When Cop is used, it is followed by the application of Ext or Red, to avoid non-determinism in the rules' application. The *Blocking Condition* is defined as follows: the new individual x_μ^θ (if it is new, then $x_\mu^\theta \notin N_O$, as in the condition) has its set of concepts $\tau(x_\mu^\theta)$ compared to the set of concepts of the previous copied individual, i.e., $\tau(x_\mu^\theta) \text{ g } \tau\left(x_{\mu-1}^\theta\right)$ [11], to test if the former is a subset of the latter.

Remark 4 ($\mathcal{ALC}$ Connection Calculus). FOL CM already copies clauses, using the indexing function μ; in $\mathcal{ALC}$ θ-CM, Cop implements blocking [1], when no alternative connection is available and cyclic ontologies are dealt. It regulates the creation of new

$$Axiom\ (Ax)\ \ \overline{\{\},M,Path}$$

$$Start\ Rule\ (St)\ \ \frac{C_1,M,\{\}}{\varepsilon,M,\varepsilon}\ \ with\ C_1 \in \alpha$$

$$Reduction\ Rule\ (Red)\ \ \frac{C,M,Path \cup \{L_2\}}{C \cup \{L_1\},M,Path \cup \{L_2\}}$$

with $\theta(L_1) = \theta(\overline{L_2})$ and the Skolem condition holds

$$Extension\ Rule\ (Ext)\ \ \frac{C_1 \backslash \{L_2\},M \backslash C_1,Path \cup \{L_1\}\quad C,M,Path}{C \cup \{L_1\},M,Path}$$

with $C_1 \in M$, $L_2 \in C_1$, $\theta(L_1) = \theta(\overline{L_2})$ and the Skolem condition holds

$$Copy\ Rule\ (Cop)\ \ \frac{C \cup \{L_1\},M \cup \{C_2^\mu\},Path}{C \cup \{L_1\},M,Path}$$

with C_2^μ is a copy of C_1, $L_2 \in C_2^\mu$, $\theta(L_1) = \theta(\overline{L_2})$ and the blocking condition holds

Fig. 3. The connection calculus $\mathcal{ALC}$ θ-CM

individuals, thus preventing non-termination. The *Skolem condition* solves the FOL cases where the combination of Skolemization and unification correctly prevents connections (see Soundness Theorem below). The Skolem condition is easy to implement: only a flag denoting if each variable/individual in any path contains an underlined concept suffices Finally, in the *Ext* and *Red rules*, θ-substitutions replace variables by variables/individuals in the whole matrix. Any individual x can have in its set of concepts $\tau(x)$ at most a single concept name with a column index in the matrix (i.e., $\forall a \left| \left\{ \underline{E^i} \in N_C | \underline{E^i}(a) \in Path \right\} \right| \leq 1$). This restriction avoids the situation in FOL matrices, where unification is tried with distinct Skolem functions (see Lemmas 1, 2).

Example 5 (*$\mathcal{ALC}$ Connection Calculus*). Figures 4 and 5 show the proof of the query from Example 1 using the matrix representation and the formal calculus, respectively.

5 Termination, Soundness and Completeness

Definition 9 (Functional Equivalence Between Decidable Inference Systems w.r.t. a Set of Formulae). An inference system A is *functionally equivalent* to a system B w.r.t. a set of formulae Σ when, for any formula α, $\Sigma \vdash_A \alpha \leftrightarrow \Sigma \vdash_B \alpha$ and $\Sigma \nvdash_A \alpha \leftrightarrow \Sigma \nvdash_B \alpha$.

Lemma 2 (Functional Equivalence Between CM and $\mathcal{ALC}$ θ-CM for Acyclic $\mathcal{ALC}$ Formulae). $\mathcal{ALC}$ θ-CM is functionally equivalent to CM w.r.t. acyclic $\mathcal{ALC}$ formulae.

Proof. Given that (i) *Cop* is not applied here; (ii) the two systems only differ on the replacement of unification by θ-substitution in the *Ext* and *Red* rules; and (iii) θ-substitution is equivalent to unification, Lemma 1 proves all cases but one: a connection with two Skolem functions. The proof for that case is inductive over the matrix structure.

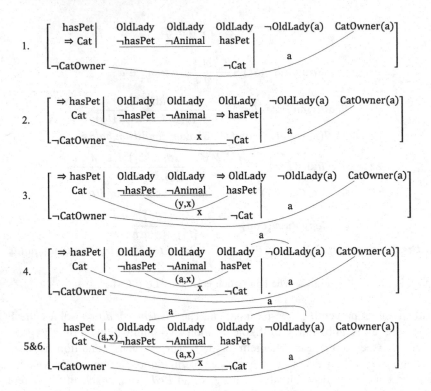

Fig. 4. The proof of the query using the graphical matrix representation. Arcs are connections whose labels are the names of the involved individual(s)/variable.

$$\frac{\overline{\{\},M,\{CO(a),\underline{C(x)},h(a,x),\neg O(a)\}}^{\ Ax}}{\{O(y)\},M,\{CO(a),\underline{C(x)},h(a,x)\}}\ Red$$
$$\frac{}{\{O(y),\underline{h(y,x)}\},M,\{CO(a),\underline{C(x)}\}}$$

$$\frac{\overline{\{\},M,\{CO(a),\underline{C(x)},O(a)\}}^{\ Ax}}{\{O(y)\},M,\{CO(a),\underline{C(x)}\}}\ Ext \qquad \overline{\{\},M,\{CO(a),\underline{C(x)}\}}^{\ Ax}\ Ext$$

$$\frac{}{\{O(y),\underline{h(y,x)}\},M,\{CO(a),\underline{C(x)}\}}$$
$$\frac{\overline{\{\},M,\{CO(a),\neg h(a,x)^1,\neg O(a)\}}^{\ Ax}}{\{O(a)\},M,\{CO(a),\neg h(a,x)^1\}}\ Red}{\{h(a,x)\},M,\{CO(a)\}}\ Red$$

$$\frac{\{h(a,x),\underline{C(x)}\},M,\{CO(a)\}}{\{CO(a)\},M,\{\}}\ Ext \qquad \overline{\{\},M,\{\}}^{\ Ax}\ Ext$$
$$\frac{\{CO(a)\},M,\{\}}{\varepsilon,M,\varepsilon}\ St$$

Fig. 5. The proof of the query using the formal connection calculus, where M is an abbreviation for $M = \{\{\underline{h,C,\neg CO}\}, \{O,h,\neg C\}, \{O,\neg h^1\}, \{O,\neg A^1\}, \{\neg O(a)\}, \{CO(a)\}\}$ (the double-ended arrow just copies the proof part to save text space).

Induction Hypothesis: $O \nvdash_{CM} \alpha$, where the only available connection is between literals with distinct Skolem functions.

Base Case: Suppose a tentative connection between literals with two distinct Skolem functions. For instance, a posed query $\exists r.A \sqsubseteq E, E \sqsubseteq \forall r.a \vDash E(a)$, represented in FOL for the CM in Fig. 6a. Unification prevents the connection (denoted by a dotted lined) in the CM. For $\mathcal{ALC}$ θ-CM, this connection is forbidden too, due to the Skolem condition. In Fig. 6b, for the new variable y, $\tau(y) = \{\underline{A^2}\}$; so it cannot contain also $\neg \underline{A^1}$, otherwise it would violate the Skolem condition $(\forall a |\{\underline{E^i} \in N_C | E^i(a) \in Path\}| \leq 1)$.

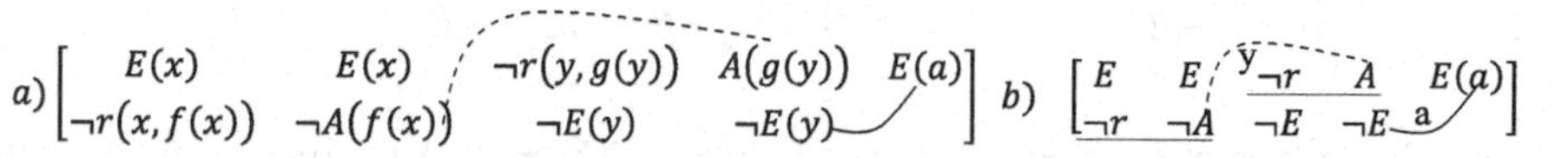

$$a) \begin{bmatrix} E(x) & E(x) & \neg r(y, g(y)) & A(g(y)) & E(a) \\ \neg r(x, f(x)) & \neg A(f(x)) & \neg E(y) & & \neg E(y) \end{bmatrix} \quad b) \begin{bmatrix} E & E & y & \neg r & A & E(a) \\ \neg r & \neg A & & \neg E & \neg E & a \end{bmatrix}$$

Fig. 6. Tentative connection proof for a) $\exists r.A \sqsubseteq E, E \sqsubseteq \forall r.A \vdash_{CM} \neg E(a)$, with $\sigma = \{y/a\}$ and b) $\exists r.A \sqsubseteq E, E \sqsubseteq \forall r.A \nvdash_{ALC\theta-CM} \neg E(a)$. Dotted lines stand for forbidden connections.

Inductive Case. Suppose an individual a or a variable x that, in CM, takes part in several connections, two of two in literals with Skolem functions. Then the same behavior from the base case will occur here: CM prevents the last connection ought to unification between two different Skolem functions, while $\mathcal{ALC}$ θ-CM avoids the connection on the basis that the Skolem condition is violated. In Fig. 7, the set of concepts of y or a cannot admit two underlined concepts, i.e. $\tau(y) = \{\underline{A^1}\}$, but $\tau(y) \neq \{\underline{A^1}, \neg \underline{C^2}\}$, again due to the Skolem condition $(|\{\underline{E^i} \in N_C | E^i(y) \in Path\}| \leq 1)$.

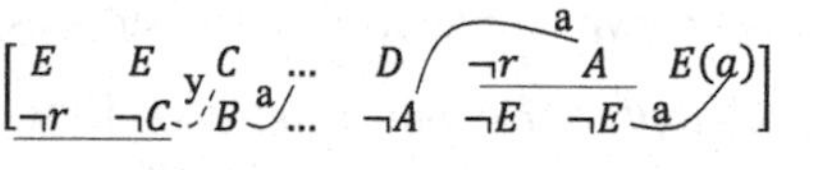

$$\begin{bmatrix} E & E & y & C & \cdots & D & \overset{a}{\frown} & \neg r & A & E(a) \\ \neg r & \neg C & B & \cdots & \neg A & \neg E & \neg E & a \end{bmatrix}$$

Fig. 7. Tentative $\mathcal{ALC}$ θ-CM connection proof for the inductive case

Since the two systems apply the same rules in the same order and the effects of unification in the FOL CM are the same as that of the θ-substitution in $\mathcal{ALC}$ θ-CM, the systems are equivalent for the set of acyclic $\mathcal{ALC}$ formulae.   ∎

Remark 5 (Cyclic Ontologies). Lemma 2 entails that termination, soundness and completeness need only to be proven for the cyclic cases. Lemma 2.22 from [1] states that in a cyclic ontology O, for an individual name $x_i \in$ ABox $\mathcal{A}$ of concept E_i (i.e., individual $E_i(x_i)$ is an assertion), then there is a unique finite sequence of roles $r_1, \ldots, r_i (i \geq 1)$, a unique finite sequence of i role instances (the so-called *role successors*) $r_1(x_0, x_1), \ldots, r_i(x_{i-1}, x_i)$, and a unique, finite sequence of individual names $x_1, \ldots, x_{i-1}$ that creates x_i for the $\mathcal{ALCN}$ tableaux system, as defined by the authors (also holds for $\mathcal{ALC}$, since it is a subset of $\mathcal{ALCN}$). For the $\mathcal{ALC}$ θ-CM, a similar corollary is valid, without needing role successors.

Lemma 3 (Uniqueness of a Generation Sequence). Suppose E_i an $\mathcal{ALC}$ concept from a query over a cyclic ontology O, x_i an individual name of concept E_i (i.e., $E_i(x_i)$). Then, there is a unique finite sequence of individual names $x_0, \ldots, x_{i-1}$, x_i in a path.

Proof. Since only one rule is applied at a time, the new concept instance $E_i(x_i)$ (with the new individual name x_i) is created in the end of the active path. Since the active path contains the unique sequence $E(x_0), \ldots, E(x_{i-1})$, the corollary holds. ∎

$$\begin{bmatrix} E & E & \overset{a}{\frown} \neg E(a) \\ \neg r & \neg E \end{bmatrix} \vdash .. \vdash \begin{bmatrix} E & \overset{a}{\overbrace{E \ b \ E \ c \ E}} & \neg E(a) \\ \neg r & \neg E & \neg E & \neg E \end{bmatrix}$$

Fig. 8. The creation of the unique sequence $\{\neg E(a), \neg E(b), \neg E(c)\}$ to arrive at $\neg E(c)$ for the unfinished query $E \sqsubseteq \exists r.E \nvdash_{ALC\theta-CM} \neg E(a)$, with $\mu(\{E, \underline{\neg E}\}) = 2$

On the other hand, it is easy to see why role successors [1, Lemma 2.22] are not needed in $\mathcal{ALC}$ θ-CM (for notation $\vdash_{ALC\theta-CM}$ and $\vdash_{CM}$ stand for deductions carried out with $\mathcal{ALC}$ θ-CM, CM respectively). Observe the generation of $\neg E(c)$ for the query described as $E \vee \exists r.E \vdash_{ALC\theta-CM} \neg E(a)$ in Fig. 8: the active path in this case is $\{\neg E(a), \neg E(b), \neg E(c)\}$. In tableaux systems, role successors $\neg r(a, b)$ and $\neg r$ (b, c) would also need to be created to arrive at $\neg E(c)$.

Theorem 1 (Termination). Given M, the matrix representing the arbitrary query $O \vdash_{ALC\theta-CM} \alpha$, and a chosen initial clause C, any rule sequence in the $\mathcal{ALC}$ θ-CM applied over the tuple "$\varepsilon, M, \varepsilon$" terminates.

Proof. Case (1) $O \vdash_{ALC\theta-CM} \alpha$, which has itself three sub-cases:

(a) $\mathcal{ALC}$ θ-CM *does not apply the Copy rule, and M is proven valid.* This case reduces to the last, since the proof is found and has no cycle; then, $\mathcal{ALC}$ θ-CM terminates.

(b) $\mathcal{ALC}$ θ-CM *uses cyclic axioms, but the Copy rule is only applied with already existent individuals.* Again, $\mathcal{ALC}$ θ-CM is equivalent to CM, because the already existent individuals are not created by *Cop*; thus, $x_\mu^\theta \in N_O$, which never meets the blocking conditions. Consequently, the indexing function μ is incremented, the cyclic columns C_2^μ are copied, and the process repeats that of the CM. Besides, after the copy, θ-substitutions work just as unifications (Lemma 1). Since CM terminates (see proof at [2, III.6.4.]), $\mathcal{ALC}$ θ-CM terminates for this case, too.

(c) $\mathcal{ALC}$ θ-CM *uses the cyclic axioms and the Copy rule (Cop) to create new individuals.* The blocking condition from *Cop* ensures termination, given that it prevents *Cop* to be applied indefinitely, thus generating infinite repetitions of finite sequence(s) of individual names $x_0, \ldots, x_{j-1}, x_j$ in the active path p that would characterize the loop. The blocking condition identifies such repetitions, by checking if the generated concept instances are new individuals names (testing if the new instance is not in the ABox already, i.e., if $x_\mu^\theta \notin N_O$), and if their set of concepts (τ) is changing (by testing whether $\tau(x_\mu^\theta)$ g $\tau\left(x_{\mu-1}^\theta\right)$, i.e., if the new instance's set of concepts is a subset of the set of the previously created instance). If both conditions are met, this active path is blocked; $\mathcal{ALC}$ θ-CM runs until a proof is found (see Completeness Theorem below) and halts.

Case (2) $O \nvdash_{ALC\theta-CM} \alpha$, which has two sub-cases:

(a) The *Copy rule is not applied or is applied with existent individuals, not created by previous Copy rule applications* (i.e., $x_\mu^\theta \in N_O$), there are open subgoals which make $O \nvdash_{ALC\theta-CM} \alpha$. This case is analogous to cases (1) (a) and (b).

(b) When the *Copy rule creates new x_i individuals*, the case is similar to (1) (c). The infinite open cycles are detected by the *Copy rule* and blocked, and the proof fails due to open subgoals. Hence, $\mathcal{ALC}$ θ-CM terminates for this case, and thus, for all cases. ∎

Theorem 2 (Soundness). An $\mathcal{ALC}$ formula in the two-lined disjunctive normal form M is valid if there is a connection proof for "$\varepsilon, M, \varepsilon$" in the $\mathcal{ALC}$ θ-CM, i.e. there exists a derivation in which all leaves are axioms.

Proof. CM is a decision procedure for $\mathcal{ALC}$, since $\mathcal{ALC}$ corresponds to the decidable FOL fragment L^2 [8]. Thus, $O \vdash_{CM} \alpha$ implies in $O \models \alpha$. Hence, it suffices to prove that $O \vdash_{ALC\theta-CM} \alpha$ implies in $O \vdash_{CM} \alpha$. For the cyclic cases and when M originally contains (Skolem) functions, the $\mathcal{ALC}$ formulae will be converted to the 2NF and 3NF, respectively $E \sqsubseteq \exists r.\hat{E}$(or $E \sqsubseteq \hat{E}$); and $\forall r.\check{D} \sqsubseteq E$ (or $\check{D} \sqsubseteq E$), E being an atomic concept, $\hat{E}$ a pure conjunction and $\check{D}$ a pure disjunction. This case is proved by the contrapositive: $O \nvdash_{CM} \alpha$ must imply in $O \nvdash_{ALC\theta-CM} \alpha$.

The contrapositive proof is by structural induction on the structure of the finite sequence of individual names $x_1, \ldots, x_{i-1}$ that generates the next individual of the cycle x_i. The cases of each of the two normal forms are proven in a similar way, as they differ only in the polarity of the class(es) involved in the existential/universal restriction. Note that in any case, the normal forms only generate columns with two elements, which facilitates the inductive proof. The proof for second normal form comes next. The set of formulae in the first, second, third normal forms is denoted by S_{1NF}, S_{2NF} and S_{3NF}.

Induction Hypothesis: $O \nvdash_{CM} \alpha$, where $a \in O$, S_{2NF} (also works for S_{3NF}).

Base Case: $O = \{E \sqsubseteq \exists r.E\} \in S_{2DNF}, O \nvdash_{CM} \alpha$, α being an arbitrary formula, e.g. $\alpha = \{\neg E(a)\}$, as shown in Fig. 9a, b. After the connection $\{E(x), \neg E(a)\}$, with $\sigma = \{x/a\}$, due to the lack of complement for $\neg E(f(x)))$, first FOL CM copies the second clause increasing this clause's μ (see Fig. 9a). Then, occurs-check blocks the new connection, and, therefore, $E \sqsubseteq \exists r.E \nvdash_{CM} \neg E(a)$.

As for the $\mathcal{ALC}$ θ-CM, the case is portrayed in Fig. 9b. The first connection is equal to that of CM (except for applying θ instead of σ). In the second clause $(\{E, \neg\underline{E^1}\})$, $\neg E(b)^1$ is built as a θ-substitution. Next (not shown in the figure), *Cop* is applied, and a new clause $\{E, \neg\underline{E^2}\}$ appears in M. Then, the connection $\{\neg E(b)^1, E(b)\}$ is settled, and instead of generating a new individual name c, b is reused in the new literal $E(b)$ (an alternative test for blocking suggested by Baader et al. [1]). A new clause copy is made, and since the connection $\{\neg E(b)^2, E(b)\}$ reappears, the process is blocked and $E \sqsubseteq \exists r.E \nvdash_{ALC\theta-CM} \neg E(a)$. The lines below the last matrix represent the fact that each new clause copy shall not represent the same individual instantiated in E, i.e., a new individual must instantiate $\neg E$. So, for the base case $a \in O$, S_{2NF}, $O \nvdash_{CM} \alpha$ implies in $O \nvdash_{ALC\theta-CM} \alpha$, q.e.d.

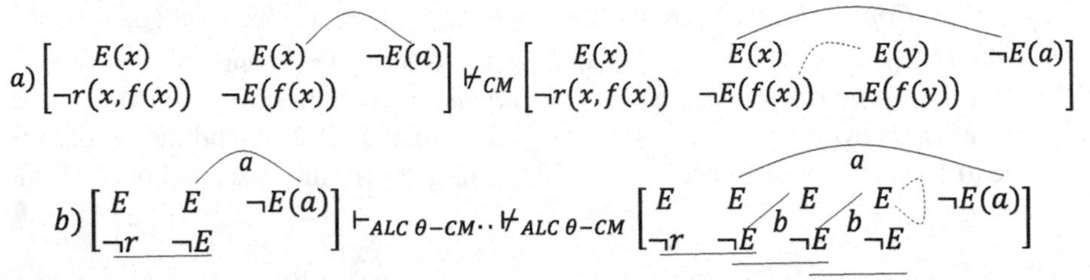

$$
a) \begin{bmatrix} E(x) & E(x) & \neg E(a) \\ \neg r(x, f(x)) & \neg E(f(x)) \end{bmatrix} \nvdash_{CM} \begin{bmatrix} E(x) & E(x) & E(y) & \neg E(a) \\ \neg r(x, f(x)) & \neg E(f(x)) & \neg E(f(y)) \end{bmatrix}
$$

$$
b) \begin{bmatrix} E & E & \neg E(a) \\ \neg r & \neg E \end{bmatrix} \vdash_{ALC\,\theta-CM} \cdot\cdot \nvdash_{ALC\,\theta-CM} \begin{bmatrix} E & E & E & E & \neg E(a) \\ \neg r & \neg E & \neg E^b & \neg E^b & \neg E \end{bmatrix}
$$

Fig. 9. Tentative connection proof, showing that a) $E \sqsubseteq \exists r.E \nvdash_{CM} \neg E(a)$, $\sigma = \{x/a\}$, with $\mu(\{E(x), \neg E(f(x))\}) = 1$, and that b) $E \sqsubseteq \exists r.E \nvdash_{ALC\theta-CM} \neg E(a)$, with $\mu(\{E, \neg E\}) = 2$

Inductive Case: Suppose $O = \{E \sqsubseteq \exists r.B, B \sqsubseteq \hat{E}\}$, $\alpha = \{\neg E(y_0)\}$, α, α an arbitrary formula), y_0 an individual name, $\hat{E}$ a pure conjunction that *uses* E. In that case, $\hat{E}$ is in one of the following forms: $E \sqcap \hat{A}$, $\exists r.(E \sqcap \hat{A})$ or $\exists r.(E \sqcap \hat{A}) \sqcap \hat{A}$, being $\hat{A}, \hat{F}$ also pure conjunctions. In either form, M contains the column $\{B(x), \neg E(f(x))\}$ for CM and $\{B(x), \neg E(y)^1\}$ for ALC θ-CM. Therefore, even after pursuing long finite sequences of individual names $x_1, \ldots, x_{i-1}$, CM fails just as in the base case, by occurs-check or looping. Similarly, ALC θ-CM generates $\neg E(y_1)^1$ after the first loop and in the next; then, the blocking condition is reached, and ALC θ-CM halts.

For the inductive case where $a \subseteq O$, S_{2NF}, $O \nvdash_{CM} \alpha$ implies in $O \nvdash_{ALC\theta-CM} \alpha$. So, the contrapositive $O \vdash_{ALC\theta-CM} \alpha$ implies in $O \vDash \alpha$ and ALC θ-CM is sound. ∎

Theorem 3 (Completeness). There is a connection proof for "$\varepsilon, M, \varepsilon$", i.e.. there exists a derivation in which all leaves are axioms, if the ALC formula F that corresponds to the matrix M is valid.

Proof. Analogously to the soundness proof, it suffices to prove that if $O \vDash \alpha$ then $O \vdash_{ALC\theta-CM} \alpha$. To show that, it is enough to demonstrate $O \vdash_{CM} \alpha$ implies in $O \vdash_{ALC\theta-CM} \alpha$, again, by the contrapositive: $O \nvdash_{ALC\theta-CM} \alpha$ must imply in $O \nvdash_{CM} \alpha$, when M contains cycles. $O \nvdash_{ALC\theta-CM} \alpha$ has two sub-cases:

(a) When the *Copy rule is not applied or is applied with already existent individual names, not created by previous Copy rule applications* (i.e., $x_\mu^\theta \in N_O$), there are open subgoals which make $O \nvdash_{ALC\theta-CM} \alpha$. This case is analogous to the last case: FOL CM fails with the same open subgoals, and $O \nvdash_{CM} \alpha$ for this case, too. Thus, $O \nvdash_{CM} \alpha$.

(b) The case when $O \nvdash_{ALC\theta-CM} \alpha$ *and the Copy rule creates new x_i instances*, can be shown by an inductive proof similar to the one of soundness. The idea is to show that FOL CM loops or finishes by occur-check when blocking takes place in $\mathcal{ALC}$ θ-CM for the base and inductive cases. Therefore, when $\mathcal{ALC}$ θ-CM fails after exhausting all possible connections and θ-substitutions, CM, by occur-checks and/or loops, is also unable to find a proof, i.e., $O \nvdash_{CM} \alpha$.

Indeed, the soundness theorem has shown that there is a functional equivalence between the two systems for the cyclic case too: $\mathcal{ALC}$ θ-CM blocks the cases that CM either loops or halts and vice-versa.

Hence, $O \nvdash_{\mathcal{ALC}\theta-CM} \alpha$ implies $O \nvdash_{CM} \alpha$, and ALC θ-CM is complete.   ∎

6 Conclusions, Ongoing and Future Work

In the current work, $\mathcal{ALC}$ θ-CM is introduced, a connection method for DL that presents two novelties: (i) it replaces Skolem functions and unification by θ-substitutions that emulate the process of creating instances in the model that is typical for DL tableaux systems; and (ii) it introduces a blocking scheme (with a new *Copy rule*) to deal with cyclic ontologies in order to assure termination.

For the inference process, it employs a normal form that minimizes redundancy in the representation and in the proof search. Moreover, termination, soundness and completeness were proven with the aid of these NFs, which restrict the more convoluted cases to matrix columns of only two literals. This facilitates to portrait the correspondence between FOL unification and θ-substitution/blocking for $\mathcal{ALC}$ θ-CM.

For future work, we will tackle cardinality restrictions ($\geq / \leq n\ r$ for $\mathcal{ALCN}$ and $\geq / \leq n\ r.C$ for $\mathcal{SHQ}$) by dealing with equality between instances. We also aim to create more sophisticated blocking schemes for dynamic and double blocking for DL constructs like inverse roles [7] or dealing with nominals.

Acknowledgements. The authors would like to thank Pernambuco's state sponsoring agency FACEPE for a grant to support the stay of Jens Otten at UFPE.

References

1. Baader, F., Calvanese, D., McGuinness, D., Nardi, D., Patel-Schneider, P. (eds.): The Description Logic Handbook. Cambridge University Press, Cambridge (2003)
2. Bibel, W.: Automated Theorem Proving. Vieweg Verlag, Wiesbaden (1987)
3. Freitas, F.: A connection method for inferencing over the description logic ALC. In: Description Logics Workshop, Barcelona, Spain (2011)
4. Melo, D., Otten, J., Freitas, F.: RACCOON: an experimental automated reasoner for description logics using a modified version of the connection method (2016). https://github.com/dmfilho/raccoon
5. Ghilardi, S., Lutz, C., Wolter, F.: Did I damage my ontology: a case of conservative extensions of description logics. In: Proceedings of 10th International Conference of Principles of KR&R (KR). AAAI Press (2006)

6. Graedel, E., Otto, M., Rosen, E.: Two-variable logic with counting is decidable. In: Proceedings of 12th IEEE Symposium on Logic in Computer Science, LICS 1997 (1997)
7. Horrocks, I., Sattler, U.: A description logic with transitive and inverse roles and role hierarchies. J. Logic Comput. **9**(3), 385–410 (1999)
8. Mortimer, M.: On languages with two variables. Zeitschrift fur mathematische Logik und Grundlagen der Mathematik **21**, 135–140 (1975)
9. Otten, J.: Restricting backtracking in connection calculi. AI Commun. **23**(2–3), 159–182 (2010)
10. Schlicht, A., Stuckenschmidt, H.: Peer-to-peer reasoning for interlinked ontologies. Int. J. Seman. Comput. 4(1), 27–58 (2010). (Special Issue on Web Scale Reasoning)
11. Schmidt, R., Tishkovsky, D.: Analysis of blocking mechanisms for description logics. In: Proceedings of the Workshop on Automated Reasoning (2007)
12. Tsarkov, D., Riazanov, A., Bechhofer, S., Horrocks, I.: Using vampire to reason with OWL. In: McIlraith, S.A., Plexousakis, D., van Harmelen, F. (eds.) ISWC 2004. LNCS, vol. 3298, pp. 471–485. Springer, Heidelberg (2004)

Representation, Reasoning, and Learning for a Relational Influence Diagram Applied to a Real-Time Geological Domain

Matthew Dirks[1]([⊠]), Andrew Csinger[2], Andrew Bamber[2], and David Poole[1]

[1] University of British Columbia (UBC), Vancouver, Canada
mcdirks@cs.ubc.ca
[2] MineSense Technologies Ltd., Vancouver, Canada

Abstract. Mining companies typically process all the material extracted from a mine site using processes which are extremely consumptive of energy and chemicals. Sorting the rocks containing valuable minerals from ones that contain little to no valuable minerals would effectively reduce required resources by leaving behind the barren material and only transporting and processing the valuable material. This paper describes a controller, based in a relational influence diagram with an explicit utility model, for sorting rocks in unknown positions with unknown mineral compositions on a high-throughput rock-sorting and sensing machine. After receiving noisy sensor data, the system has 400 ms to decide whether to divert the rocks into either a keep or discard bin. We learn the parameters of the model offline and do probabilistic inference online.

1 Introduction

This paper considers the problem of sorting rocks, separating valuable, high-grade rocks from low-grade rocks as they pass over an array of electromagnetic sensors. By sorting more effectively ahead of the mill, we reduce costs and help preserve the environment because the amount of material sent to further downstream mining processes is reduced [1,7]. The rock sorting machine, known as SortOre[TM], has been deployed in field pilot situations in Ontario, Canada and Guatemala in addition to the 60 tonnes per hour unit available in lab.

We have developed Rock Predictor Sorting Algorithm (RPSA) based on a rock sorting machine on which we have performed training and evaluation. The machine, with schematics shown in Fig. 1(b), passes rocks on a conveyor belt, moving downward on the y-axis, over a **sensor array** of 7 electromagnetic (EM) coils. From each sensor we read the change in voltage caused by the rock disrupting the magnetic field above the EM coil every millisecond (ms), which we call **sensor readings**. The rocks travel for about 400 ms after the sensor array to the end of the conveyor belt where they fall off onto a **diverter array** which may have one or more diverters **activated**, displacing rocks into a **keep bin** or else leaving them to fall into a **discard bin**. Both the sensing and diverting are imperfect.

© Springer International Publishing Switzerland 2016
R. Khoury and C. Drummond (Eds.): Canadian AI 2016, LNAI 9673, pp. 257–262, 2016.
DOI: 10.1007/978-3-319-34111-8_31

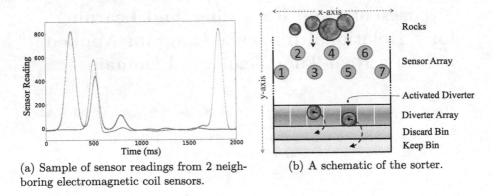

(a) Sample of sensor readings from 2 neighboring electromagnetic coil sensors.

(b) A schematic of the sorter.

Fig. 1. Rock sorting machine.

An example of sensor readings from 2 neighboring sensors of an unknown number of rocks is shown in Fig. 1(a). We can see there is a rock passing over the red sensor at around 250 ms, and one over the green sensor at 1800 ms. At 500 ms, most likely a rock passed between the sensors, slightly closer to the green sensor, or there were two rocks. If there was only one rock, because it passed between the sensors, its signal is lower than if it had passed directly over a sensor, and we should adjust for this. We need to be able to reason about the inherent uncertainty about the world based on the sensor readings.

Much research has been done in electromagnetic (EM) sensing for other applications, such as locating unexploded ordnance beneath the ground [5]. Lowther simulates electromagnetic devices in a full 3D setting in which he very accurately approximates electromagnetic physics [11]. However, computation time ranges from 1 s to 1 h per simulation typically – whereas our work requires a very fast response. In a lab setting, Mesina et al. have explored the use of EM for sorting scrap metal, using simple thresholds to classify particles [12]. Their work differs from ours in that they do not attempt to locate the particles or determine how much metal is present. Our work appears to be the first to apply artificial intelligence to the task of sensing and sorting individual rocks.

2   The Model

The system can be described using a relational influence diagram, as seen in Fig. 2. Influence diagrams, developed by Howard and Matheson [8], are an extension to belief networks and consist of 3 node types: random variables (ovals), decision variables (rectangles), and utility nodes (diamonds). Arrows going into random variables represent probabilistic dependence: the value of a random variable is probabilistically dependent on its parents. Arrows going into decision variables represent the information available when making a decision. Arrows going into a utility node represent deterministic dependence on the value of its parents.

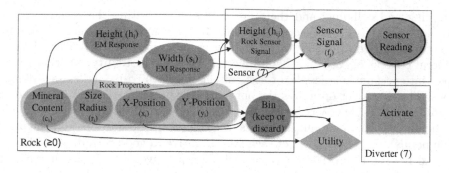

Fig. 2. Relational influence diagram

We extend influence diagrams further with the idea of relations between variables as in relational probabilistic models [13]. A similar technique is used under the name relational decision diagram [9]. We use plate notation [3] to avoid redundancy when a node, or set of nodes, is repeated more than once. A plate is drawn as a box around one or more nodes, and is labelled with the name of a class. For instance, we have zero or more rocks which are represented by the plate labelled "Rock (≥ 0)". The nodes within a plate are duplicated, along with any arrows connected to them, for each instance of the class. We chose to use a relational influence diagram because it explains the model and reveals explicitly what is and isn't being modeled, the assumptions made, and the areas where future work may need to take place.

There is an unknown number of rocks, denoted with an i subscript (1 through n), which may or may not be overlapping. We assume rocks are circles and we do not model mineral type (iron versus copper bearing minerals, etc.), instead we focus on a scalar mineral content which is tailored to the rocks we are looking for. We model each rock as a 2-dimensional symmetric Gaussian function, with height and width. Together these form the **EM response** of the rock (see Fig. 2 and Table 1). The rocks do not necessarily pass directly over a sensor, but may pass between sensors or over multiple sensors. Sensors are denoted with a j subscript. The **rock sensor signal** node is within two plates (rock and sensor) in Fig. 2 because at each time, for each rock, there is one rock sensor signal for each of the 7 sensors. A rock's sensor signal is the signal of the rock over a particular sensor, which we model as a 1-dimensional Gaussian. The height of this Gaussian depends on the EM response of the rock. For each sensor, the **sensor signal**, $f_j(t)$, sums each rock sensor signal from the rocks passing (partially or wholly) over the sensor, where t is continuous time. Finally, a **sensor reading** is a noisy sample from the sensor signal at discrete times (milliseconds). Given only the sequence of sensor readings of all 7 sensors, a decision is made to activate or not activate each diverter for each millisecond in time by comparing each rock's mineral content to a threshold (T in Table 1). The diverters guide each rock into either the keep bin or the discard bin. However, rocks can end up in the wrong bins because the diverter array is a chaotic system, with rocks tumbling and colliding, and there are delays in activating the diverters. The **bin** node simulates this chaotic behaviour. Rocks can also pass between sensors unnoticed and can be close together or even on top of another. The bin and mineral content of each rock determine the value of the **utility** function which takes into account misclassified rocks, misclassification costs, and class distributions. The controller optimizes the normalized expected cost (NEC), as shown at the end of Table 1. Misclassification costs depend on various factors outside of the model (e.g. environmental costs, the cost of transportation, and the price of minerals). C_{FN} is the cost of each false negative (a good rock diverted to the discard bin) and C_{FP} is the cost of each false positive (a bad rock diverted to the keep bin). Class distributions are the probability of a good and bad rock. A mine operator will input the misclassification costs and class distributions under which the mine operates which affect how the sorting algorithm behaves and the performance therein. In Table 1, FN_r is the false negative rate ($\frac{FN}{TP+FN}$), and FP_r is the false positive rate ($\frac{FP}{FP+TN}$). FN is the count of false negative rock classifications, and similarly for FP, TN, TP.

Table 1. Definitions of relation influence diagram nodes.

Node Name	Definition
Height – EM Response	$h_i = c_i m$
Width – EM Response	$s_i = a r_i + b$
Height – Rock Sensor Signal	$h_{ij} = h_i e^{-1/2 \frac{\left(x_j - x_i\right)^2}{s_i^2}}$
Sensor Signal	$f_j(t) = \sum_{i=1}^{n} h_{ij} e^{-1/2 \frac{(t - y_i)^2}{s_i^2}}$
Sensor Reading	$reading \sim \mathcal{N}(f_j(t), \sigma)$
Activate	$\begin{cases} yes & \text{if } c_i \geq T \\ no & \text{otherwise} \end{cases}$
Utility	$NEC = \frac{FN_r \cdot P(+) \cdot C_{FN} + FP_r \cdot P(-) \cdot C_{FP}}{P(+) \cdot C_{FN} + P(-) \cdot C_{FP}}$

3　Inference

There are two computational problems. First is the online reasoning that occurs in the approximately 400 ms between the time a rock passes over the sensors and the time it reaches the diverters. The second is the offline computation to learn the parameters of the model.

Online Inference. At runtime RPSA samples rock parameters – the number of rocks, and the position, size, and mineral content of each rock – then generates a sensor signal for each sensor conditioned on the rock parameters and compares the sensor signal to the observed sensor readings. Sampling repeats until time runs out, at which point the maximum a posteriori (MAP) hypothesis is chosen. Each rock's mineral content from this hypothesis is compared to a threshold parameter (T in Table 1). If the mineral content exceeds this threshold, the rock is classified as good and the appropriate diverters are activated. In most cases only one diverter is activated, but if the rock position is predicted to be on the boundary between two diverters then both diverters are activated.

MAP takes into account both the prior and the likelihood. The prior information we use is an exponential distribution over the number of rocks and also reflects that it is unlikely that rocks overlap (they tend to fall off each other). The likelihood specifies how well the hypothesis fits the observed data. RPSA is an anytime algorithm [15]; the search can be stopped at any time in which case the algorithm returns the best result it has found. We experimented with a number of search algorithms including exhaustive search, MCMC [4], gradient-based methods [14], and random search [2]. An exhaustive search over all possible rock properties cannot be done in a reasonable amount of time. MCMC and gradient-based methods are also too computationally intensive for our timing requirements, plus they have a tendency to get stuck in local optima. What worked best is a mix between random and exhaustive search with a coarse discretization of some of the parameters.

Offline Learning. A number of random restarts of the whole online procedure ensure RPSA has searched enough of the space. The number of restarts were automatically chosen offline, as well as the values for a, b, m, σ, and T from Table 1. We learn the model offline using recorded sensor data and video. We ran about 480 rocks through the machine while diverting using our algorithm (with hand-picked parameters since the model was not optimized at data collection time). Generating data and manually labelling every rock's position, approximate size, and mineral content (either good or bad) is expensive, but it allows us to train and evaluate the sorting machine in a simulated setting. With this data we automatically configure our model's parameters using a state-of-the-art algorithm configuration method called sequential model-based algorithm configuration (SMAC) [10]. SMAC searches over the space of parameters optimizing a given cost function — in our case we optimize for our utility function, NEC.

Conclusion

Sorting rocks using electromagnetic sensors is a challenging real-world planning problem where there is an unknown number of objects and we have to act in real-time. We created a new sorting algorithm, RPSA, by modeling the problem using a relational influence diagram and automatically configuring its parameters offline. RPSA does online inference and beats the previously used algorithm, VBSA, by 9 % on average (see [6]), while still being able to make decisions online within the real-time constraint.

References

1. Bamber, A., Klein, B., Pakalnis, R., Scoble, M.: Integrated mining, processing and waste disposal systems for reduced energy and operating costs at xstrata nickel's sudbury operations. Min. Technol. **117**(3), 142–153 (2008)
2. Bergstra, J., Bengio, Y.: Random search for hyper-parameter optimization. J. Mach. Learn. Res. **13**, 281–305 (2012). http://dl.acm.org/citation.cfm?id=2188395
3. Buntine, W.L.: Operations for learning with graphical models. J. Artif. Intell. Res. **2**, 159–225 (1994)
4. Christian Robert, G.C.: A short history of markov chain monte carlo: subjective recollections from incomplete data. Stat. Sci. **26**(1), 102–115 (2011). http://www.jstor.org/stable/23059158
5. Das, Y., McFee, J.E., Toews, J., Stuart, G.C.: Analysis of an electromagnetic induction detector for real-time location of buried objects. IEEE Trans. Geosci. Remote Sens. **28**(3), 278–288 (1990)
6. Dirks, M.: Sensing and Sorting Ore Using a Relational Influence Diagram. Master's thesis, University of British Columbia (2014). http://circle.ubc.ca/handle/2429/49998
7. Hitch, M., Bamber, A., Oka, P.: Presorting of high grade molybdenum ore-a case for enhanced small mine development. J. Eng. Appl. Sci. **2**(5), 129–135 (2015)
8. Howard, R.A., Matheson, J.E.: Influence diagrams. Decis. Anal. **2**(3), 127–143 (2005). http://pubsonline.informs.org/doi/abs/10.1287/deca.1050.0020
9. Hsu, W., Joehanes, R.: Relational decision networks. In: Proceedings of the ICML Workshop on Statistical Relational Learning, p. 122 (2004)
10. Hutter, F., Hoos, H.H., Leyton-Brown, K.: Sequential model-based optimization for general algorithm configuration. In: Coello, C.A.C. (ed.) LION 2011. LNCS, vol. 6683, pp. 507–523. Springer, Heidelberg (2011)
11. Lowther, D.: The development of industrially-relevant computational electromagnetics based design tools. IEEE Trans. Magn. **49**, 2375–2380 (2013)
12. Mesina, M.B., de Jong, T.P.R., Dalmijn, W.L.: Improvements in separation of non-ferrous scrap metals using an electromagnetic sensor. Phys. Sep. Sci. Eng. **12**(2), 87–101 (2003). http://dx.doi.org/10.1080/1478647031000139079
13. Poole, D.: Logic, probability and computation: foundations and issues of statistical relational AI. In: Delgrande, J.P., Faber, W. (eds.) LPNMR 2011. LNCS, vol. 6645, pp. 1–9. Springer, Heidelberg (2011)
14. Snyman, J.: Practical Mathematical Optimization: An Introduction to Basic Optimization Theory and Classical and New Gradient-Based Algorithms. Applied Optimization. Springer, New York (2005). http://opac.inria.fr/record=b1132592
15. Zilberstein, S.: Using anytime algorithms in intelligent systems. AI Mag. **17**(3), 73–83 (1996). http://rbr.cs.umass.edu/shlomo/papers/Zaimag96.html

Nearly Counterfactual Revision

Aaron Hunter[✉]

British Columbia Institute of Technology, Burnaby, Canada
aaron_hunter@bcit.ca

Abstract. We consider belief revision involving conditional statements
where the antecedent is almost certainly false. In order to represent such
statements, we use Ordinal Conditional Functions that may take infinite values. In this manner, we are able to capture the intuition that
the antecedent can not be verified by a finite number of observations.
We define belief revision in this context through basic ordinal arithmetic, and we propose an approach to conditional revision in which only
the right hypothetical levels are revised by conditional information. We
compare our approach to existing work on conditional revision and belief
improvement.

1 Introduction

In the literature on belief change, the focus is typically on an agent incorporating
new information that is given as a propositional formula. In this paper, we are
concerned with situations where an agent needs to revise by a conditional where
the antecedent is almost certainly false. More precisely, we consider antecedents
that will not be believed given any finite amount of "regular" supporting evidence. We represent the degree of belief in such formulas using Ordinal Conditional Functions that may take infinite values, and we provide an approach to
conditional revision based on basic ordinal arithmetic.

1.1 Motivating Example

Consider the following basic claims:

1. *heavy*: Your dog is overweight.
2. *hollow*: Your dog has hollow bones.
3. *fly*: Your dog can fly.

We are interested in revision by conditionals of the following form:

4. ¬*hollow* | *heavy*: If your dog is overweight, then it has solid bones.
5. *hollow* | *fly*: If your dog can fly, then it has hollow bones.

These conditional statements have a similar form, but we claim the purpose of
each statement is quite different. Statement (4) has an antecent that can easily
be verified, and it allows us to rule out an unlikely claim in the actual world.

© Springer International Publishing Switzerland 2016
R. Khoury and C. Drummond (Eds.): Canadian AI 2016, LNAI 9673, pp. 263–269, 2016.
DOI: 10.1007/978-3-319-34111-8_32

On the other hand, statement (5) has a nearly impossible antecedent. The purpose of this claim is therefore not to inform an agent about some condition about the actual world. Instead, the purpose of this claim is to force revision of one's beliefs about a hypothetical world.

Note that the perceived "impossibility" of (3) does not mean that (5) is free of content; revision by (5) should change an agent's beliefs in a counterfactual sense. But if ever the notion of flying dogs becomes believable, then this report will take on significance at the level of factual beliefs. In this paper, we refer to claims such as (3) as *nearly counterfactual*.

2   Preliminaries

2.1   Belief Revision

Assume an underlying propositional signature $\mathbf{P}$. An interpretation over $\mathbf{P}$ is called a *state*, while a logically closed set of formulas over $\mathbf{P}$ is called a *belief set*. A *belief revision operator* is a function that takes an initial belief set along with a formula for revision, and it returns a new belief set. Formal approaches to belief revision typically require an agent to have some form of *ordering* or *ranking* that gives the relative plausibility of possible states [1,6]. For *iterated belief revision*, we need to explicitly describe how the ordering changes, rather than just the belief set [2,4].

Belief revision can also be defined in terms of Ordinal Conditional Functions (OCFs). An OCF is a function that maps every state to an ordinal, with the restriction that the pre-image of 0 is non-empty [9,10]. A function mapping states to ordinals that does not necessarily take the value 0 is called a *free OCF* [7]. In an OCF, we interpret lower ordinal values to indicate greater plausibility. If r is an OCF, we write $Bel(r) = \{x \mid r(x) = 0\}$. The *degree of strength* of a plausibility function r is the least n such that $n = r(v)$ for some $v \notin Bel(r)$. Hence, the degree of strength is a measure of how difficult it would be for an agent to abandon the currently believed set of states.

2.2   Levels of Implausibility

An ω^2-CF is an OCF where all plausibility values are in ω^2 [3]. Roughly, an ω^2-CF consists of an ordered sequence of finite valued OCFs, which allows us to represent infinite jumps in plausibility.

The following definition from [3] gives a transformation on free OCFs that shifts the minimum plausibility value to 0.

Definition 1. *Let r be a free OCF over ω^2 with $\min(r) = \omega \cdot k + c$. Define $\bar{r}$ as follows. Let s be a state with $r(s) = \omega \cdot m + p$.*

1. *If $m > k$, then $\bar{r}(s) = \omega \cdot (m - k) + c$.*
2. *If $m = k$, then $\bar{r}(s) = (p - c)$.*

We call $\bar{r}$ the *finite zeroing* of r. Using this operation, we can define a form of addition on ω^2-CFs as follows:

$$r_1 \bar{+} r_2 = \overline{r_1 + r_2}.$$

Essentially, this is just addition of ω^2-CFs followed by a shift to ensure the result is also an ω^2-CF. However, we can not do this with simple substraction, because subtraction of ordinals in not well defined. It can be shown that AGM revision and Spohn's conditionalization are both special cases of this operation.

It is useful to introduce some notation that lets us explicitly refer to the levels of ω^2. Every ordinal $\alpha \in \omega^2$ can be written as $\alpha = \omega \cdot k + c$. In this case, let $deg(\alpha) = k$; we refer to this value as the *degree* of α.

3   Nearly Counterfactual Revision

3.1   Intuition

Our goal is to give a precise meaning to expressions of the form $r * (B|A)$. One important feature that is typically taken as a requirement for conditional reasoning is the so-called Ramsey Test. In the context of revision by conditional statements, Kern-Isberner formulates the Ramsey Test as follows: when revising by a conditional, one would like to ensure that revision by $(B|A)$ followed by a revision by A should guarantee belief in B [5]. We suggest that this formulation needs to be refined in order to be used in the case where infinite ranks are possible.

In the case of the flying dog, one is quite likely to accept the conditional $(hollow|fly)$ based on a single report with finite strength. However, a single report of fly with finite strength will not be believed. If the antecedent of the conditional is "very hard" to believe, then we should not expect the Ramsey Test to hold without some condition on the strength of the subsequent report. The problem is that the notion of believing a conditional is quite different than the notion of believing a fact. In order to believe $(hollow|fly)$, we simply need to keep some kind of record of this fact for the unlikely case where we discover that flying dogs happen to exist. On the other hand, in order to believe fly, we really need to make a significant change in our current world view.

3.2   Formalization

We say that a formula A is *nearly counterfactual* with respect to r if $r(A) \geq \omega$. For any formula B and any $n \in \omega$, we let r_B^n denote the OCF where $r(s) = 0$ if $s \models B$ and $r(s) = n$ otherwise.

To proceed, we require the following notion of *possibility* which captures the idea that a certain state is *believed* at a fixed level of plausibility if it is minimal at that level.

Definition 2. *Let r be an ω^2-CF and let A be a formula. Define $poss(r, A)$ to be the set of natural numbers k such that there is some state s with $deg(r(s)) = k$, $s \models A$ and $r(s)$ is minimal among all states t with $deg(r(t)) = k$.*

Using this notion, we can define a form of strengthening for conditionals.

Definition 3. *Let r be an ω^2-CF and let B, A be formulas where A is nearly counterfactual with respect to r. Let $n \in \omega$.*

$$r * (n, B|A)(s) = \begin{cases} r(s), \text{ if } deg(r(s)) \notin poss(A) \\ r \bar{+} r_B^n(s) \text{ otherwise} \end{cases}$$

We call this function the *n-stengthening* of B conditioned on A. This function essentially finds all levels of r where A is "believed," and then strengthens B at only those levels.

Example 1. Let r be the plausibility function

$$r(s) = \begin{cases} \omega \text{ if } s \models fly \\ 10 \text{ if } s \models heavy \wedge \neg fly \\ 0 \text{ otherwise} \end{cases}$$

Now suppose that we extend the vocabulary to include the predicate symbol *hollow*. Define a new function r' as follows:

$$r'(s) = \begin{cases} r(s), \text{ if } s \not\models hollow \\ r(s) + 1, \text{ if } s \models hollow \end{cases}$$

This just says that we initally believe our dog does not have hollow bones; however, it is not particularly implausible. It follows that:

- $r'(s) = \omega$ if $s \models fly \wedge \neg hollow$.
- $r'(s) = \omega + 1$ if $s \models fly \wedge hollow$.

From these results, it follows that:

- $r' * (2, hollow|fly)(s) = r'(s)$, if $s \not\models fly$.
- $r' * (2, hollow|fly)(s) = \omega$, if $s \models fly \wedge hollow$.
- $r' * (2, hollow|fly)(s) = \omega + 1$, $s \models fly \wedge \neg hollow$.

So, roughly speaking, after strengthening by $(hollow|fly)$, we now believe that hollow bones are more plausible in all hypothetical situations where we believe flying dogs are possible.

The preceding example illustrates an important feature of our approach, that is given by the following condition.

Proposition 1. *Let A be a formula, let r be an OCF, and let s be a state such that $deg(r(s)) < deg(r(t))$ for all t such that $t \models A$. Then $r(s) = r * (n, B|A)(s)$ for all n and B.*

Hence, the plausibility of a state is only changed at levels where A is possible. Since the definition is only applied to nearly counterfactual conditionals, this means that only hypothetical states are affected by the strengthening.

It remains to move from conditional strengthening to conditional revision. For any ω^2-CF r and natural number k, let $r_k = \{r(s) \mid deg(r(s)) = k\}$. So r_k is the set of ordinal values assigned by r at plausibility level k. If $min(r_k) = \omega \cdot k + c$, then let $fin(r_k) = c$. Note that r_k must have a minimum value, since it is a set of ordinals.

Definition 4. *Let r be an ω^2-CF and let B, A be formulas where A is nearly counterfactual with respect to r.*

$$r * (B|A)(s) = \begin{cases} r(s), & \text{if } deg(s) \notin poss(A) \\ r \bar{+} r_B^n(s) & \text{where } n = fin(r_{deg(r(s))}), \text{ otherwise} \end{cases}$$

Hence, for revision, we strengthen belief in B by the least value that will ensure B is believed at level k. This satisfies a modified form of the Ramsey Test.

Proposition 2. *Let r be an ω^2-CF and let s be a state with $r(s) = \omega \cdot k + c$. If r' is an ω^2-CF with degree of strength larger than k and $Bel(r') \models A$, then $Bel((r * (B|A)) * r') \models B$.*

Hence, if we revise by $(B|A)$ followed by an OCF with "sufficiently strong" belief in A, then B will be believed.

4 Relation to Existing Work

4.1 Infinite Plausibility Values

There has been related work on the use of infinite valued ordinals in OCFs [7]. However, in this work, different "levels" are used to represent beliefs that are independent; the lowest level represents an agent's actual beliefs about the world, whereas higher levels are used to represent integrity constraints. Our approach is different in that we explicitly use the ordering on limit ordinals to represent infinite leaps in plausibility.

In our framework, it is possible to define a correspondence between sequences of orderings and ordinals in ω^2. Therefore, belief change by normalized addition on ω^2-CFs can really just be seen as a finite collection of improvements at each level. The important point, however, is that our approach does not define an *improvement operator* in the sense of [8], because no finite sequence of improvements at level d will ever impact the actual beliefs at different higher level of implausibility.

4.2 Conditional Belief Revision

Conditional belief revision was previously addressed by Kern-Isberner, who proposes a set of rationality postulates as well as a concrete approach to conditional revision [5]:

$$r * (B|A)(s) = \begin{cases} r(s) - r(B|A), & \text{if } s \models A \wedge B \\ r(s) + \alpha + 1, & \text{if } s \models A \wedge \neg B \\ r(s), & \text{if } s \models \neg A \end{cases}$$

where $\alpha = -1$ if $r(\{A, B\}) < r(\{A\})$, and $\alpha = 0$ otherwise. This operation satisfies all of the postulates for conditional revision, and it also satisfies the Ramsey Test. However, that this definition does not work if we allow infinite values for plausibilities, because it involves subtraction of ordinals. Hence, our approach is not equivalent to Kern-Isberner's approach, nor do we want it to be. Conditional revision involving statements that are "almost certainly" false requires a different treatment.

5   Conclusion

In this paper, we have explored the use of infinite ordinals for reasoning about revision by nearly counterfactual conditionals. In the present framework, counterfactual revision can be seen as a tool for keeping a sort of "memory" about unlikely situations, in order to incorporate this information later if necessary. This memory is essentially maintained by keeping a series of finite-valued OCFs that capture beliefs under different counterfactual conditions.

We suggest that this model may also be useful for reasoning by analogy. When we revise by counterfactual information, we modify some portion of our OCF that is related to hypothetical states. However, in some circumstances, a hypothetical world may be isomorphic to some "part" of the actual world. When this is the case, it may then be reasonable to use our hypothetical beliefs to draw conclusions about the actual world. For example, our beliefs about flying dogs might be useful if we encounter a new flying mammal that shares many physical traits with dogs. This is a form of *ampliative reasoning* that we intend to explore in future work.

References

1. Alchourrón, C.E., Gärdenfors, P., Makinson, D.: On the logic of theory change: partial meet functions for contraction and revision. J. Symbolic Logic **50**(2), 510–530 (1985)
2. Darwiche, A., Pearl, J.: On the logic of iterated belief revision. Artif. Intell. **89**(1–2), 1–29 (1997)
3. Hunter, A.: Infinite ordinals and finite improvement. In: van der Hoek, W., Holliday, W.H., Wang, W.-F. (eds.) LORI 2015. LNCS, vol. 9394, pp. 416–420. Springer, Heidelberg (2015)
4. Jin, Y., Thielscher, M.: Iterated belief revision, revised. Artif. Intell. **171**(1), 1–18 (2007)
5. Kern-Isberner, G.: Postulates for conditional belief revision. In: Proceedings of IJCAI, pp. 186–191 (1999)
6. Katsuno, H., Mendelzon, A.O.: Propositional knowledge base revision and minimal change. Artif. Intell. **52**(2), 263–294 (1992)
7. Konieczny, S.: Using transfinite ordinal conditional functions. In: Sossai, C., Chemello, G. (eds.) ECSQARU 2009. LNCS, vol. 5590, pp. 396–407. Springer, Heidelberg (2009)
8. Konieczny, S., Péréz, R.P.: Improvement operators. In: Eleventh International Conference on Principles of Knowledge Representation and Reasoning (KR 2008), pp. 177–186 (2008)

9. Spohn, W.: Ordinal conditional functions. a dynamic theory of epistemic states. In: Harper, W.L., Skyrms, B. (eds.) Causation in Decision, Belief Change, and Statistics, vol. II, vol. 42, pp. 105–134. Kluwer Academic Publishers, Netherlands (1988)
10. Williams, M.A..: Transmutations of knowledge systems. In: Proceedings of the Fourth International Conference on the Principles of Knowledge Representation and Reasoning (KR 1994), pp. 619–629 (1994)

Improving Conversation Engagement Through Data-Driven Agent Behavior Modification

Michael Procter[1(✉)], Fuhua Lin[1], and Robert Heller[2]

[1] School of Computing and Information Systems,
Athabasca University, Athabasca, Canada
mikeprocter@shaw.ca, oscarl@athabascau.ca
[2] Faculty of Humanities and Social Sciences,
Athabasca University, Athabasca, Canada
bobh@athabascau.ca

Abstract. E-learning systems based on a conversational agent (CA) provide the basis of an intuitive, engaging interface for the student. The goal of this paper is to propose an agent-based framework for providing an improved interaction between students and CA-based e-learning applications. Our framework models both the student and the CA and uses agents to represent data sources for each. We describe an implementation of the framework based on BDI (Belief-Desire-Intention) architecture and results of initial testing.

1 Introduction and Background

Conversational agents (CAs) are designed to provide users with a natural language interface to an application, with the intention of mimicking the experience of speaking with another human. CAs in e-learning applications have the potential to provide an intuitive, user-friendly interface that engages the student. Educational applications of CA technology include animated pedagogical agents (APA) [1, 2], intelligent tutoring systems (ITS) [3], and collaborative learning [2]. CAs can play an important role in game-based learning systems (GBL) [4–6].

Two models can be developed to improve the student's experience when interacting with the CA. (1) Modeling human characteristics, such as personality and emotion, possibly expressed through an embodied agent, have been shown to help users adopt a more positive attitude towards the agent [7]. (2) Modeling student characteristics, including learning goals and requirements, as well as an awareness of student personality and affective state [8] provides opportunities for effective interventions and performance tailored to the individual's needs. The information that is available about a student may depend on several factors, including what measurement devices (camera, biometric and physiological) are available. The type of student data that is relevant and useful to a particular CA also varies.

Much of the research which focuses on improving the student's perception of the CA concentrates on methods for providing human characteristics, such as emotion, personality, intelligence and goals [5, 9]. This suggests the need to model the CA in software. Research associated with improving the agent's understanding of the student

© Springer International Publishing Switzerland 2016
R. Khoury and C. Drummond (Eds.): Canadian AI 2016, LNAI 9673, pp. 270–275, 2016.
DOI: 10.1007/978-3-319-34111-8_33

examines methods for detecting user characteristics such as emotion, personality, and goals [10]. This suggests the need to model the student.

Modeling the student carries the additional challenges associated with collecting information about the user, particularly when it changes over time, as in the case of affective data. Often this data is provided by devices, such as cameras, eye-trackers, EEG sensors, and heart-rate monitors. These devices can be obtrusive, and simply may not be available to the typical student.

A system that improves the interaction between a student and a CA should ideally be able to adapt to whatever student model data is accessible. It should also be able to adapt to the capabilities of the CA, the student information it can recognize, and how it can respond to it.

We propose an agent-based framework supporting both of these models and describe an implementation that focuses on the second, i.e. the CA modifies its behavior based on assessing student engagement in real time, using conversational quality.

2 The Proposed Agent-Based Framework

We address the dynamic nature of modeling both the student and the CA using autonomous intelligent agents, which, by their nature, are designed to adapt to changes in the environment. Each source of student data and CA behavior are to be represented by an agent, providing information to a central agent responsible for maintaining a model of the student and the CA.

A high level overview of the agent framework is shown in Fig. 1. The student and the CA, shown at the bottom, communicate through the agent framework above them. The architecture consists of three layers. The Representation layer is responsible for providing an interface between the student and the CA, with an agent representing each. The Model layer maintains information about the state of each of the participants, again with agents assigned to each. The Model Sources layer provides information to the Model layer agents. Multiple data source agents (DSAs) process data from devices and provide one or more information channels to the model.

1. **Student Representation (ST-REP)** Goal: Represent the student by: Communicating student input to CA (via CA-REP); Communicating CA response to student; Provides feedback to student based on data provided by CA-REP; Provides student model data from ST-MODEL to CA-REP.
2. **CA Representation (CA-REP)** Goal: Represent the CA by: Sending student input (from ST-REP) to CA; Sending CA response to student (via ST-REP); Recommending conversation strategies based on student data from ST-MODEL; Analyze student responses and provide feedback to ST-REP; Send data to embodied agents to support lip-sync, gestures, expressions.
3. **Student Modeling (ST-MODEL)** Goal: Maintain data about student by: Subscribing to data source agents (DSAs) based on CA requirements; Receiving data from DSAs (affect, personality, engagement, goals); Sending updates to representation layer agents integrating data where appropriate.

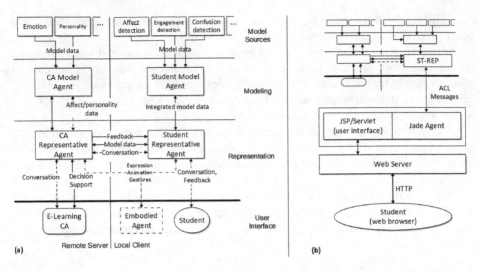

Fig. 1. (a) Agent architecture; (b) Remote user access

4. **CA Modeling (CA-MODEL)** Goal: Maintain information about CA by: Subscribing to data source agents (DSAs) based on CA requirements; Updating CA-REP from data provided by DSA models (e.g. emotion, goals, personality, CA performance assessment) to support embodied agents and conversation strategy recommendations.
5. **Student and CA Data Source Agents (DSA)** Goal: Provide a source of data about the student (or CA) by: Provides information channel using a publish/subscribe mechanism (e.g. affect detected from facial expressions, engagement detected from conversational analysis).

3   Implementation

A version of the described system has been implemented using Jason, a Java-based interpreter of an improved version of AgentSpeak(L), which supports multi-agent systems (MAS) based on the BDI (Belief-Desire-Intention) agent paradigm [11]. The BDI architecture is commonly used in the development of cognitive agents and has been used successfully in agent-based pedagogical applications [12]. This system implements the two representation layer agents (ST-REP and CA-REP) and the student model agent (ST-MODEL) as BDI agents. ST-MODEL employs a simple integration strategy that combines multiple sources of a data type using a weighted average based on the accuracy rating provided by the DSA. The CA-MODEL agent and associated DSAs will be developed as part of a future research phase.

Core DSAs - Conversation Text Classifiers. The DSAs are the key components to providing a dynamically configured system capable of adapting to whatever student information is available. However, as this is a CA application, a core set of conversation-based DSAs should always be available. Currently two DSAs have been created using text classifiers to provide measures of user engagement. The classifiers were trained on manually annotated conversational records from a previous study [1]. They rate user input on whether it is conversational in nature, and appropriate to the CA's statements. An article describing this approach to measuring student engagement based on conversational analysis is in development.

External Communication – Connecting Users and Devices to the System. Jason supports a Jade (jade.tilelab.com) environment to provide distributed MAS. This was used to provide remote access for users and devices. A Java servlet connected to a Jade agent supplies the student interface (Fig. 1b). User input through a web page is sent from the

Table 1. Agent activity phases.

Initialization
1. User connects to JSP agent which requests new start from Invoker agent.
2. Invoker creates a set of agents (ST-REP, CA-REP, ST-MODEL, core DSAs).
3. DSAs announce **data types available**. CA-REP announces **data types needed**.
4. ST-MODEL resolves needed and available data, subscribes to DSA data streams.
5. ST-MODEL records cases of more than one source of a data type (typically from two or more DSAs) and stores these for integration.
Communication
1. ST-REP receives user input (text). Sends to CA-REP.
2. CA-REP receives user input. Sends to CA.
3. CA-REP receives CA response (text).
4. CA response may trigger a **Decision Support action**.
5. Student data update may trigger an **Intervention action**.
6. CA-REP sends CA response to ST-REP. ST-REP display response to student.
Intervention (CA-REP initiated)
1. CA-REP receives student data update, executes **Student Update action.**
2. Determines if an **Intervention plan** should be executed.
3. Gets CA output from intervention (e.g. "Would you like to change topics?").
4. CA-REP combines CA intervention output with response to student input.
Decision Support (DS) (CA initiated)
1. CA-REP receives CA response indicating decision request, triggering a **DS plan.**
2. CA-REP requests student data from ST-MODEL if needed.
3. CA-REP executes **DS plan** (e.g. select topic), sends input to CA.
4. CA sends new response. CA-REP sends CA response to ST-REP.
Student data update
1. DSA **processes incoming data** (e.g. wearable device, camera, new text log entry)
2. DSA sends data update to subscribers for each information streams <C,S,V,A>
3. ST-MODEL receives student data message and executes **integration plan** if other sources for the same data type are received.
4. ST-MODEL sends data update to CA-REP (via ST-REP)

Jade agent to the ST-REP agent, using Agent Communication Language (ACL) messages. This allows for a relatively thin client and a mechanism for devices to connect to DSAs through a Jade agent installed on the user's system.

System Data and Execution. Student and CA data are represented as a tuple <C,S,V,A> (Category, Sub-category, Value, Accuracy). For example <Affect, Engaged, 2, 0.6>. ST-MODEL maintains a list of known data types in the form <C,S,_,_> and current student data.

Agent interactions and system behavior are described in terms of six activity phases in Table 1. Bold text identifies components to be provided or customized by the developer to meet the needs of the CA or student data to be used. With the exception of Initialization, activity phases occur asynchronously. CA-REP and ST-REP are responsible for combining resulting messages for the user into the conversation.

4 Evaluation

Our search of the literature did not find any frameworks that provide models for both the student and the tutor system, and none were specifically tailored to CAs. The system most similar to part of what we propose is ABE [13]. Although they share the same approach to representing sources of user information (specifically affect data) by individual agents, their framework stops at the model level. It is not oriented specifically to CA-based learning applications, nor does it address modeling the ITS.

Performance. The implemented framework was tested to demonstrate how information is passed from the DSA to ST-MODEL to CA-REP, supporting decision making regarding CA responses. In addition to the two core DSAs, two additional DSAs were created to provide simulated student data at regular intervals to add load to the system.

We used Freudbot [1] to act as the CA for test purposes. User input from logs of previous studies was fed to the CA via the agent framework, with the classifier DSAs providing data for each conversational exchange pair (CA statement and user response), and 2 test DSAs sending simulated data on a total of 5 information channels.

Tests were carried out on a relatively low-powered desktop computer (3 GHz Intel Core Duo CPU, 3 GB memory, 32-bit Windows 7).

To measure efficiency of the agent system, we collected response times by introducing code to log timestamps at user input and CA response events and subtracting the response times reported by the CA server. For 142 conversational exchanges (student input/CA response pairs), the mean delay was 393.2 ms (sd = 179.7). CPU usage was very low (below 1 % after startup).

5 Conclusions and Future Work

The implementation of the proposed framework demonstrates the ability of a multi-agent system to adapt to a variety of different student data sources and use the student data in real-time to drive the behavior of the CA. Initial observations of response times are encouraging, but load testing with Java profiling tools using multiple DSA configurations is underway to confirm the scalability of the system.

We plan to develop other DSAs that analyze the conversation to detect or predict Student affect and engagement. Controlled experiments will compare how students interact with a CA when the framework is added.

References

1. Heller, R., Procter, M.: Animated pedagogical agents: the effect of visual information on a historical figure application. Int. J. Web-Based Learn. Teach. Technol. **4**, 54–65 (2009)
2. Kumar, R., Rosé, C.P.: Architecture for building conversational agents that support collaborative learning. IEEE Trans. Learn. Technol. **4**, 21–34 (2011)
3. D'Mello, S., Craig, S., Witherspoon, A., McDaniel, B., Graesser, A.: Automatic detection of learner's affect from conversational cues. User Model. User-Adap. Inter. **18**, 45–80 (2008)
4. McClure, G., Chang, M., Lin, F.: MAS controlled NPCs in 3D virtual learning environment. In: 2013 International Conference on Signal-Image Technology and Internet-Based Systems (SITIS), pp. 1026–1033. IEEE (2013)
5. Löckelt, M.: Design and implementation issues for convincing conversational agents. Conversational Agents and Natural Language Interaction: Techniques and Effective Practices: Techniques and Effective Practices. p. 156 (2011)
6. Bellotti, F., Berta, R., De Gloria, A., Lavagnino, E.: Towards a conversational agent architecture to favor knowledge discovery in serious games. In: Proceedings of the 8th International Conference on Advances in Computer Entertainment Technology, pp. 1–7. ACM, New York, NY, USA (2011)
7. Nunamaker Jr., J.F., Derrick, D.C., Elkins, A.C., Burgoon, J.K., Patton, M.W.: Embodied conversational agent-based kiosk for automated interviewing. J. Manag. Inf. Syst. **28**, 17–48 (2011)
8. Chrysafiadi, K., Virvou, M.: Student modeling approaches: a literature review for the last decade. Expert Syst. Appl. **40**, 4715–4729 (2013)
9. Callejas, Z., López-Cózar, R., Ábalos, N., Griol, D.: Affective conversational agents: the role of personality and emotion in spoken interactions. In: Perez-Marin, D., Pascual-Nieto, I. (eds.) Conversational Agents and Natural Language Interaction, pp. 203–222. IGI Global, Hershey (2011)
10. Calvo, R.A., D'Mello, S.: Affect detection: an interdisciplinary review of models, methods, and their applications. IEEE Trans. Affect. Comput. **1**, 18–37 (2010)
11. Bordini, R.H., Hübner, J.F., Wooldridge, M.: Programming Multi-agent Systems in AgentSpeak using Jason. Wiley, New York (2007)
12. Soliman, M., Guetl, C.: Experiences with BDI-based design and implementation of intelligent pedagogical agents. In: 2012 15th International Conference on Interactive Collaborative Learning (ICL), pp. 1–5 (2012)
13. Gonzalez-Sanchez, J., Chavez-Echeagaray, M.E., Atkinson, R., Burleson, W.: ABE: An agent-based software architecture for a multimodal emotion recognition framework. In: 2011 9th Working IEEE/IFIP Conference on Software Architecture (WICSA), pp. 187–193 (2011)

Active Recruitment Mechanisms for Heterogeneous Robot Teams in Dangerous Environments

Geoff Nagy and John Anderson[✉]

Autonomous Agents Laboratory, University of Manitoba,
Winnipeg, MB R3T 2N2, Canada
{geoffn,andersj}@cs.umanitoba.ca

Abstract. Using teams of autonomous, heterogeneous robots to operate in dangerous environments means increased cost-effectiveness and the ability to spread skills among team members. The high risk of loss in these domains is a challenge to team management. Teams must be able to recruit the help of other robots in the environment, while balancing searching with performing immediately useful work. This paper describes additions to a framework for dynamic team management in dangerous domains in order to support various levels of active search for useful agents while balancing useful work in the domain.

1 Introduction

Heterogeneity is useful in multi-robot teams because it may not be cost-effective to provide every team member with the most expensive equipment, and doing so would only increase the robots' design and control complexity [1]. Losses in dangerous environments can limit a team's skill set, and since many or even all robots are likely to eventually be damaged or lost in such environments, it is sensible and necessary to release replacement robots periodically. To accomplish this, teams must also be prepared to integrate new robots (possibly including previously lost units) and decide how much effort to expend into searching for them. More active approaches involving physical searches will yield faster results at the expense of immediately useful work. Passive approaches that rely on chance encounters will result in more useful work being performed by individual robots, but will lessen the chances of successfully locating desired robots [8]. Strategies in between these extremes would contain elements of both, such as simply calling for help wirelessly without physically searching. We have made additions to an existing framework [8] enabling teams of heterogeneous robots to use various recruitment strategies, and this paper focuses only on a specific subset of these. We examine in simulation how our framework performs in a dangerous domain: Urban Search and Rescue (USAR), the search for human victims inside a partly collapsed structure, under conditions of partial communication failure and robot loss.

© Springer International Publishing Switzerland 2016
R. Khoury and C. Drummond (Eds.): Canadian AI 2016, LNAI 9673, pp. 276–281, 2016.
DOI: 10.1007/978-3-319-34111-8_34

2   Previous Work

Existing work does not employ a wide range of recruitment strategies and many works assume that robots are always in communication range of each other [2,3,11], or focus on domains [4,5,12,14] that do not reflect real-world conditions that our work takes into account (e.g., communication failures or robot loss).

Krieger et al. [10] modelled foraging tasks using recruitment among homogeneous robots, but did not consider factors such as multiple robot or task types, or unreliable communication. Pitonakova et al. [13] endeavoured to discover under what conditions recruitment is beneficial for more complex foraging but did not explore elements of robotic heterogeneity or specialization despite the presence of different resource types. Kiener and Von Stryk [9] presented a task allocation framework for teams of highly heterogeneous and specialized robots, demonstrated with a small range of tasks. It contained no mechanisms to support the addition of new robots or the possibility of larger roles.

Gunn and Anderson [8] developed a framework allowing teams of heterogeneous robots to operate in complex simulated USAR environments, including unreliable communication and possible loss of robots. Teams of robots were assigned tasks by a leader to explore the environment and locate human victims. Losses were countered by the ability to rebalance roles (including leadership) upon losing or encountering robots. We have made additions to this framework allowing robots to search for useful skills as needed (as opposed to relying on chance encounters) and balance this activity with performing useful work. The next section describes relevant portions of the original framework and follows with the new mechanism we have added.

3   Methodology

3.1   Framework

In order to describe our recruitment mechanisms, the main components of our existing framework for dynamic team management and task allocation must be described. Space limits the background that can be presented here; the interested reader is directed to [8].

Robots. Our framework is designed to support heterogeneous robots that will not be equally suitable for all types of work. In our work, robots are able to determine how well-suited they are for a particular job and every robot maintains a priority queue of outstanding jobs. Robots can be assigned tasks by a team leader and can also discover them on their own. Newly discovered tasks can be executed by the robot itself or can be passed to a team leader for reassignment elsewhere if the robot is ill-equipped or too busy.

Tasks, Roles, Team Leaders, and Task Allocation. We define a *task* as a single piece of work that can be completed by a robot. Every robot in our work fills a *role*, which includes a description of the types of tasks associated with it.

As the milieu of team members changes due to losses or robot encounters, robots use their own (limited and possibly inconsistent) team knowledge and may switch roles in order to fill gaps in team abilities [8]. A leader robot, although not always reachable (and thus without the most up-to-date knowledge of the team), brings global perspective to a team to the degree that communication is available. The leader should be the most computationally-capable unit on a team, and in the original framework, is responsible for task assignments, updating known victim locations, and keeping a limited model of the current team.

The team leader uses roles as a heuristic to indicate default assignments to team members, but communication limitations (such as a delayed report of a role change) and robot losses can cause this to fail. Task broadcast is used as a fall-back, and members respond with their ability to complete the task irrespective of their role [7]. A robot receiving a request to complete a task may not always be able to oblige due to limited capabilities or an overfull task queue. While the original framework supported task assignments only from the team leader, our recruitment extensions add a mechanism whereby other team members unable to accomplish a task can recruit others to complete it.

3.2   Recruitment Strategies

Recruitment strategies vary in terms of how much effort robots expend towards locating others. Here we describe two mechanisms that represent points on a spectrum of more active recruitment mechanisms.

Concurrent recruitment increases the chances of encountering other robots while also completing useful work. Robots will perform tasks as normal while simultaneously broadcasting messages asking for help with other tasks. Other robots in wireless range, even if on another team, will accept the task if they are able, or offer to recruit for the task if they are not. A recruiter will examine these responses and attempt to assign a task to the most suitable robot who is able to complete the task themselves. If no such robot is available, the recruiter will attempt to offload the task to one of the other responding robots for recruitment. If no responses are received or the assignment fails (due to robot failures or unreliable communication) the recruiter continues to broadcast the request.

With *active recruitment*, a robot will perform a physical search for a robot that can complete a specific task, rather than completing pending tasks of its own. Physically searching for another robot has the advantage of providing increased opportunities for encountering others, at the cost of spending time away from useful work. In situations where a robot is given one task too many, or has been assigned a task for which it is poorly-equipped, it will enqueue the task and mark it to be completed via recruitment. This ensures that the recruit-ment task is weighed against other pending tasks. When recruiting, the robot performs a physical search to increase the chances of encountering another robot, and broadcasts a request for help and the task details. As with concurrent recruit-ment, the recruiter assigns the task to the most suitable robot who responds to the message. Once the task has been successfully assigned, the recruiter returns to its team's last known location and resumes normal tasks.

4   Experimental Evaluation and Design

We evaluated a subset of our framework using Stage [6]. We constructed two 60 m × 60 m USAR settings, each containing significant debris, human victims, and several objects appearing to be human but requiring more capable robots to verify. The robots' goal was to maximize area coverage and correctly identify as many true victims as possible within 30 min. Like previous work [8] we abstracted heterogeneity to three types of robots: a MinBot (intended for exploration and potential victim discovery, but unsuitable as leaders); a MidBot, with advanced victim sensors to correctly identify victims, that can also potentially serve as leaders, and; a MaxBot, having only basic victim sensors but larger memory and computational abilities making them ideal for leadership roles. Teams were composed of 1 MaxBot, 2 MidBots, and 4 MinBots.

We evaluated all three recruitment strategies against three communication success rates (100 %, 60 %, and 20 %), and three probability levels (moderate, low, or no chance) that any robot could experience a random temporary (between 3–4 min.) or permanent failure. To offset these losses, replacement robots (10 MinBots, 2 MidBots, and 1 MaxBot) were released into the environment at the 10-min mark of every trial. Each trial was repeated 50 times in each of our two environments (2700 runs total).

5   Results and Discussion

5.1   Environment Coverage

Results of area coverage (Fig. 1) indicate that poor communication rates encourage greater environmental coverage in active recruitment settings. Failed task assignment attempts lead robots to undertake active searches to find suitable robots to complete tasks which could not be assigned by a leader, resulting in greater coverage. Poor communication further results in failures to recruit other robots, resulting in more searches. Since poor communication results in

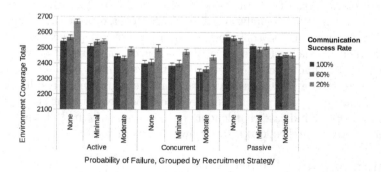

Fig. 1. Sum of all robots' environment coverage, averaged over 50 trials in each configuration. Error bars show standard error.

increased recruitment, exploration tasks are more likely to be assigned to members of different teams as well. This effect is also visible in concurrent settings when communication success rates are low: as recruitment tasks are created to compensate for the lack of assignment acceptances, members of other teams can respond to these requests, resulting in greater coverage of the environment.

5.2 Victims Found

We evaluate our framework against the number of true victims known by a team leader (Fig. 2), since in reality, robots must eventually communicate victim locations to human rescue teams. Team leaders are the logical choice since they are likely to have the most knowledge.

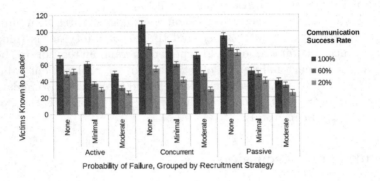

Fig. 2. Sum of all true victims identified and communicated to a team leader, averaged over 50 trials in each configuration. Error bars show standard error.

Concurrent recruitment performs better than other configurations when communication is reliable and robot failures do not occur, since robots continually attempt to recruit others whenever required. This results in a greater number of task assignments overall. In active settings, physical searches for a task are only initiated when a recruitable task is pulled from the robot's task queue. Depending on the priority of the task, there may be a significant delay before recruitment begins. Additionally, active searches take a robot away from its team and leader, resulting in less communication and more lost potential work.

Active recruitment performed poorly in terms of communicating victim locations to a leader. The number of verified victims recorded by individual robots (as opposed to only a leader) indicated that active recruitment results in a high number of victims located when compared to other recruitment strategies, but the results in Fig. 2 indicate that leaders are not well-informed about these victims, since active searches lead a robot away from its leader.

6 Summary and Future Work

We have described a specific subset of recruitment strategies implemented as part of a framework for managing teams of robots in dangerous domains, and evaluated our framework in a simulated USAR environment. Future work will include examining priorities of certain tasks to determine what kind of recruitment strategy, if any, should be used to complete them, as well as the possibility of maintaining copies of useful information over multiple robots to help mitigate the effects of robot failures or insufficient communication with a leader. Information-passing itself could be implemented as a task that would help to improve the effectiveness of the approaches outlined in this paper.

References

1. Brooks, R.A.: A robust layered control system for a mobile robot. IEEE J. Robot. Autom. **2**(1), 14–23 (1986)
2. Costa, E.D., Shiroma, P.M., Campos, M.F.: Cooperative robotic exploration and transport of unknown objects. In: SBR-LARS 2012, pp. 56–61. IEEE (2012)
3. Dos Santos, F., Bazzan, A.L.: Towards efficient multiagent task allocation in the robocup rescue: a biologically-inspired approach. Auton. Agent. Multi-Agent Syst. **22**(3), 465–486 (2011)
4. Dutta, P.S., Sen, S.: Forming stable partnerships. Cogn. Syst. Res. **4**(3), 211–221 (2003)
5. Gage, A., Murphy, R.R.: Affective recruitment of distributed heterogeneous agents. In: AAAI, pp. 14–19 (2004)
6. Gerkey, B., Vaughan, R.T., Howard, A.: The player/stage project: tools for multi-robot and distributed sensor systems. In: Proceedings of the 11th International Conference on Advanced Robotics, vol. 1, pp. 317–323 (2003)
7. Gunn, T., Anderson, J.: Effective task allocation for evolving multi-robot teams in dangerous environments. In: Proceedings of the IAT 2013. IEEE Computer Society Press (2013)
8. Gunn, T., Anderson, J.: Dynamic heterogeneous team formation for robotic urban search and rescue. J. Comput. Syst. Sci. **81**(3), 553–567 (2015)
9. Kiener, J., Von Stryk, O.: Cooperation of heterogeneous, autonomous robots: a case study of humanoid and wheeled robots. In: Proceedings of the IROS 2007, pp. 959–964. IEEE (2007)
10. Krieger, M.J., Billeter, J.B.: The call of duty: self-organised task allocation in a population of up to twelve mobile robots. Robot. Auton. Syst. **30**(1), 65–84 (2000)
11. Mathews, N., Christensen, A.L., O'Grady, R., Rétornaz, P., Bonani, M., Mondada, F., Dorigo, M.: Enhanced directional self-assembly based on active recruitment and guidance. In: Proceedings of the IROS 2011, pp. 4762–4769. IEEE (2011)
12. Pinciroli, C., O'Grady, R., Christensen, A.L., Dorigo, M.: Self-organised recruitment in a heteregeneous swarm. In: ICAR 2009, pp. 1–8. IEEE (2009)
13. Pitonakova, L., Crowder, R., Bullock, S.: Understanding the Role of Recruitment in Collective Robot Foraging. MIT Press, Cambridge (2014)
14. Van De Vijsel, M., Anderson, J.: Coalition formation in multi-agent systems under real-world conditions. In: Proceedings of the AAAI (2004)

Streams and Distributed Computing

Compression of General Bayesian Net CPTs

Yang Xiang$^{(\boxtimes)}$ and Qian Jiang

University of Guelph, Guelph, Canada
yxiang@uoguelph.ca

Abstract. Non-Impeding Noisy-AND (NIN-AND) Tree (NAT) models offer a highly expressive approximate representation for significantly reducing the space of Bayesian Nets (BNs). They can also significantly improve efficiency of BN inference, as shown for binary NAT models. To enable these advantages for general BNs, advancements on three technical challenges are made in this work. We overcome the limitation of well-defined Pairwise Causal Interaction (PCI) bits and present a flexible PCI pattern extraction from general target Conditional Probability Tables (CPTs). We extend parameter estimation for binary NAT models to constrained gradient descent for compressing target CPTs into multi-valued NAT models. The effectiveness of the compression is demonstrated experimentally. A novel framework is also developed for PCI pattern extraction when persistent leaky causes exist.

1 Introduction

A discrete BN quantifies causal strength between each effect and its n causes by a CPT whose number of parameters is exponential in n. Common Causal Independence Models (CIMs), e.g., noisy-OR [4], reduce the number to being linear in n, but are limited in expressiveness. As members of CIM family, NAT models [8,9,11,12] express both reinforcing and undermining as well as their recursive mixture using only a linear number of parameters. Thus, NAT models offer a highly expressive approximation for significantly reducing the space of BNs.

CIMs are not directly operable by common BN inference algorithms, e.g., the cluster tree method [1]. Several techniques exist to overcome the difficulty [2,5,6,14]. By applying multiplicative factorization to binary NAT models and compiling NAT modeled BNs for lazy propagation [3], it has been shown that efficiency of exact inference with BNs can also be improved significantly [8].

The above efficiency gain was shown with binary NAT models. However, binary NAT models are not sufficiently general. The ultimate goal of this research is to achieve similar efficiency improvement for inference in general BNs compressed into multi-valued NAT models [9]. Advancing from binary to multi-valued NAT models encounters several challenges. In this work, we investigate the following. To gain efficiency with both space and inference time through NAT modeling, each (target) CPT in BNs is approximated (compressed) into a NAT model. The first step is to find a small set of candidate NAT structures to focus subsequent parameter search. A NAT can be uniquely identified by a function

© Springer International Publishing Switzerland 2016
R. Khoury and C. Drummond (Eds.): Canadian AI 2016, LNAI 9673, pp. 285–297, 2016.
DOI: 10.1007/978-3-319-34111-8_35

that specifies interactions between each pair of causes, termed a PCI pattern [10]. Therefore, we extract a PCI pattern from the target CPT, which yields the candidate NATs. Since a target CPT is generally not a NAT model, how to extract a PCI pattern that provides good approximation of its causal interaction structure is a challenge. The first contribution of this work is a scheme that meets this challenge.

Once the candidate NATs are obtained, probability parameters of corresponding NAT models must be assessed. The second contribution of this work is to extend the framework for doing so with binary NAT models to multi-valued NAT models. We present a constrained gradient descent as the key component of the extension. Although the general idea of constrained gradient descent already exists, this work investigates specific constraints for compressing multi-valued CPTs. CIMs allow both explicit causes and implicit causes, termed leaky causes. Leaky causes may be persistent or non-persistent. We analyze implications of both types of leaky causes to NAT compression of CPTs. We show that persistent leaky causes raise another challenge. The third contribution of this work is a framework for PCI pattern extraction with persistent leaky causes.

Section 2 briefly introduces the background. Contribution on PCI pattern extraction from general target CPTs is presented in Sect. 3. Constrained gradient descent for compressing multi-valued CPTs is covered in Sect. 4. Their effectiveness is shown through experimental study in Sect. 5. Contribution on PCI extraction with persistent leaky causes is presented in Sect. 6.

2　Background

Consider an effect e and the set of all causes $C = \{c_1, ..., c_n\}$ that are multi-valued and graded. That is, e has domain $D_e = \{e^0, ..., e^\eta\}$ ($\eta \geq 1$), where e^0 is *inactive*, $e^1, ..., e^\eta$ are *active*, and a higher index signifies higher intensity. The domain of c_i is $D_i = \{c_i^0, ..., c_i^{m_i}\}$ ($m_i > 0$). An active value may be written as e^+ or c_i^+. A causal event is a *success* or *failure* depending on whether e is rendered active at certain intensity, is *single-causal* or *multi-causal* depending on the number of active causes, and is *simple* or *congregate* depending on the range of effect values. $P(e^k \leftarrow c_i^j) = P(e^k | c_i^j, c_z^0 : \forall z \neq i)$ ($j > 0$) is the probability of a *simple single-causal success*. $P(e \geq e^k \leftarrow c_1^{j_1}, ..., c_q^{j_q}) = P(e \geq e^k | c_1^{j_1}, ..., c_q^{j_q}, c_z^0 : c_z \in C \setminus X)(j > 0)$ is the probability of a *congregate multi-causal success*, where $X = \{c_1, ..., c_q\}$ ($q > 1$). It is also denoted $P(e \geq e^k \leftarrow \underline{x}^+)$.

A NAT consists of two types of NIN-AND gates, each over disjoint sets of causes $W_1, ..., W_q$. An input event of a *direct* gate is $e \geq e^k \leftarrow \underline{w}_i^+$ and the output event is $e \geq e^k \leftarrow \underline{w}_1^+, ..., \underline{w}_q^+$. An input of a *dual* gate is $e < e^k \leftarrow \underline{w}_i^+$ and the output event is $e < e^k \leftarrow \underline{w}_1^+, ..., \underline{w}_q^+$. Probability of the output event of a gate is the product of probabilities of its input events. Interactions among causes may be reinforcing or undermining.

Definition 1. *Let e^k be an active effect value, $R = \{W_1, W_2, ...\}$ be a partition of a set $X \subseteq C$ of causes, $R' \subset R$, and $Y = \cup_{W_i \in R'} W_i$. Sets of causes in R*

reinforce each other relative to e^k, *iff* $\forall R'\ P(e \geq e^k \leftarrow \underline{y}^+) \leq P(e \geq e^k \leftarrow \underline{x}^+)$.
They undermine each other iff $\forall R'\ P(e \geq e^k \leftarrow \underline{y}^+) > P(e \geq e^k \leftarrow \underline{x}^+)$.

A direct gate models undermining and a dual gate models reinforcing. A NAT organizes multiple gates into a tree and expresses mixture of reinforcing and undermining recursively. A NAT specifies interaction between each pair of c_i and c_j, denoted by *PCI bit* $pci(c_i, c_j) \in \{u, r\}$ with u for undermining. The collection of PCI bits is the *PCI pattern* of the NAT. A NAT can be uniquely identified by PCI pattern [12]. Given a NAT and probabilities of input events, called *single-causals*, the probability of its output event can be obtained. From the single-causals and all derivable NATs [7], the CPT $P(e|C)$ is uniquely defined [9].

3   Extracting PCI Patterns from General CPTs

To compress a target CPT over e and C into a NAT model, we need to determine a NAT over C. This can be achieved by searching for a PCI pattern relative to each e^k and determine the NAT by the best pattern over all k. By Definition 1, given c_i and c_j, $pci(c_i, c_j)$ is *well defined* relative to e^k when one of the following conditions holds for all active values of c_i and c_j.

$$(c_i, c_j) = \begin{cases} u\ :\ P(e \geq e^k \leftarrow c_i^+, c_j^+) < min(P(e \geq e^k \leftarrow c_i^+), P(e \geq e^k \leftarrow c_j^+)), \\ r\ :\ P(e \geq e^k \leftarrow c_i^+, c_j^+) \geq max(P(e \geq e^k \leftarrow c_i^+), P(e \geq e^k \leftarrow c_j^+)). \end{cases} \quad (1)$$

As shown experimentally (Sect. 5.1), in a general CPT, neither condition may hold for a significant number of cause pairs. For such a CPT, very few PCI bits are well defined, resulting in a *partial* PCI pattern. A partial pattern of a few bits is compatible with a large candidate set of NATs, making subsequent search costly. Hence, a best pattern has the most bits. Below, we develop a scheme to overcome the difficulty where the best PCI pattern has too few well defined bits.

We aim to extract a partial PCI pattern that approximates causal interactions in a target CPT. For a partial pattern, PCI bit of a given cause pair may be u, r, or undefined. For uniformity, we expand the domain of a PCI bit into $\{u, r, nul\}$ with nul for unclassified.

For a well-defined bit, one condition in Eq. (1) must hold for all active cause value pairs. Consider interaction for one value pair first. To indicate the e^k value, we denote the interaction as $pci(e^k, c_i^+, c_j^+) \in \{u, r, nul\}$. To simplify notation, we denote $P(e \geq e^k \leftarrow c_i^+)$, $P(e \geq e^k \leftarrow c_j^+)$, and $P(e \geq e^k \leftarrow c_i^+, c_j^+)$ as p, q, and t, respectively. A well-defined interaction is extracted by the following rule.

Rule 1 (Well-Defined). *If* $t \notin [min(p, q), max(p, q)]$, *then*

$$pci(e^k, c_i^+, c_j^+) = \begin{cases} u\ :\ t < min(p, q), \\ r\ :\ t > max(p, q). \end{cases}$$

A well-defined interaction satisfies $t \notin [min(p, q), max(p, q)]$. Rules below relax this requirement. When $t \in [min(p, q), max(p, q)]$, $pci(e^k, c_i^+, c_j^+)$ is deemed nul only if $|p - q|$ is too small, e.g., less than a threshold $\tau_0 = 0.2$.

Rule 2 (Tight Enclosure). *If* $t \in [min(p,q), max(p,q)]$ *and* $|p-q| \leq \tau_0$, *then* $pci(e^k, c_i^+, c_j^+) = nul$, *where* $\tau_0 \in (0,1)$ *is a given threshold.*

Rational of the rule is the following. Under tight enclosure, both u and r may well approximate interaction between c_i and c_j. Hence, NATs compatible with either should be included in the candidate set, which is what value nul entails.

We refer to condition $t \in [min(p,q), max(p,q)]$ and $|p-q| > \tau_0$ as *loose enclosure*, where we compute the ratio $R = \frac{t-0.5(p+q)}{|p-q|}$. Ratio $R \in [-0.5, 0.5]$ and the bounds are reached when t equals p or q. When $R < 0$, t is closer to $min(p,q)$. When $R > 0$, t is closer to $max(p,q)$. When $R = 0$, t is equally distant from p and q. We refer to R as *normalized deviation* and specify the interaction as follows, where a possible value for τ_1 is 0.4. Its rational follows from the above analysis.

Rule 3 (Sided Loose Enclosure). *Given thresholds* $\tau_0, \tau_1 \in (0,1)$, *if* $t \in [min(p,q), max(p,q)]$ *and* $|p-q| > \tau_0$, *then*

$$pci(e^k, c_i^+, c_j^+) = \begin{cases} nul & : & |R| \leq \tau_1, \\ u & : & R < -\tau_1, \\ r & : & R > \tau_1. \end{cases}$$

Given e^k, the above determines $pci(e^k, c_i^+, c_j^+)$ for a pair c_i^+ and c_j^+. If each cause has $m+1$ values, there are m^2 pairs of active values for c_i and c_j. The next rule determines the PCI bit $pci(e^k, c_i, c_j)$ by majority of value based interactions. A possible value for threshold τ_2 may be 0.51.

Rule 4 (Majority Value Pairs). *Let* M *be the number of active cause value pairs* (c_i^+, c_j^+), M_u *be the number of interactions where* $pci(e^k, c_i^+, c_j^+) = u$, *and* M_r *be the number of interactions where* $pci(e^k, c_i^+, c_j^+) = r$. *For a given threshold* $\tau_2 \in (0.5, 1)$,

$$pci(e^k, c_i, c_j) = \begin{cases} u & : & M_u > \tau_2 M, \\ r & : & M_r > \tau_2 M, \\ nul & : & Otherwise. \end{cases}$$

After PCI bit $pci(e^k, c_i, c_j)$ is extracted for each pair (c_i, c_j), a set $pci(e^k)$ of PCI bits relative to e^k is defined. From η such sets, the next rule selects the best as the PCI pattern, where a possible value for threshold τ_3 may be 0.8.

Rule 5 (Partial PCI Pattern). *Let* n *be the number of causes of* e, *the set of PCI bits relative to* e^k *(k > 0) be* $pci(e^k) = \{pci(e^k, c_i, c_j) \mid \forall_{i,j} \ c_i \neq c_j\}$, *and* N_k *be the number of PCI bits in* $pci(e^k)$ *such that* $pci(e^k, c_i, c_j) \neq nul$. *Let* $N_x = max_k N_k$ *and* $\tau_3 \in (0.5, 1)$ *be a given threshold. Then select* $pci(e^x)$ *as the partial PCI pattern if* $N_x > \tau_3 C(n, 2)$.

If $N_x \leq \tau_3 C(n, 2)$, the above rule is inconclusive. We require the search procedure to relax thresholds τ_0 through τ_3 until a PCI pattern is selected. Effectiveness of the procedure for reducing NAT space while extracting good NAT candidates is shown in Sect. 5.2.

4   Parameter Estimation with Constrained Descent

Once a partial PCI pattern is extracted, the set of candidate NATs compatible with the pattern can be determined [12]. For each candidate NAT, single-causals can be estimated from target CPT through gradient descent. From resultant NAT models, the best NAT model can be selected. These steps parallel those for compression of binary CPTs into binary NAT models [11]. In this section, we extend gradient descent to compression of multi-valued CPTs.

A NAT and a set of single-causals define a NAT model M. We measure similarity of a target CPT P_T and the CPT P_M of M by Kullback–Leibler divergence,

$$KL(P_T, P_M) = \sum_i P_T(i) log \frac{P_T(i)}{P_M(i)},$$

where i indexes probabilities in P_T and P_M. Gradient descent estimates the set of single-causals of M such that $KL(P_T, P_M)$ is minimized. In experimental study (Sect. 5), average Euclidean distance $ED(P_T, P_M) = \sqrt{\frac{1}{K} \sum_{i=1}^K (P_T(i) - P_M(i))^2}$ is also obtained, where K counts parameters in P_T.

During descent, the point descending the multi-dimensional surface is a vector of single-causals. For a binary NAT model with n causes, the vector has n parameters and each can be specified independently. For multi-valued NAT models, where $|D_e| = \eta + 1$ and $|D_i| = m + 1$ for $i = 1, ..., n$, the descent point is a $\eta m n$ vector. Each parameter is a $P(e^+ \leftarrow c_i^+)$. Unlike the binary case, the $\eta m n$ parameters are not independent. We consider below constraints that they must observe during descent. First, each parameter $P(e^+ \leftarrow c_i^+) > 0$. That is, each parameter is lower bounded by 0, but cannot reach the bound since otherwise c_i^+ no longer causes e^+. Second, in the binary case, each parameter is upper bounded by 1, but cannot reach the bound since otherwise c_i is no longer an uncertain cause. In the multi-valued case, this constraint is replaced by a more strict alternative. For each c_i^+, $\sum_{j=1}^{\eta} P(e^j \leftarrow c_i^+) < 1$ must hold. If violated, the resultant parameters $P(e^1 \leftarrow c_i^+), ..., P(e^{\eta} \leftarrow c_i^+)$ will not be valid single-causals of an uncertain cause. This amounts to mn constraints, each governing η parameters. To satisfy these constraints, we extend gradient descent for binary NAT models below.

At the start of each round of descent, each group of η single-causals under the same constraint are initialized together as follows. Generate $\eta + 1$ random numbers in the range$[\delta, 1 - \delta]$, where $\delta > 0$ is a small real. Let S be their sum and $0 < \gamma < 1$ be a real close to 1. Drop one number arbitrarily, multiply the remaining η numbers by γ/S, and assign results as initial single-causals. Proposition 1 summarizes properties of the initialization, whose proof is omitted due to space.

Proposition 1. *Let* $P(e^1 \leftarrow c_i^+), ..., P(e^{\eta} \leftarrow c_i^+)$ *be initial values of parameters with the same active cause value* c_i^+. *The following hold.*

1. *For each parameter,* $P(e^j \leftarrow c_i^+) \geq \delta$, $j = 1, ..., \eta$.
2. *For the subset of parameters,* $\sum_{k=1}^{\eta} P(e^k \leftarrow c_i^+) \leq \gamma$.

Each step of gradient descent updates the ηmn parameters in sequence. To ensure that both conditions of Proposition 1 continue to hold, we constrain descent as follows. After each $P(e^j \leftarrow c_i^+)$ is updated, check if $P(e^j \leftarrow c_i^+) \geq \delta$. If not, set $P(e^j \leftarrow c_i^+) = \delta$ and stop $P(e^j \leftarrow c_i^+)$ from further descent. Otherwise, check if $S = \sum_{k=1}^{\eta} P(e^k \leftarrow c_i^+) \leq \gamma$ holds. If not, set $P(e^j \leftarrow c_i^+)$ to $P(e^j \leftarrow c_i^+) + \gamma - S$ and stop $P(e^j \leftarrow c_i^+)$ from further descent. If both tests succeed, commit to the updated value of $P(e^j \leftarrow c_i^+)$ and allow it to continue descent. Proposition 2 summarizes properties of the method, whose proof is omitted due to space.

Proposition 2. *Let* $P(e^1 \leftarrow c_i^+), ..., P(e^\eta \leftarrow c_i^+)$ *be current values of a subset of parameters with the same active cause value c_i^+, such that the following hold.*

1. *For each parameter,* $P(e^j \leftarrow c_i^+) \geq \delta$, $j = 1, ..., \eta$.
2. *For the subset of parameters,* $\sum_{k=1}^{\eta} P(e^k \leftarrow c_i^+) \leq \gamma$.

After each $P(e^j \leftarrow c_i^+)$ is updated during descent, the above conditions still hold.

By Proposition 1, each round of descent starts with valid single-causals. By Proposition 2, for each step of descent, after each parameter is updated, the entire set of single-causals is still valid. Hence, the constrained gradient descent terminates with valid single-causals.

5 Experimental Results

5.1 Necessity of Flexible PCI Extraction

This experiment reveals difference between general and NAT CPTs and need for flexible PCI extraction. Two batches of CPTs are simulated each over $n = 5$ causes with all variable domain sizes being $k = 4$. The 1st batch consists of 100 random CPTs and the 2nd 100 NAT CPTs (of randomly selected NATs and single-causals).

Given a target CPT, for each pair of causes, Eq. (1) is applied relative to each of e_1, e_2, and e_3. With $n = 5$, there are $C(5, 2) = 10$ cause pairs. For each pair, there are $3 * 3 = 9$ active value pairs. For each pair, the PCI bit is well-defined if and only if one condition of Eq. (1) holds for all 9 value pairs. A target CPT has between 0 and 10 well-defined PCI bits.

In the 1st batch, 97 CPTs have 0 well-defined PCI bit extracted. For each of the 3 remaining CPTs, one well-defined PCI bit is extracted relative to e_1, one relative to e_2, and one relative to e_3. Hence, the rate of well-defined PCI bits is 0.003 for each of e_1, e_2, and e_3. In the 2nd batch, 10 well-defined PCI bits are extracted from each CPT. This shows that general CPTs and NAT CPTs differ significantly and the flexible PCI pattern extraction presented in Sect. 3 is necessary.

5.2   Performance of Flexible PCI Extraction and Descent Search

This experiment examines compression error and efficiency gain from the flexible PCI extraction of Sect. 3, as well as the effectiveness of constrained gradient descent of Sect. 4. A 3rd batch of 100 random CPTs with $n = 4$ and $k \leq 4$ are generated. Each CPT is compressed by flexible PCI extraction and constrained descent, referred to as NAT-Com. It is also compressed by descent search exhaustively (hence optimally) for each of 52 NATs of $n = 4$, referred to as NAT-Opt. The choice $n = 4$ is made as NAT-Opt is much more costly for $n = 5$ with a total of 472 NATs.

Table 1 compares their performance, where ED refers to $ED(P_T, P_M)$, KL refers to $KL(P_T, P_M)$, RT refers to Runtime in seconds, and SR refers to Space Reduction. SR is the ratio of numbers of independent parameters between target CPT and NAT CPT. For instance, if $n = 4$ and $k = 3$ for all variables, the ratio is $(3^5 - 1)/(3 * 3 * 4) = 6.75$. SR and ED of NAT-Opt show that NAT compression by constrained descent is effective with significant space reduction (14.67) while incurring reasonable error (0.189 ED). NAT-Com has 8 % larger ED (0.204) (same space reduction), but is 9 times faster. This shows that flexible PCI extraction trims NAT space significantly while retaining good NAT candidates. This efficiency gain is expected to grow exponentially with n as will be shown below.

Table 1. Performance summary of NAT-Com and NAT-Opt

	NAT-Com		NAT-Opt			NAT-Com		NAT-Opt	
	Mean	Stdev	Mean	Stdev		Mean	Stdev	Mean	Stdev
ED	0.204	0.043	0.189	0.036	SR	14.670	6.908	14.670	6.908
KL	22.189	32.590	17.287	20.822	RT	4.456	3.849	41.258	29.420

5.3   Comparison Between NAT and Noisy-MAX Compression

This experiment compares effectiveness of NAT compression with the well-known noisy-MAX as a baseline [13]. A 4th batch of 100 random CPTs with $n = 5$ and $k \leq 4$ and a 5th batch of 100 random CPTs with $n = 6$ and $k \leq 4$ are generated and are processed together with the 3rd batch ($n = 4$ and $k \leq 4$). Each CPT is compressed by NAT-Com, as well as by NMAX-Com where each target CPT is compressed into a noisy-MAX model.

Table 2 compares their performance. From the SR row, as n grows, space reduction by both method grows significantly (from 14.67 to 89.96). Since NAT-Com searches through multiple NATs while NMAX-Com processes a single causal model, NMAX-Com is about 10 times faster.

At the same time, as n grows, ED distance by NMAX-Com increases from 0.23 to 0.45, while ED distance by NAT-Com increases from 0.20 to 0.32. On average, NAT-Com reduces distance to target CPTs by 13 %, 27 %, and 29 %, respectively. Since target CPTs are randomly generated, the experiment is conducted at the most general (worst) condition. It is expected that target CPTs

Table 2. Performance summary of NAT-Com and NMAX-Com

	NAT-Com ($n=4$)		NMAX-Com ($n=4$)		NAT-Com ($n=5$)		NMAX-Com ($n=5$)		NAT-Com ($n=6$)		NMAX-Com ($n=6$)	
	Mean	Stdev	Mean	Stdev	Mean	Stdev	Mean	Stdev	Mean	Stdev	Mean	Stdev
ED	0.20	0.04	0.23	0.07	0.27	0.08	0.37	0.10	0.32	0.09	0.45	0.06
KL	22.19	32.59	41.96	70.15	126.39	119.45	283.07	248.31	453.82	362.47	866.06	547.53
SR	14.67	6.91	14.67	6.91	36.60	20.73	36.60	20.73	89.96	53.32	89.96	53.32
RT	4.46	3.85	0.47	0.32	14.45	16.54	1.49	1.36	54.12	54.41	4.86	3.85

from real BNs display more regularity [13] and compression accuracy by NAT-Com will be further reduced. We leave this to future work.

6   PCI Pattern Extraction with Persistent Leaky Causes

A leaky cause in a causal model integrates all causes that are not explicitly named. In the following, we assume that a leaky cause exists. We denote it by c_0 and denote other causes as $c_1, ..., c_n$. A leaky cause may be persistent or non-persistent. A *non-persistent* leaky cause is not always active. Hence, it can be modeled the same way as other causes. A target CPT in the form $P(e|c_0, c_1, ..., c_n)$ is fully specified with $P(e^0|c_0^0, c_1^0, ..., c_n^0) = 1$ and $P(e^+|c_0^0, c_1^0, ..., c_n^0) = 0$.

On the other hand, a *persistent* leaky cause (PLC) c_0 is always active, and a target CPT has the form $P(e|c_0^+, c_1, ..., c_n)$. This has two implications. First, for all active e^+, we have $P(e^+|c_0^+, c_1^0, ..., c_n^0) > 0$. Second, parameters corresponding to $P(e|c_0^0, c_1, ..., c_n)$ are unavailable. This raises an issue when we compress the target CPT into a NAT model. Since $P(e|c_0^0, c_1, ..., c_n)$ is undefined, the target CPT appears as $P'(e|c_1, ..., c_n) = P(e|c_0^+, c_1, ..., c_n)$.

Example 1. *A target CPT $P(e|c_0, c_1)$ over binary e, c_0 and c_1, where c_0 is a PLC, is shown below (left). Since it is only partially defined and c_0 is an implicit cause, it may be viewed as $P'(e|c_1)$ (right) which is fully defined.*

c_0	c_1	e	$P(e\|c_0,c_1)$		c_0	c_1	e	$P(e\|c_0,c_1)$		c_1	e	$P'(e\|c_1)$
c_0^0	c_1^0	e^0	*undefined*		c_0^1	c_1^0	e^0	0.85		c_1^0	e^0	0.85
c_0^0	c_1^0	e^1	*undefined*		c_0^1	c_1^0	e^1	0.15		c_1^0	e^1	0.15
c_0^0	c_1^1	e^0	*undefined*		c_0^1	c_1^1	e^0	0.32		c_1^1	e^0	0.32
c_0^0	c_1^1	e^1	*undefined*		c_0^1	c_1^1	e^1	0.68		c_1^1	e^1	0.68

Should we define the NAT model over $\{e, c_1, ..., c_n\}$ to match $P'(e|c_1, ..., c_n)$? We reject this option for two reasons. First, the CPT of a NAT model thus defined has $P_M(e^+|c_1^0, ..., c_n^0) = 0$. From the first implication above, we have $P'(e^+|c_1^0, ..., c_n^0) = P(e^+|c_0^+, c_1^0, ..., c_n^0) > 0$, which leads to an inherent modeling error. Second, c_0 can undermine or reinforce another cause. There are 2^n possible causal interactions between c_0 and $c_1, ..., c_n$. They do not approximate target CPT equally well. It is impossible to parameterize the NAT model according to the most suitable causal interaction, unless c_0 is explicitly represented. In the remainder, we assume that the NAT model is defined over family $\{e, c_0, c_1, ..., c_n\}$.

To compress target CPT $P(e|c_0^+, c_1, ..., c_n)$ into a NAT model, we need to extract a PCI pattern. By Eq. (1), PCI bit $pci(c_i, c_j)$ is defined based on comparison among

$$P(e^+ \leftarrow c_i^+), P(e^+ \leftarrow c_j^+), \text{ and } P(e^+ \leftarrow c_i^+, c_j^+).$$

Two difficulties arise when c_0 is a PLC. First, when $i = 0$, the target CPT contains $P(e^+ \leftarrow c_0^+)$ and $P(e^+ \leftarrow c_0^+, c_j^+)$, but not $P(e^+ \leftarrow c_j^+)$. Second, for $i, j > 0$, none of $P(e^+ \leftarrow c_i^+)$, $P(e^+ \leftarrow c_j^+)$, and $P(e^+ \leftarrow c_i^+, c_j^+)$ is specified. In summary, when c_0 is a PLC, no PCI bit can be extracted based on Eq. (1), or based on rules in Sect. 3. One alternative is to extract $pci(c_i, c_j)$ for $i, j > 0$ from comparison among

$$P(e^+ \leftarrow c_0^+, c_i^+), P(e^+ \leftarrow c_0^+, c_j^+), \text{ and } P(e^+ \leftarrow c_0^+, c_i^+, c_j^+).$$

Unfortunately, although they are available from target CPT, it can be shown that causal interaction between c_i and c_j does not uniquely correspond comparison among the three. To meet this challenge, we investigate another alternative. For simplicity in presentation, we assume binary variables, i.e., $D_e = \{e^-, e^+\}$ and $D_i = \{c_i^-, c_i^+\}$.

Given c_0, c_i and c_j where $i, j > 0$, there are 8 causal interaction relations as Fig. 1. We refer to the 8 NATs as T_a through T_h.

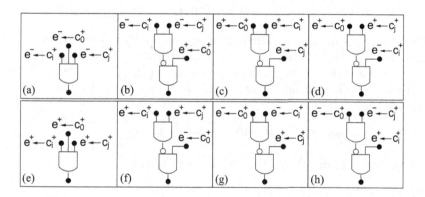

Fig. 1. NATs over c_0, c_i and c_j

Their PCI patterns are summarized in Table 3.

A target CPT $P(e|c_0^+, c_1, ..., c_n)$ specifies the following parameters that involve only c_0, c_i and c_j,

$$P(e^+ \leftarrow c_0^+), P(e^+ \leftarrow c_0^+, c_i^+), P(e^+ \leftarrow c_0^+, c_j^+), \text{ and } P(e^+ \leftarrow c_0^+, c_i^+, c_j^+).$$

We write them compactly as $P(^+0), P(^+0i), P(^+0j)$, and $P(^+0ij)$. If each NAT above can be identified by comparing these parameters, PCI bits in Table 3 will be obtained. We investigate this possibility below.

Table 3. PCI patterns of NATs

	$pci(c_0,c_i)$	$pci(c_0,c_j)$	$pci(c_i,c_j)$		$pci(c_0,c_i)$	$pci(c_0,c_j)$	$pci(c_i,c_j)$
T_a	r	r	r	T_e	u	u	u
T_b	u	u	r	T_f	r	r	u
T_c	u	r	r	T_g	r	u	u
T_d	r	u	r	T_h	u	r	u

For T_a, since any disjoint subsets of $\{c_0, c_i, c_j\}$ reinforce each other (Fig. 1(a)),

$$P(^+0ij) > P(^+0i), P(^+0ij) > P(^+0j), \text{ and } P(^+0ij) > P(^+0).$$

For T_e, since any disjoint subsets of $\{c_0, c_i, c_j\}$ undermine each other (Fig. 1(e)),

$$P(^+0ij) < P(^+0i), P(^+0ij) < P(^+0j), \text{ and } P(^+0ij) < P(^+0).$$

For T_b (Fig. 1(b)), $P(^+0i)$ results from interaction between c_0 and c_i. Since $P(^+0ij)$ results from interaction between c_0 and group $\{c_i, c_j\}$ where c_i is reinforced by c_j, it follows that $P(^+0ij) > P(^+0i)$. From symmetry between c_i and c_j, we derive $P(^+0ij) > P(^+0j)$. Since c_0 undermines the group $\{c_i, c_j\}$, it follows that $P(^+0ij) < P(^+0)$. From the dual relation between T_b and T_f (Fig. 1 (f)), we derive

$$P(^+0ij) < P(^+0i), P(^+0ij) < P(^+0j), \text{ and } P(^+0ij) > P(^+0).$$

For T_c (Fig. 1(c)), since c_j reinforces group $\{c_0, c_i\}$, it follows that $P(^+0ij) > P(^+0i)$. Since $P(^+0ij)$ results from interaction between c_j and group $\{c_0, c_i\}$ where c_0 is undermined by c_i, it follows that $P(^+0ij) < P(^+0j)$. Since T_d (Fig. 1(d)) is obtained from T_c by switching between c_i and c_j, we derive for T_d the following,

$$P(^+0ij) < P(^+0i) \text{ and } P(^+0ij) > P(^+0j).$$

To compare $P(^+0ij)$ and $P(^+0)$ for T_d, we analyze

$$P(^-0) - P(^-0ij) = [1 - P(^+0)] - [1 - P(^+0)P(^+j)]P(^-i)$$
$$= 1 - P(^+0) - P(^-i) + P(^+0)P(^+j)P(^-i) = P(^+i) - P(^+0)[1 - P(^+j)P(^-i)].$$

If $P(^+i)$ is close to 1, the sum is about $1 - P(^+0) > 0$. If $P(^+i)$ is close to 0, the sum is about $-P(^+0)P(^-j) < 0$. Hence, comparison of $P(^+0ij)$ and $P(^+0)$ is non-deterministic for T_d. Due to relation between T_d and T_c, the same holds for T_c.

From dual relation between T_d and T_h, for T_h, we have $P(^+0ij) > P(^+0i)$, $P(^+0ij) < P(^+0j)$, and non-deterministic comparison of $P(^+0)$ and $P(^+0ij)$. Since T_g results from switching c_i and c_j in T_h, we derive for T_g $P(^+0ij) < P(^+0i)$, $P(^+0ij) > P(^+0j)$, and non-deterministic comparison of $P(^+0)$ and $P(^+0ij)$.

Table 4. Causal probability comparison

	$P(^+0ij)$ $-P(^+0i)$	$P(^+0ij)$ $-P(^+0j)$	$P(^+0ij)$ $-P(^+0)$	$P(^+0i)$ $-P(^+0)$	$P(^+0j)$ $-P(^+0)$	$P(^+0i)$ $-P(^+0j)$
T_a	> 0	> 0	> 0			
T_b	> 0	> 0	< 0			
T_e	< 0	< 0	< 0			
T_f	< 0	< 0	> 0			
T_d	< 0	> 0	$+/-$	> 0	< 0	> 0
T_g	< 0	> 0	$+/-$	> 0	< 0	> 0
T_c	> 0	< 0	$+/-$	< 0	> 0	< 0
T_h	> 0	< 0	$+/-$	< 0	> 0	< 0

The first 4 columns of Table 4 summarize the above. It can be seen that T_a, T_b, T_e and T_f can be uniquely identified by the comparisons, and hence all three PCI bits in Table 3. The group of T_d and T_g can be identified from two comparisons, and so can the group of T_c and T_h. This allows specification of $pci(c_0, c_i)$ and $pci(c_0, c_j)$. Since the two NATs in each group cannot be differentiated, $pci(c_i, c_j)$ cannot be specified.

There are $C(4, 2) = 6$ pairs of comparisons among $P(^+0), P(^+0i), P(^+0j)$, and $P(^+0ij)$, with the remaining shown in the last three columns of Table 4. Comparisons between $P(^+0i)$, $P(^+0j)$ and $P(^+0)$ in col. 5 and 6 are derived from Table 3. Col. 7 compares $P(^+0i)$ and $P(^+0j)$. For T_d and T_g, col. 5 and 6 imply $P(^+0i) > P(^+0) > P(^+0j)$. For T_c and T_h, col. 5 and 6 imply $P(^+0i) < P(^+0) < P(^+0j)$. As can be seen, col. 5, 6 and 7 do not improve differentiation.

We conclude the following from this analysis. If target CPT is an unknown NAT model with a PLC, a partial PCI pattern can be extracted by comparing $P(^+0)$, $P(^+0i), P(^+0j)$, and $P(^+0ij)$ for each pair of $i, j > 0$. In particular, $pci(c_0, c_j)$ is extractable for all $j > 0$. For $i, j > 0$, 50 % of bits $pci(c_i, c_j)$ are extractable on average. This result can be extended to general CPTs by applying the technique in Sect. 3 to probability comparison, which we do not elaborate here due to space.

Given $c_0, c_1, ..., c_n$, there are $C(n+1, 2) = (n+1)n/2$ PCI bits. Among them, $pci(c_i, c_j)$ where $i, j > 0$ counts $C(n, 2) = (n-1)n/2$ bits and $pci(c_0, c_j)$ counts n bits. Hence, the proposed framework allows extraction of $n(n+3)/4$ PCI bits on average. It follows that, as n grows from 4 (a total of 5 causes) to 12, the expected percentage of extractable PCI bits changes from 70 % to 58 %. In the next section, we outline future research regarding the remaining PCI bits.

7 Conclusion

The first contribution of this work is a flexible PCI pattern extraction that obtains a partial PCI pattern with a sufficient number of bits from general target CPTs. Experiment in Sect. 5.1 demonstrates the necessity of such a flexible extraction. Experiment in Sect. 5.2 shows that the extraction significantly reduces the number of candidate NATs for subsequent parameter estimation while incurs only minor loss of accuracy. The second contribution extends gradient descent for compression of binary NAT models to constrained descent for compression of multi-valued NAT models. Experiment in Sect. 5.3 shows that compression of random CPTs into NAT models achieves better accuracy than compression into noisy-MAX models. Further research will examine effectiveness of compression in real BN CPTs. The impact of compression errors to BN inference will also be evaluated.

The above compression assumes non-PLCs. If applied to CPTs with PLCs, modeling errors occur when all explicit causes are absent. The third contribution is a framework for extracting PCI pattern when PLCs exist, which significantly differs from PCI pattern extraction with non-PLCs. Further research is needed to answer the following open questions. With PLC presence, if a $pci(c_i, c_j)$ is unclassified, does assigning $pci(c_i, c_j) = u$ or r matter to accuracy of the resultant NAT model? If it does, can $pci(c_i, c_j)$ be inferred from higher order conditional probabilities? Answers to these questions will enable development of a compression algorithm for target CPTs with PLCs.

Acknowledgement. Financial support from NSERC Discovery Grant is acknowledged. We thank anonymous reviewers. We apologize for not moving explanations of figures and tables from text to captions, as it does not appear feasible to us.

References

1. Jensen, F.V., Lauritzen, S.L., Olesen, K.G.: Bayesian updating in causal probabilistic networks by local computations. Comput. Stat. Q. **4**, 269–282 (1990)
2. Madsen, A.L., D'Ambrosio, B.: A factorized representation of independence of causal influence, lazy propagation. Int. J. Uncertainty Fuzziness Knowl.-Based Syst. **8**(2), 151–166 (2000)
3. Madsen, A.L., Jensen, F.V.: Lazy propagation: A junction tree inference algorithm based on lazy evaluation. Artif. Intell. **113**(1–2), 203–245 (1999)
4. Pearl, J.: Probabilistic Reasoning in Intelligent Systems: Networks of Plausible Inference. Morgan Kaufmann, Burlington (1988)
5. Savicky, P., Vomlel, J.: Exploiting tensor rank-one decomposition in probabilistic inference. Kybernetika **43**(5), 747–764 (2007)
6. Takikawa, M., D'Ambrosio, B.: Multiplicative factorization of noisy-max. In: Proceedings of 15th Conference Uncertainty in Artificial Intelligence, pp. 622–630 (1999)
7. Yang, X.: Acquisition and computation issues with NIN-AND tree models. In: Myllymaki, P., Roos, T., Jaakkola, T., (eds.) Proceedings of 5th European Workshop on Probabilistic Graphical Models, Finland, pp. 281–289 (2010)
8. Yang, X.: Bayesian network inference with NIN-AND tree models. In: Cano, A., Gomez-Olmedo, M., Nielsen, T.D. (eds.) Proceedings 6th European Workshop on Probabilistic Graphical Models, Granadapp, pp. 363–370 (2012)

9. Xiang, Y.: Non-impeding noisy-and tree causal models over multi-valued variables. Int. J. Approx. Reason. **53**(7), 988–1002 (2012)
10. Xiang, Y., Li, Y., Zhu, Z.J.: Towards effective elicitation of NIN-AND tree causal models. In: Godo, L., Pugliese, A. (eds.) SUM 2009. LNCS, vol. 5785, pp. 282–296. Springer, Heidelberg (2009)
11. Xiang, Y., Liu, Q.: Compression of Bayesian networks with NIN-AND tree modeling. In: van der Gaag, L.C., Feelders, A.J. (eds.) PGM 2014. LNCS, vol. 8754, pp. 551–566. Springer, Heidelberg (2014)
12. Xiang, Y., Truong, M.: Acquisition of causal models for local distributions in Bayesian networks. IEEE Trans. Cybern. **44**(9), 1591–1604 (2014)
13. Zagorecki, A., Druzdzel, M.J.: Knowledge engineering for Bayesian networks: How common are noisy-MAX distributions in practice? IEEE Trans. Syst. Man Cybern. Syst. **43**(1), 186–195 (2013)
14. Zhang, N., Poole, D.: Exploiting causal independence in Bayesian network inference. J. Artif. Intell. Res. **5**, 301–328 (1996)

Sampling Graphical Networks via Conditional Independence Coupling of Markov Chains

Guichong Li[1,2(✉)]

[1] Department of Computer Science, University of Ottawa, Ottawa, Canada
jli136@site.uottawa.ca
[2] Data and Systems Integration, NPD Group, Port Washington, NY, USA
Guichong.Li@npd.com

Abstract. Markov Chain Monte Carlo (MCMC) methods have been used for sampling Online SNs. The main drawbacks are that traditional MCMC techniques such as the Metropolis-Hastings Random Walk (MHRW) suffer from slow mixing rates, and the resulting sample is usually approximate. An appealing solution is to adapt the MHRW sampler to probability coupling techniques for perfect sampling. While this MHRW coupler is theoretically advanced, it is inapplicable for sampling large SNs in practice. We develop a new coupling algorithm, called Conditional Independence Coupler (CIC), which improves existing coupling techniques by adopting a new coalescence condition, called Conditional Independence (CI), for efficient coalescence detection. The proposed CIC algorithm is outstandingly scalable for sampling large SNs without any bias as compared to previous traditional MCMC sampling algorithms.

Keywords: Sampling online social networks · Markov Chain Monte Carlo · Metropolis-Hastings algorithm · Coupling of Markov Chains · Conditional independence

1 Introduction

Online Social Networks (OSNs) are large graph networks consisting of nodes and links, which are denoted as users and friend or follow relationships between users. Recent statistics show that the number of active users on Twitter is more than 270 millions, and the number of active users in Facebook is more than 1.3 billion. Mining these large graphical networks becomes a big challenge. Recent research has focused on Markov Chain Monte Carlo (MCMC) techniques [1, 2, 10] for sampling OSNs to find a relatively small representative sample suitable for Social Network Analysis (SNA) [4–6].

Metropolis-Hastings Random Walk (MHRW) [4] is a tyical MCMC sampler, which employs the well-known Metropolis-Hastings algorithm [9] for producing an unbiased sample. This is achieved by customizing the transition kernel into the symmetrical transition probability such that the chain has a uniform distribution

---

G. Li—This research is funded by NSERC Engage and SME4SME.

R. Khoury and C. Drummond (Eds.): Canadian AI 2016, LNAI 9673, pp. 298–303, 2016.
DOI: 10.1007/978-3-319-34111-8_36

on the state space Ω. The main issues are that MHRW is subject to significant burn-in lengths and correlation with the initial node choice. Resulting sample can be approximately obtained by using convergence diagnotics [3].

A possible alternative to convergence diagnostics in MHRW is to develop MHRW coupler [11], which adapts the base MHRW sampler to perfect sampling techniques such as Coupling From The Past (CFTP). In MHRW coupler convergence is achieved by coalescence to a single state, which becomes an exact sample from π. Therefore, issues of selecting proper convergence diagnostics are discarded [8,12].

There are two crucial issues for this development. Firstly, defining an effective update function for the coupler is not trivial. Secondly, the whole state space is unachievable from large SNs. A grand coupling over the whole state space is prohibitive. Researchers still face many technical and theoretical challenges for application of state of the art coupling techniques.

The significance of this work is that we propose a new forward coupling algorithm, which is regarded as a perfect sampler, to tackle current challenges. The key issue which is distinct from traditional coupling techniques is that CIC adopts a new coalescence condition, called *Conditional Independence (CI)*, for efficient and effective coalescence detection.

2   Preliminary

Two related coupling techniques are introduced as follows.

2.1   Couple from the Past

CFTP, a well-known coupling technique developed by Propp and Wilson [12], allows perfect (exact) sampling from a desired distribution. The algorithm (omitted due to space limitations) performs backward coupling for exact sampling. The coalescence condition is called *geometrical coalescence*, as defined in (1).

$$T = \inf\{t | F^0_{-t} \text{ is a constant function}, t \in \mathbb{N}\}, \tag{1}$$

where $F^0_{-t}(x) = (\Phi_0 \circ \Phi_{-1} \circ \Phi_{-t+1} \circ \Phi_{-t})(x), x \in \Omega$, and Φ_{-t} is the update function, defined by

$$\Phi_{-t} : X_{-t} = \Phi(X_{-t-1}, R^{(-t)}),$$

where $R^{(-t)} \sim U(0,1)$, and $t \in \mathbb{N}$.

2.2   Metropolis Hastings Random Walk Coupler

Provided an ergodic Markov chain $\{X_t\}_0^\infty$ with the transition probability P on a finite and discrete state space Ω, the stationary distribution π is defined as

$$\forall y \in \Omega, \pi(y) = \sum_{x \in \Omega} \pi(x)P(x,y). \tag{2}$$

The Metropolis-Hastings (MH) algorithm [9] is an important MCMC method to simulate the Markov chain for sampling from π. In particular, the Metropolis-Hastings Random Walk (MHRW) recently has been used for sampling OSNs [4]. The symmetric transition probability P can be given by

$$P(x_i, x_{ij}) = \begin{cases} \min(\frac{1}{k_i}, \frac{1}{k_{ij}}) & \text{if } x_{ij} \text{ is a neighbor of } x_i \\ 1 - \sum_j P(x_i, x_{ij}) & \text{if } x_{ij} = x_i \\ 0 & \text{otherwise} \end{cases} \quad (3)$$

where k_i and k_{ij} are the degrees of x_i and x_{ij}, respectively.

Multiple MCMC chains can be useds for simulating chains from different initial states in the way in which large collections of states are all updated to a single new state [11]. For instance, the Metropolis-Hastings (MH) coupler has been designed by adapting the MH sampler to perform coupling from the past from a continuous state space [11].

Given the base MHRW sampler on Ω, the adapted MHRW coupler runs multiple copies of the Markov chain $\{X_t\}_{-T}^0$ for coupling from the past. At time t we are in the state X_t, and wish to generate X_{t+1} using a deterministic update function Φ, which is rewritten here as

$$X_{t+1} = \Phi(X_t, R^{(t+1)}), \quad (4)$$

where $R^{(t+1)} \sim U(0, 1)$ and $t \in \mathbb{Z}$. Given P in (3), let the proposal density be

$$q(y|x_i) = \frac{1}{k_i}, \quad (5)$$

where $x_i \in X_t$. We, thus, define a function $\psi(x_i, R^{(t+1)})$ such that a proposal distribution y has the density $q(. \mid .)$, i.e.,

$$y = x_{ij} = \psi(x_i, R^{(t+1)}), \quad (6)$$

where $j = \lfloor R^{(t+1)} \times k_i \rfloor$. That is, y is always one of the neighbors of x_i.

3   Conditional Independence Coupling

We develop a new coupling technique for efficiently sampling OSN as follows.

3.1   Generating Small State Spaces

The grand coupling on a large state space in CFTP is prohibitive at the cost of computational time and space. It is also impossible in OSNs because the whole state space is unavailable and undefined at runtime.

Note that a number of states $x \in \Omega$ might be independent of the target state y such that $\exists \Omega' \subset \Omega$, y is only dependent of Ω', i.e., $y \perp \Omega \setminus \Omega' | \Omega'$. We can rewrite the definition of the stationary distribution π in (2) as follows.

$$\pi(y) = \sum_{x \in \Omega} \pi(x) P(x, y) = \sum_{x \in \Omega'} \pi(x) P(x, y). \quad (7)$$

Algorithm 1. Small State Space algorithm

Require: X_0: initial states; N: the size of the small state space, with the minimum
  $N = 2$;
Ensure: Ω': a generated small state space w.r.t. X_0
  1: $\Omega' = X_0, t = 0$
  2: **while** $t < \log N$ **do**
  3:    $R^{(t+1)} \sim U(0, 1)$
  4:    $X_t = \Omega'$
  5:    $X_{t+1} = Update(X_t, R^{(t+1)})$
  6:    $\Omega' = \Omega' \cup X_{t+1}, t = t + 1$
  7: **end while**

This implies that the stationary value y is associated with a small local state space $\Omega' \subset \Omega$ w.r.t. y.

The related idea behind Ω' has been introduced in the literature [10,13]. Eventually, we propose a random process, called *Small State Space (SSS)*, for generating a small state space Ω'. SSS produces Ω' incrementally given initial X_0 using the update function in (4), which is implemented by Update (omitted due to space limitations).

3.2 Conditional Independence

By rewriting (1) for MHRW coupler, the mixing (coupling) time T of a grand coupling is given by

$$T = \inf\{t | |X_t| = 1, t \in \mathbb{N}\}. \tag{8}$$

Bounding mixing time for a random walk on the hypercube can be analyzed by coupling [7],

$$T(\varepsilon) = O(n \log(n/\varepsilon)). \tag{9}$$

Unfortunately, T is subject to infinity in a large social network. We, thus, re-consider the equation in (7). As we know, it describes that y is dependent of Ω', i.e., $y \perp \Omega \setminus \Omega' | \Omega'$. It exactly means that a valid sample y from π can be drawn by using a grand coupling of some Ω' if y is only dependent of Ω'. On the other hand, the coalescence to y can be detected by examining possible independency of y from Ω' if any.

Heuristically, we develop a new coalescence condition, called Conditional Independence (CI), beyond the geometrical coalescence for a fast mixing time. We rewrite (8) by introducing CI as follows:

$$T = \inf\{t | |X_t| = 1 \vee \text{CI}, t \in \mathbb{N}\}, \tag{10}$$

where $CI = \exists! y \in X_t, \forall x \in X_t \setminus \{y\}, x \in X_0$. That is, the single value y obtained due to CI is also a perfect sample because some chain copies starting from $X_0' \subseteq X_0$ coalesce to y at time $t = T$ while other chain copies starting from $X_0 \setminus X_0'$ just stay at the past at time $t = 0$, and then $y \perp X_0 \setminus X_0' | X_0'$.

Algorithm 2. Conditional Independence Coupler

Require: Ω': a small state space; Ω: whole state space; τ_0: the minimal coupling time, $\tau_0 = 1$ by default

Ensure: X_0: singleton state

1: $W = \Omega', V = \{i|x_i \in \Omega'\}, T = \tau_0$
2: **repeat**
3:     $R^{(T)} \sim U(0,1)$
4:     $t = 0, X_t = W$
5:     **while** $t < T$ **do**
6:         $X_{t+1} = \text{Update}(X_t, R^{(t+1)})$
7:         $t = t + 1$
8:     **end while**
9:     **if** $|X_0| = 1$ **then**
10:         **return** X_0
11:     **end if**
12:     $X_0 = X_0 - \{x_i|x_i \in X_0 \wedge i \in V\}$
13:     **if** $|X_0| > 0$ **then**
14:         **if** $|X_0| = 1$ **then**
15:             **return** X_0
16:         **end if**
17:         $W = X_0, V = V \cup \{i|x_i \in X_0\}$
18:         loop
19:     **end if**
20:     $W = \Omega', V = \{i|x_i \in \Omega'\}, T = T + 1$
21: **until** true

3.3 Conditional Independence Coupler

We develop a new sampling algorithm, called *Conditional Independence Coupler (CIC)* in Algorithm 2, for perfect sampling according to (10). The technical core of CIC is to create two set variables: W, called *working set* containing the new initial states for forward coupling, and V, called *visited set* containing all states accessed at time $t = 0$. This lets us effectively monitor the evolution of the Markov chain within one tour for regeneration.

Furthermore, it is easy to know that V is just X_0 as a small set in (10) when the coalescence occurs because V, also called *regenerated small sets of states*, contains all starting states, and $|V| \approx 2|\Omega'|$, see (11). As a result, T is the learned coupling time according to (10) if the algorithm returns at Steps 10 or 15. Finally, at Step 20 the algorithm increases T with increment of 1 for forward coupling, and re-uses the previously generated random numbers due to Step 3.

The algorithm actually explores a binary random tree on Ω' through the iterations from Steps 2 to 18. As a result, the expected time depth is $O(\log N)$, and the expected size of V is given by

$$N\left(1 + \frac{1}{2} + \frac{1}{4} + \dots\right) = O(2N), \tag{11}$$

where $N = |\Omega'|$. That is, the expected size of V is twice as large as the size of Ω'. Moreover, both the expected time and space complexities are $O(N \log N)$.

References

1. Andrieu, C., de Freitas, N., Doucet, A., Jordan, M.: An introduction to MCMC for machine learning. Mach. Learn. **50**(2003), 5–43 (2003)
2. Brooks, S.P.: Markov Chain Monte Carlo method and its application. Statistician **47**, 69–100 (1998)
3. Geweke, J.: Evaluating the accuracy of sampling-based approaches to the calculation of posterior moments. In: Bayesian Statistics, pp. 169–193 (1992)
4. Gjoka, M., Kurant, M., Butts, C.T., Markopoulou, A.: Walking in Facebook: a case study of unbiased sampling of OSNs. In: INFOCOM 2010 Proceedings of the 29th Conference on Information Communications, pp. 2498–2506 (2010)
5. Kwak, H., Lee, C., Park, H., Moon, S.: What is Twitter, a social network or a news media? In: Proceedings of the 19th International Conference on World Wide Web, pp. 591–600 (2010)
6. Leskovec, J., Faloutsos, C.: Sampling from large graphs. In: Proceedings of the 12th ACM SIGKDD International Conference on Knowledge Discovery and Data Mining, pp. 631–636 (2006)
7. Levin, D.A., Peres, Y., Wilmer, E.L.: Markov Chains and Mixing Times. AMS, Providence (2009)
8. Meng, X.L.: Towards a more general ProppWilson algorithm: multistage backward coupling. Monte Carlo Methods - Fields Inst. Commun. **26**, 85–93 (2000)
9. Metropolis, N., Rosenbluth, A.W., Rosenbluth, M.N., Teller, A.H., Teller, E.: Equations of state calculations by fast computing machines. J. Chem. Phys. **21**(6), 1087–1092 (1953)
10. Meyn, S.P., Tweedie, R.L.: Markov Chains and Stochastic Stability. Springer, New York (1993)
11. Murdoch, D.J., Green, P.J.: Exact sampling from a continuous state space. Scand. J. Stat. **25**(3), 483–502 (1998)
12. Propp, J.G., Wilson, D.B.: Exact sampling with coupled Markov Chains and applications to statistical mechanics. Random Struct. Algorithms **9**, 223–252 (1996)
13. Robert, G., Rosenthal, J.S.: Small and pseudo-small sets for Markov chains. Commun. Stat. Stoch. Models **17**, 121–145 (2000)

Threaded Ensembles of Supervised and Unsupervised Neural Networks for Stream Learning

Yue Dong[1]([✉]) and Nathalie Japkowicz[2]

[1] Department of Mathematics and Statistics, University of Ottawa,
Ottawa, ON, Canada
ydong029@uottawa.ca
[2] School of Electrical Engineering and Computer Science, University of Ottawa,
Ottawa, ON, Canada
nat@site.uottawa.ca

Abstract. Most existing model-based approaches to anomaly detection in streaming data are based on decision trees due to their fast construction speed [1]. This paper proposes two fast anomaly detectors based on ensembles of neural networks for evolving data streams. One model is a supervised online learning algorithm involving an ensemble of threaded multilayer perceptrons (MLP). The other model is a one-class learning algorithm with an ensemble of threaded autoencoders. The latter model only requires data from the positive class for training and is accurate even when anomalous training data are rare. The models feature an ensemble of multilayer perceptrons or autoencoders from multi-threads which evolve with data streams. Using multi-threads makes the methods highly efficient because both methods process data streams in parallel. Our analysis shows that both methods in streaming data have constant small time complexity and constant memory requirement. When compared with Very Fast Decision Trees (VFDT), a state-of-the-art algorithm, our methods performed favorably in terms of detection accuracy and training time for the datasets under consideration.

Keywords: Stream mining · Anomaly detection · Neural networks · Multilayer perceptrons · Autoencoder · Ensemble learning · Supervised learning · One-class learning · Ensemble

1 Introduction

Anomalies or outliers are rare events that are statistically significant deviations of data from expected normal behavior. Anomaly detection is the problem of identifying outliers from normal data. Without being correctly detected, abnormal events could lead to devastating consequences in many practical domains, such as safety surveillance, network security, and fraud detection. Moreover, with the advances in technology, many datasets are no longer static and are continuously updated, sometimes with millions or even billions of new records per day.

© Springer International Publishing Switzerland 2016
R. Khoury and C. Drummond (Eds.): Canadian AI 2016, LNAI 9673, pp. 304–315, 2016.
DOI: 10.1007/978-3-319-34111-8_37

In most cases, this data arrives in streams and if it is not processed immediately or stored, it will be lost forever. In addition, the data arrives so rapidly that it is usually not feasible to store it all. To extract useful knowledge from this high-speed real-time data, data stream mining became popular.

As [2] mentioned, there are many special characteristics in detecting anomalies in streaming data. Firstly, data streams arrive continuously without stopping, so any off-line batch learning algorithm that requires storing the entire stream for analysis will run out of memory space. Secondly, data streams for anomaly detection usually contain mostly normal data instances (positive instances), where abnormal events are rare and may not be available for training. In this case, any multi-classes classifiers that require fully labeled data from both normal (positive) and abnormal (negative) classes will not be suitable. Thirdly, streaming data often evolves over time with concept drift. Therefore, the model must adapt to different parts of the stream and learn continuously in order to maintain high detection accuracy.

In many streaming data applications, analysis can only be performed in one pass over the data due to the sheer size of the input. Because of this, traditional batch learning algorithms are often unsuitable for data stream mining. We thus need algorithms that can summarize information about prior data instances without keeping all of them in memory.

Several attempts have been made to modify the existing batch learners for data stream mining. According to Gama et al. [3], most of these attempts use decision trees "that continuously evolve over time, taking into account that the environment is non-stationary and computational resources are limited." Examples of these algorithms are Very Fast Decision Trees (VFDT) [4] and Concept-adapting Very Fast Decision Trees (CVFDT) [5].

However, it is not natural to use decision trees for data stream mining because they are very inefficient at dealing with concept drift — an important issue in data streams. Decision trees work top-down in a way involving choosing the best attributes to split data at each level. To grow one level, decision tree algorithms need to scan the whole dataset once. Then, once the attributes are chosen, it is costly to change the shape of the tree or to change the nodes which have already been chosen. In the worst case scenario, if the new training data requires the root attribute to be changed, then all the nodes have to be chosen again. Since the characteristics of streaming data require the algorithm to constantly adjust to the new data, decision trees are not the best for data stream mining.

Unlike algorithms that alter decision trees' structure dynamically as streaming data arrives such as Hoeffding Trees [4], it is easy to see that neural networks (NN) are suitable for dealing with concept drift. In fact, they naturally handle well the task of incremental learning. A typical feed-forward neural network has one input layer, one output layer and several hidden layers. It learns by adjusting the weights between neurons in iterations. When continuous, high-volume, open-ended data streams come, it is necessary to pass the data in smaller groups (mini-batch learning) or one at a time (online learning) before updating the weights. In this manner, neural networks can learn by seeing each example only once and therefore do not require examples to be stored.

This paper proposes two anomaly detection algorithms based on ensembles of neural networks with multi-threads and mini-batches. The proposed methods have several features that distinguish themselves from other existing techniques. Firstly, they process data in one pass and only require small constant time per data instance. Secondly, they use fixed amounts of main memory to process potentially endless streaming data or massive datasets. Thirdly, streaming autoencoders is a one-class anomaly detector, which is useful when a data stream contains a significant amount of normal data. Fourthly, both algorithms have the ability to deal with concept drift because they learn the data continuously with incremental updates. Last but not least, the models separate streaming data into multiple threads to speed up the training process, then form an ensemble of multilayer perceptrons (MLP) or autoencoders from those threads to improve the detection accuracy.

Our experimental study shows that ensembles of streaming multi-threaded MLP or autoencoders lead to robust and accurate anomaly detectors. In order to have a good initial weight, we collect the first n samples as a window in each thread and run multiple epochs on it. Then it learns incrementally from the rest of the streaming data. This provides an anytime learner. In particular, it can still perform well early on, when there is not much data yet.

The rest of this paper is organized as follows: in Sect. 2, we discuss the related works. In Sect. 3, we propose our methods for mining streaming data. In Sect. 4, we discuss the datasets and the experimental design. In Sect. 5, we present the results and compare them with the state-of-the-art stream mining algorithm — VFDT. In Sect. 6, we conclude with remarks and discussions.

2   Related Works

In the literature, there are already numerous studies devoted to anomaly detection in static datasets. Typical examples include the statistical methods [6], classification-based methods [7], clustering-based methods [8], distance-based methods [9], One-Class Support Vector Machine (SVM) [10] and Isolation Forest [11]. These batch learning methods require loading the entire dataset into the main memory for training [2]. Therefore they are not suitable for processing streaming data.

Data stream mining has become prevalent in the last two decades. Based on [12], several classification methods suitable for data streams have been developed. Examples are Very Fast Decision Trees (VFDT) [4] based on Hoeffding trees, Online Information Network (OLIN) [13], On-demand Classification [14] based on micro-clusters found with data streams, and the clustering algorithm CluStream [15]. There are many other popular algorithms, such as Ultra Fast Forest Trees (UFFT) [1] and Concept Based Decision Tree (CBDT) [16].

As we can see, most of these stream mining algorithms are based on decision trees. However, it is not natural to use decision trees when learning has to be continuous, which it frequently does due to concept drift. When concept drift is present, the decision tree structure is unstable in the setting of stream mining and it is costly to adjust the trees' shape with new data.

We study VFDT in this paper as an illustration for the disadvantages of tree based models. VFDT is a popular supervised learning method for anomaly detection in data streams. It is an incremental anytime decision tree based algorithm for classifying high-speed data streams. VFDT requires positive as well as negative class labels to be available for training, which is sometimes not a realistic assumption because anomalous data is usually rare or not available for training. However, the biggest problem for VFDT is its slow training time and inability to adjust to concept drift.

VFDT builds each node of the decision tree based on a small amount of data instances from a data stream. According to Domingos and Hulten [4], VFDT works as follows: "given a stream of examples, the first ones will be used to choose the root test; once the root attribute is chosen, the succeeding examples will be passed down to the corresponding leaves and used to choose the appropriate attributes there, and so on recursively."

To decide how many data instances are necessary for splitting a node, the Hoeffding bound is used. It works as follows: consider r as a real valued random variable, n as the number of independent observations of r, and R as the range of r, then the mean of r is at least $r_{avg} - \epsilon$, with probability $1 - \delta$:

$$P(\bar{r} \geq r_{avg} - \epsilon) = 1 - \delta \qquad (1)$$

where

$$\epsilon = \sqrt{\frac{R^2 ln(1/\delta)}{2n}} \qquad (2)$$

With the Hoeffding bound, VFDT can guarantee that the performance of Hoeffding trees converges to that of a batch learning decision tree [4]. However, this data structure is not "stable" because this convergence only works when there is no concept drift. Subtle changes of the data distribution can cause Hoeffding trees to fail as demonstrated in the HTTP + SMTP dataset from our experiment.

To mitigate the issue of concept drift, the same authors introduced Concept-adapting Very Fast Decision Trees (CVFDT). CVFDT adapts better to concept drift by monitoring changes of information gain for attributes and generating alternate subtrees when needed. However, it still can't solve the issue completely because the algorithm only monitors the leaves. Due to this disadvantage of decision tree based models, other researchers tried to stay away from having to regrow decision trees at all.

Many researchers have proposed algorithms based on incremental neural networks, such as the papers [17–20]. These algorithms pass the data points one by one and update the weights every time after scanning a data point. The shortcoming of these algorithms is that the results are highly dependent on the random initial weights. In batch learning, we pass the data in many epochs to avoid the effect of random initial weights. This is impossible in incremental neural networks since the data only gets passed once and the previous data is discarded. Moreover, updating the weights with backpropagation can be slow. If the data arrives too quickly, the data might be lost before it can be processed.

The models proposed in this paper overcome the disadvantages of incremental neural networks by the following three means: (i) using weights obtained from a mini-batch neural network to minimize the effect of random initial weights; (ii) dividing the data into multiple threads and processing them in parallel to speed up the training; (iii) forming ensembles of neural networks (MLP or autoencoders) to improve the detection accuracy.

3   The Proposed Methods

The proposed method is an ensemble of online multi-threaded MLP or autoencoders with a mini-batch window. Each MLP or autoencoder is built from a particular subspace (one thread) of the data stream. To facilitate a better set of initial weights, we collect n data points in a mini-batch window at the beginning of each thread. An MLP or autoencoder is then trained on the corresponding data with multiple epochs. After that, the final weights are passed as the initial weights for training a subsequent neural network on each evolving data thread. The system then operates with multiple threads of incremental MLP or autoencoders for anomaly detection.

3.1   Streaming Neural Networks

The first model we are introducing for data stream mining is called Streaming Multilayer Perceptrons (SMLP). Essentially, this algorithm involves an ensemble of multilayer perceptrons from multiple threads. Each multilayer perceptron uses the initial weights passed from a neural network trained on the first n data points in the corresponding data thread.

The model is trained with the following steps:

1. Initialize m threads for parallel processing of data streams.
2. The data point x_i goes to the thread $i \bmod m$.
3. For each thread, collect n data samples at the beginning of the stream. Then train an MLP with the n data points in the window for up to t epochs. Obtain final weights to pass as the initial weights for the subsequent incremental multilayer perceptron.
4. Train an incremental multilayer perceptron with the initial weights obtained from step 3 for each thread. Update the weights every time after one data instance passes. Thus, we train an MLP by feeding the data one by one in order with only one epoch.

After passing all the training data, SMLP is ready for classification. When a testing instance arrives, the m neural networks trained from m threads will vote for the output. If the labels of testing data arrive later, we can use them to update the model.

Time and Space Complexities: The key operation in SMLP is updating the weights each time a training data point passes. Training a neural network with backpropagation can be slow. However, with multi-threaded processing and online learning where we train neural networks with only one epoch in parallel, the speed of processing data improves dramatically. With this fast data processing, we can safely discard the data points which have been processed, so only the weights of the m neutral networks are being stored in memory.

In terms of accuracy, SMLP optimizes incremental neural networks since the initial weights are adjusted by training from the first n data instances. This minimizes the problem brought by the random initial weights. Therefore, the performance of SMLP is less random than that of incremental neural networks. Moreover, having ensembles of MLP from different segments of the data stream improves the performance of SMLP. As shown in the paper [23], ensembles of classifiers usually have better performances than a single classifier as long as individual classifiers are diverse enough. In SMLP, each MLP in the ensemble is trained on a unique subset of the data stream. This guarantees individual MLP in the ensemble are diverse enough to give a better performance than a single classifier.

3.2 Streaming Autoencoders

The second model we are introducing is called Streaming Autoencoders (SA), which is based on ensembles of autoencoders with similar settings as the first model. Since the autoencoders are only trained on normal instances in the dataset, this is an unsupervised learning model. This model has the advantage of dealing with the class imbalance problem, namely when the abnormal instances are rare or not available for training [21].

Architecturally, an autoencoder is a feedforward, non-recurrent neural network with an input layer, an output layer and one or more hidden layers connecting them. The output layer has the same number of nodes as the input layer. Autoencoders have the goal of reconstructing their own inputs X.

An autoencoder consists of two parts, the encoder and the decoder, which can be defined as two maps ϕ and ψ, such that the goal is to minimize the reconstruction errors of inputs X:

$$\phi : \mathcal{X} \to \mathcal{F} \tag{3}$$
$$\psi : \mathcal{F} \to \mathcal{X} \tag{4}$$
$$\arg\min_{\phi,\psi} \|X - (\Psi \circ \Phi)X\|^2 \tag{5}$$

In our study, we consider the case when there is only one hidden layer. An autoencoder with one hidden layer takes the input $\mathbf{x} \in \mathbb{R}^d$ and maps it onto $\mathbf{z} \in \mathbb{R}^p$ (p is the number of hidden units in the hidden layer) with a non-linearity activation function σ_1 such as the sigmoid:

$$\mathbf{z} = \sigma_1(\mathbf{Wx} + \mathbf{b}) \tag{6}$$

The latent representation $\mathbf{z}$ is then mapped back (with a decoder) to $\mathbf{x}'$ which is a reconstruction of $\mathbf{x}$ with the same shape as $\mathbf{x}$. The mapping happens through a similar transformation:

$$\mathbf{x}' = \sigma_2(\mathbf{W}'\mathbf{z} + \mathbf{b}')$$

The reconstruction error can be measured in many ways based on the distributional assumptions on the input. Autoencoders are trained to minimize reconstruction errors (such as squared errors as follows):

$$\mathcal{L}(\mathbf{x}, \mathbf{x}') = \|\mathbf{x} - \mathbf{x}'\|^2 = \|\mathbf{x} - \sigma_2(\mathbf{W}'(\sigma_1(\mathbf{W}\mathbf{x} + \mathbf{b})) + \mathbf{b}')\|^2$$

If the hidden layer of an autoencoder has fewer nodes than that of the input layer, the feature vector $\phi(\mathbf{x})$ can be regarded as a compressed representation of the input $\mathbf{x}$. If the hidden layers are larger than the input layer, an autoencoder can potentially become useless by learning the identity function. This can be avoided by using denoising autoencoders or autoencoders with dropout [22]. In our model, we use autoencoders with dropout since it gives the best results in our experiment setting.

In order to be used for anomaly detection, the autoencoders are trained on normal (positive) instances. Once trained, they have the ability to recognize abnormal instances in new data (those instances with a large reconstruction error). Therefore, the reconstruction error is used as the measurement to profile the degree of anomaly for autoencoders.

To decide the threshold that discriminates normal and abnormal instances, we need to pass all the training data from both classes. The threshold is then set empirically as the reconstruction error which gives the best separation of the two classes as discussed in paper [21].

3.3   Dealing with Concept Drift

When dealing with data streams, the underlying data distribution can change over time. This phenomenon is called concept drift. Concept drift can cause predictions to become less accurate over time [24]. Our study is restricted to one type of concept drift where the conditional distribution of the output changes, but the distribution of the input may or may not change.

When the conditional distribution changes in the output, there are two solutions to prevent deterioration of prediction accuracy [25]. Active solutions rely on triggering mechanisms which explicitly detect concept drift in the data generating process. On the contrary, passive solutions build models which are continuously updated. This can be achieved by retraining the model on the most recently observed samples [24] or enforcing an ensemble of classifiers [26].

The two models proposed in this paper are passive predictive models with adaptive learning. They use blind adaptation strategies [26] without any explicit detection of concept drift. As [26] explained, the blind approaches forget old concepts at a constant rate whether changes are happening or not. As time goes on, the newly arrived data tends to erase away the prior patterns.

In neural networks, learning is inevitably associated with forgetting. The newly arrived data updates the weights of neural networks, which makes the effects of old data in the stream becomes less and less important. Since streaming data arrives at a high speed, the weights in the models get updated rapidly and the model adjusts to the concept drift quickly.

Moreover, the use of threaded ensemble learning makes our models less sensitive to noise and more stable in adapting to concept drift. The ability to continuously learn from data streams while retaining previously learned knowledge is known as the stability–plasticity dilemma [27]. This dilemma happens because there is a trade-off between handling noise stably and learning new patterns. Since we split the data into multiple threads, the chance of noise affecting all threads is small compared to a single neural network. If a concept drift is truly happening, it will show up in multiple threads and be detected and adapted by the majority vote of the neural networks in the ensemble.

4   Experimental Setup

4.1   Dataset

Columns 2 to 4 of Table 1 summarize the five large datasets used in this study. SMTP and HTTP (from KDD Cup 99) are streaming data involving distinguishing between "bad" connections, called intrusions or attacks, from "good" normal connections. HTTP has sudden surges of anomalies in some streaming segments. SMTP possibly exhibits some distribution changes (concept drift) within the streaming sequence without having surges of anomalies [2].

To test our algorithms on a dataset where a distribution change has indeed occurred within a stream, we created the dataset, SMTP + HTTP, containing the SMTP data instances followed by the HTTP data instances as demonstrated in paper [2]. This dataset has a distribution change when the communication protocol is switched from SMTP to HTTP.

COVERTYPE is a UCI dataset commonly used in data stream research. It has some distribution changes since there are multiple different cover-types in the normal class [2]. We used the smallest class Cottonwood/Willow with 2747 instances as the anomaly class. The last dataset we used is the SHUTTLE dataset from UCI repository. This dataset has no distribution change.

In VFDT, all datasets are processed as streams with a chunk size of 1000. In SMLP and SA, the window size for the mini-batch learning is 100 and the model is updated incrementally (one by one). The order of the data instances is preserved in each data stream. We split the data stream into 80 % for training and 20 % for testing.

We are aware that our datasets are small for today's standards of data streams, which arrive very rapidly in a large volume. However, SMLP and SA are fully distributed streaming algorithms that have the potential to process big data streams. In fact, our models are implemented by the h2o deep learning package, which processes data point-by-point in each core processor.

4.2   Experimental Settings

The parameter settings for our two models are 8 threads with window sizes of 100 for initial weights. For streaming autoencoders, we use input dropout of 0.1 and Tanh function with 50 % dropout as the activation function.

Our methods are implemented in R with h2o deep learning package. The total memory allocated for the h2o cluster is 1.78 GB. We use VFDT from the RMOA package for the comparison. All experiments were conducted on a 2.7 GHz Intel Core i5 CPU with 8 GB RAM.

The area under the curve (AUC) is used as the evaluation metric. It is computed based on anomaly scores. We use the probability that a point is predicted normal as the anomaly score for SMLP and VFDT. Thus, a true anomaly instance generally has a low prediction probability to be a normal point, while a normal point has a high probability. For SA, we use the inverse of the reconstruction error as the anomaly score, because a test data point with a smaller reconstruction error is more likely to be a normal instance. This serves as a ranking measure for the tasks of anomaly detection.

Once the anomaly scores (probabilities) for all instances were obtained, the instances were ranked based on their anomaly scores. From this ranking and the true labels, we then computed the AUC to measure the performance of all anomaly detectors. AUC is the main indicator reported in this paper as in Table 1.

In all experiments, we conducted 10 independent runs of each algorithm on each dataset and then computed the average results. A t-test at 5 % level of significance was used to compare performance levels of the algorithms. Significant results of our models (results which are significantly better than that of VFDT) are highlighted in bold in Table 1.

5   Experimental Results

We report the experiments' results in this section. First, we compute the AUC of our models and compare them with the AUC of VFDT. VFDT and SMLP are both multiclass classifiers. They are given the advantage of using the actual positive and negative class labels for training, and this is done immediately after each new instance is scored. In contrast, Streaming Autoencoders use only "normal" data for training, by which we mean data from the positive class. We expect VFDT and Streaming Multilayer Perceptrons (SMLP) to produce better results than streaming autoencoders. Since we use VFDT as the baseline for comparison, our streaming data anomaly detectors will be deemed competitive if their performance can be shown to be comparable with the performance of VFDT.

Interestingly, Table 1 shows that SA actually gives higher AUC scores than VFDT and SMLP on three (i.e., SMTP, SMTP + HTTP, and SHUTTLE) out of five datasets tested. Moreover, when there is a strong concept shift as in SMTP + HTTP, SA performs significantly better than the other two approaches.

VFDT has the ability to adapt its tree structure because the nodes of the tree are developed from different segments of the data stream. When incrementally inducing a decision tree from a data stream, it uses the Hoeffding Bound to decide the number of data instances needed to split the attributes. However,the need to monitor and modify the tree causes VFDT's runtime to be four to six

Table 1. Average AUC scores for Streaming Multilayer Perceptrons (SMLP), Streaming Autoencoders (SA), and Very Fast Decision Trees (VFDT). Using VFDT as a reference, scores significantly higher than VFDT are printed in boldface.

Dataset	Data size	Dimensionality	Anomaly	AUC			Training time (seconds)		
				SMLP	SA	VFDT	SMLP	SA	VFDT
HTTP	623091	41	6.49 %	0.988	0.993	0.996	70.03	17.88	348.75
SMTP	96554	41	12.25 %	**0.996**	**0.997**	0.993	24.17	6.64	52.25
SMTP + HTTP	719645	41	7.26 %	0.630	**0.997**	0.857	31.11	29.47	394.24
COVERTYPE	581012	54	0.47 %	0.994	0.977	0.999	110.26	10.38	353.26
SHUTTLE	14500	10	5.97 %	0.984	**0.994**	0.990	4.89	2.75	7.02

times slower than SMLP and eight to ten times slower than SA, as shown in the last three columns of Table 1. More interestingly, we found that streaming autoencoders are always much faster than SMLP. This is because it is faster to update the autoencoders when the normal data instances are very similar to each other.

Unlike VFDT, SMLP and SA do not need to modify or extend the tree structure during the streaming process. The backpropagation in multiple threads is very efficient because it requires neither evaluation for dimensions nor selections for splitting points as in VFDT.

In our experiments, we found that the performance of our models is not too sensitive to the number of hidden units we choose in the hidden layers. Suppose the dimensionality (the number of attributes) of the datasets is n. Varying the number of hidden units from $2/3 * n$ to $2 * n$ give similar results. Parameter settings diminish as the ensemble size grows. This is due to the power of ensemble learning — while individual base learners may be weakened by non-optimal parameter settings for a problem at hand, the combination of these weak learners still produces reasonably good results.

6   Conclusions

The proposed two anomaly detection algorithms, Streaming Multilayer Perceptrons and Streaming Autoencoders, satisfy the key requirements for mining evolving data streams: (i) they pass data only once to learn without the need of storing the entire dataset, which implies they are capable of processing infinite data streams; (ii) they have the ability to deal with concept drift by learning continuously and adaptively. Moreover, SA has the ability to learn different anomalies even if they have not been observed in the training data.

Our empirical studies show that both models are robust in evolving data streams. In terms of detection accuracy, both algorithms are comparable to VFDT — a state-of-the-art algorithm. In terms of training time, both models outperforms VFDT. Moreover, the streaming autoencoder is very fast compared to SMLP or VFDT. It can be trained using only data instances from the positive class, which is an advantage when the abnormal instances are rare or not even available for training.

References

1. Gama, J., Medas, P., Rodrigues, P.: Learning decision trees from dynamic data streams. In: Proceedings of the 2005 ACM Symposium on Applied Computing. ACM (2005)
2. Tan, S.C., Ting, K.M., Liu, T.F.: Fast anomaly detection for streaming data. In: IJCAI Proceedings of the International Joint Conference on Artificial Intelligence, vol. 22, No. 1 (2011)
3. Gama, J., Rodrigues, P.P., Sebastio, R.: Evaluating algorithms that learn from data streams. In: Proceedings of the 2009 ACM Symposium on Applied Computing. ACM (2009)
4. Domingos, P., Hulten, G.: Mining high-speed data streams. In: Proceedings of the 6th ACM SIGKDD International Conference on Knowledge Discovery and Data Mining. ACM (2000)
5. Hulten, G., Spencer, L., Domingos, P.: Mining time-changing data streams. In: Proceedings of the 7th ACM SIGKDD International Conference on Knowledge Discovery and Data Mining. ACM (2001)
6. Vic, B., Lewis, T.: Outliers in Statistical Data, 3rd edn. Wiley, Hoboken (1994)
7. Abe, N., Zadrozny, B., Langford, J.: Outlier detection by active learning. In: Proceedings of the 12th ACM SIGKDD International Conference on Knowledge Discovery and Data Mining. ACM (2006)
8. He, Z., Xiaofei, X., Deng, S.: Discovering cluster-based local outliers. Pattern Recogn. Lett. **24**(9), 1641–1650 (2003)
9. Bay, S.D., Schwabacher, M.: Mining distance-based outliers in near linear time with randomization and a simple pruning rule. In: Proceedings of the Ninth ACM SIGKDD International Conference on Knowledge Discovery and Data Mining. ACM (2003)
10. Heller, K., et al.: One class support vector machines for detecting anomalous windows registry accesses. In: Workshop on Data Mining for Computer Security (DMSEC), Melbourne, FL, 19 November 2003
11. Liu, F.T., Ting, K.M., Zhou, Z.-H.: Isolation forest. In: 9th IEEE International Conference on Data Mining, ICDM2008. IEEE (2008)
12. Hahsler, M., Bolanos, M., Forrest, J.: Introduction to stream: an extensible framework for data stream clustering research with R
13. Last, M.: Online classification of nonstationary data streams. Intell. Data Anal. **6**, 129–147 (2002). ISSN 1088-467X
14. Aggarwal, CC., Han, J., Wang, J., Yu, P.S.: On demand classification of data streams. In: Proceedings of the 10th ACM SIGKDD International Conference on Knowledge Discovery and Data Mining, KDD 2004, pp. 503–508. ACM, New York, NY, USA (2004)
15. Aggarwal, C.C., Han, J., Wang, J., Yu, P.S.: A framework for clustering evolving data streams. In: Proceedings of the International Conference on Very Large Data Bases (VLDB 2003), pp. 81–92 (2003)
16. Hoeglinger, S., Pears, R., Koh, Y.S.: CBDT: A concept based approach to data stream mining. In: Theeramunkong, T., Kijsirikul, B., Cercone, N., Ho, T.-B. (eds.) PAKDD 2009. LNCS, vol. 5476, pp. 1006–1012. Springer, Heidelberg (2009)
17. Polikar, R., et al.: Learn++: An incremental learning algorithm for supervised neural networks. IEEE Trans. Syst. Man Cybern. Part C: Appl. Rev. **31**(4), 497–508 (2001)

18. Carpenter, G.A., et al.: Fuzzy ARTMAP: A neural network architecture for incremental supervised learning of analog multidimensional maps. IEEE Trans. Neural Netw. **3**(5), 698–713 (1992)
19. Shen, F., Ogura, T., Hasegawa, O.: An enhanced self-organizing incremental neural network for online unsupervised learning. Neural Netw. **20**(8), 893–903 (2007)
20. Shen, F., Hasegawa, O.: Self-organizing incremental neural network and its application. In: Diamantaras, K., Duch, W., Iliadis, L.S. (eds.) ICANN 2010, Part III. LNCS, vol. 6354, pp. 535–540. Springer, Heidelberg (2010)
21. Japkowicz, N., Myers, C., Gluck, M.: A novelty detection approach to classification. In: IJCAI (1995)
22. Bengio, Y.: Learning deep architectures for AI. Found. Trends Mach. Learn. **2**(1), 1–127 (2009)
23. Wang, H., et al.: Mining concept-drifting data streams using ensemble classifiers. In: Proceedings of the 9th ACM SIGKDD International Conference on Knowledge Discovery and Data Mining. ACM (2003)
24. Widmer, G., Kubat, M.: Learning in the presence of concept drift and hidden contexts. Mach. Learn. **23**(1), 69–101 (1996)
25. Gama, J., et al.: A survey on concept drift adaptation. ACM Comput. Surv. (CSUR) **46**(4), 44 (2014)
26. Ryan, E., Polikar, R.: Incremental learning of concept drift in nonstationary environments. IEEE Trans. Neural Netw. **22**(10), 1517–1531 (2011)
27. Carpenter, G.A., Grossberg, S., John, H.R.: ARTMAP: Supervised real-time learning and classification of nonstationary data by a self-organizing neural network. Neural Netw. **4**(5), 565–588 (1991)

A Density-Grid Based Clustering Algorithm on Data Stream Using Resilient Distributed Datasets

Yuan Zhang[✉] and Jiongmin Zhang

Department of Computer Science and Technology,
East China Normal University, 200241 Shanghai, China
ecnu_zy@hotmail.com

Abstract. To find the clusters of arbitrary shapes rapidly from the sustainable growth of data stream, this paper proposes GDRDD-Stream algorithm. To capture the evolving characteristics of the data stream, this paper defines the effective time for the data points and design the eliminating strategy based on the effective time to remove the historical data. Secondly, we design the partitioning method based on resilient distributed datasets to balance the computing load between different nodes. Finally, we improve the traditional DBSCAN algorithm in order to compute in parallel between different partitions. The experimental results show that the proposed algorithm can cluster data stream distributed in arbitrary shape rapidly, capture the evolving behaviors of data stream, and its performance and quality are better than the CluStream algorithm.

Keywords: Data stream · Clustering · RDD · Spark · DBSCAN

1 Introduction

With the rapid development of Internet applications, an enormous amount of data stream arrived in real time, like stock data, network intrusion monitoring data and etc. Different with the static data, data stream [1] is massive, diverse, continuous, and rapid [2]. So the clustering algorithms on static data are no longer applicable to data stream. There have been some algorithms [3] which improved the traditional clustering algorithms in view of the characteristics of data stream: the clustering algorithms proposed in the articles [5, 6] are based on k-means which cluster data stream by single scanning. CluStream [7] is a framework, in which clustering process is divided into the online stage and the offline stage. And it could capture the evolution. But all of these algorithms are based on k-means, so which can only mine spherical clusters. D-Stream [8] applies the two-stage framework of the CluStream so as to capture the evolving process and is a density-based clustering algorithm to find the clusters of arbitrary shape. But its efficiency will decreased significantly for data stream of high dimension or rapid growth. So it is an urgent need for efficient clustering algorithm. Distributed cluster provides a revolution for large-scale data stream clustering. However, the clustering algorithms based on MapReduce [4] store the intermediate results on disks, and I/O operation is frequent with a certain delay. In order to improve the capability of

© Springer International Publishing Switzerland 2016
R. Khoury and C. Drummond (Eds.): Canadian AI 2016, LNAI 9673, pp. 316–322, 2016.
DOI: 10.1007/978-3-319-34111-8_38

parallel computing and the throughput of the cluster, Matei et al. proposed Resilient Distributed Datasets (RDD) [9]. RDD is a shared memory model without the need for frequent access to the disk so as to improve the computing performance of the cluster significantly.

This paper proposes a density-grid based clustering algorithm (GDRDD-Stream), which is based on the DBSCAN [10] algorithm on the Spark platform.

2   Design and Implementation of GDRDD-Stream

GDRDD-Stream algorithm achieves parallelization by storing data in resilient distributed datasets. Firstly, the whole data space is divided into the grids with the same size that is not less than 2*ε using the traditional spatial partitioning algorithm. Then these arrived data points are matched to the corresponding grids, and the elimination algorithm based on the effective time is implemented to filter out the grids and the data points. And then the RDD partitioning algorithm based on the number of the data points is implemented to partition. Finally, this algorithm improves the traditional DBSCAN algorithm to achieve parallelization, and adopts the basic idea of "parallel DBSCAN in partitions – DBSCAN in border points – merging results" to generate the clustering results. The flow chart of GDRDD-Stream algorithm is shown in Fig. 1.

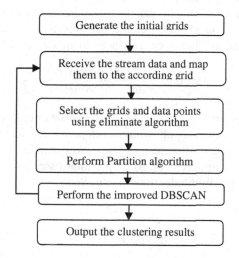

Fig. 1. The flow chart of GDRDD-Stream algorithm

2.1   Eliminate Algorithm Based on the Effective Time

The timeliness of the data affects the accuracy of clustering result. If clustering only the recent data stream, it cannot reflect the evolution of the data stream. Therefore, the concept of the effective time of data point is proposed in GDRDD-Stream algorithm. Data objects in the effective time will be clustered, weakening the impact of historical data and reflecting the evolution of data stream.

Definition 4 **Effective Time of the data point ($T_{effective}$):** Refers to a period of time that the data point impacts on the clustering results, that is, the longest time that the data points in memory can survive.

In order to further optimize the algorithm, GDRDD-Stream designs the elimination algorithm based on the effective time, which can reduce the amount of computation, and timely release of memory. The elimination rules are as follows:

- Record the update time of grid. If the update time of the grid over the effective time, delete its historical data and this grid does not participate in the cluster calculation.
- Detect every data point in the grid that if they exceed the valid time.

2.2 RDD Partitioning Algorithm

The grids selected by Elimination algorithm easily result in the uneven distribution of the data points in different partitions. GDRDD-Stream algorithm adopts the partitioning method based on the number of data points in grid. It balances the computing load between different nodes by merging the adjacent grids to ensure that the data points in each grid are relatively average.

- The minimum number that could be processed per partition can be calculated by the formula (1). The parameter processedPts is the points selected by Elimination algorithm. The parameter defaultParallelism, parallel degree of Spark, is set in Spark. Because the distribution of points in the partition is not even, use defaultParallelism*2 to make the number of the partition floating around the average value.

$$MinNum = \frac{count(processedPts)}{defaultParallelism \times 2} \qquad (1)$$

- Merge the adjacent grids according to the minimum number of partition. Spark provides Partitioner interface, allowing developers to customize partitioning rule. Reallocate the grid IDs for the merged grids and generate the objects that inherits from the Partitioner interface. The RDD partition based on grid IDs is generated using the MapPartitionWithIndex interface in Spark.

2.3 The Improved DBSCAN Algorithm

GDRDD-Stream, based on DBSCAN, can discover clusters of arbitrary shapes and is of high clustering quality. Because the traditional DBSCAN algorithm needs to calculate the distances between all of data points, its time complexity is high. But the position of point in the data space is fixed so only the data points in the ε-neighbor need to be calculated. We could partition the data by the positions of them in dimension space. The data points in the same partition are of the close distance that most of the points in the same partition only find the ε-neighbor in this partition. And the border points(ε from the partition boundary) need to find the ε-neighbor from other partitions. Because the number of the border points is small and the clustering calculation in partitions could be executed in parallel, the efficiency is significantly improved compared with the original algorithm. Based on the partitioning method, we improve the DBSCAN as following:

(1) DBSCAN algorithm is executed to cluster the data in every partition in parallel.
(2) Then, record the border points(ε from the partition boundary) and cluster the border points between different partitions.
(3) Finally, merge the clustering results.

It is important to note that the cluster identifier may be duplicate between different partitions. To avoid this situation, the identifier of the first point is allocated as the identifier of the new cluster because of the point identifier with the global uniqueness. For any two points in the border points, if they are not in the same grid and do not belong to the same cluster but their distance is less thanε, create a tuple (clusterId1, clusterId2) using their cluster identifiers and then merge them into the same cluster; if their distance is less than$\varepsilon$and one of them does not belong to any cluster, allocate this point with the cluster identifier of another point. And then merge all of the tuples and reassign the cluster identifiers for all data points.

If the total number of data points is n, the complexity of the traditional DBSCAN is $O(n^2)$. This algorithm improves the efficiency of clustering through parallel computing. If the partition number is m, the complexity is $O((n/m)^2)$, the efficiency may improve m^2 times nearly.

3   Experimental Results

We evaluate the evolution, quality and efficiency of GDRDD-Stream and compare it with CluStream. Our experiments are conducted on four PCs with 2.2 GHz CPU and 1 G memory. We evaluate GDRDD-Stream on CentOS 6.5 with Spark 1.3.1 and Scala 2.10.4 and use $T_{effective} = 40$ s, eps = 30, Mpt = 8. We use two testing datasets. The first one is DS1 with 2-dimension which are synthetic. Its size is 80000 (the original distribution of the dataset is shown in Fig. 2). The second one is a real dataset used by KDD-CUP-99 [11] which contains 41 attributes and 5 categories.

3.1   Evolution Testing

We simulate the data stream based on DS1 at the speed of 1000 points per second to evaluate the evolution of GDRDD-Stream. We check the clustering results at four different times, including t1 = 40 s and t2 = 60 s. The clustering results are shown from Figs. 3 to 4. Since this algorithm is based on the effective time of data point and parallel in-memory computing, it can mine dynamically the evolution of the data stream and discover clusters of arbitrary shapes. The grids and data points are detected and eliminated periodically to release the memory resources in time.

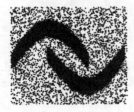

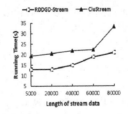

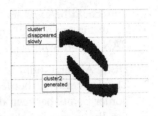

Fig. 2. Original distribution of the dataset DS1

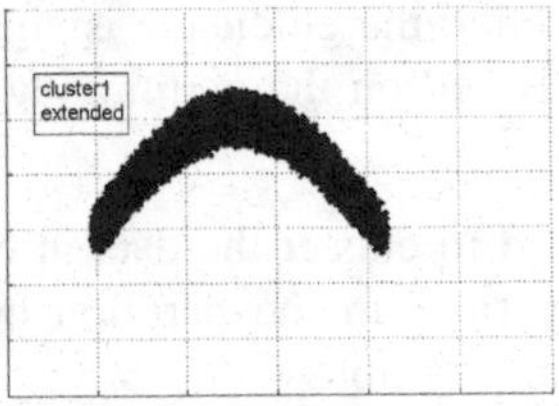

Fig. 3. Clustering result at t1 = 40 s

Fig. 4. Clustering result at t2 = 60 s

3.2 Correct Rate Comparison

The cluster purity is used to measure the quality of the clustering algorithm.

$$purity = \frac{1}{k}\sum_{i=1}^{k} \frac{|C_i^r|}{|C_i|} \tag{2}$$

In the formula (2), k is the number of clusters, $|C_i|$ is the number of the data points in the ith cluster, $|C_i^r|$ is the number of the data points that are clustered rightly in the ith cluster. The cluster purity of GDRDD-Stream and CluStream are compared on KDD CUP-99, shown in Fig. 5. The cluster purity of GDRDD-Stream is apparently higher than CluStream's at the time of 40 s, 60 s, and 80 s. It is caused by that CluStream can only discover the spherical clusters and need to know the value of k in advance while GDRDD-Stream can discover non-convex clusters. The clusters of arbitrary shape are generated and the number of clusters is unknown with flowing of data stream. GDRDD-Stream can allocate the network intrusion detection type to the right cluster while CluStream will allocate the different intrusion types into the same cluster. So the cl-u-s-ter p-u-ri-ty of

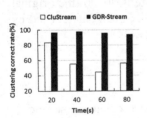

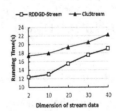

Fig. 5. Correct rate comparison on KDD CUP-99 (1000points/s)

Fig. 6. Running time with changes of length of data stream

Fig. 7. Running time with changes of dimension of data stream

GDRDD-Stream is higher than CluStream.

3.3 Efficiency Comparison

The efficiency comparison of GDRDD-Stream and CluStream with the growth of data stream length on KDD CUP-99 is shown in Fig. 6. The processing time of CluStream increases apparently with the extension of data stream length while the processing time of GDRDD-Stream increases slowly. The efficiency comparison of GDRDD-Stream and CluStream is shown in Fig. 7 with dimension from 2 to 40. The efficiency of GDRDD-Stream is higher than CluStream's. It is caused by that CluStream, based on complex distance calculation, will increase the calculation load with the growth of length and dimension of data stream. While GDRDD-Stream improves the efficiency through parallel computing between the RDD partitions based on grids.

4   Conclusion

In this paper, the GDRDD-Stream algorithm is proposed that are based on DBSCAN algorithm and resilient distributed datasets on Spark in view of the drawbacks of the current clustering algorithms of data stream. Experimental results show that GDRDD-Stream could capture the evolution of data stream and discover the clusters of arbitrary shapes in real time. And it has high efficiency without affecting the clustering results because of parallel computing using RDD. The algorithm makes high speed data stream clustering feasible without degrading the clustering quality.

References

1. Amineh, A., Teh, Y.W., Mahmoud, R., et al.: A study of density-grid based clustering algorithms on data streams. In: Proceeding of Eighth International Conference on Fuzzy Systems and Knowledge Discovery (FSKD), pp. 1652–1656 (2011)
2. Jonathan, A.S., Elaine, R.F., Rodrigo, C.: Data stream clustering: a survey. ACM Comput. Surv. **46**(1), 13:1–13:31 (2013)
3. Shifei, D., Fulin, W., Jun, Q., Hongjie, J., Fengxiang, J.: Research on data stream clustering algorithms. Artif. Intell. Rev. **43**, 593–600 (2015)
4. Jeffrey, D., Sanjay, G.: MapReduce: simplified data processing on large clusters. Commun. ACM **51**(1), 107–113 (2008)
5. Guha, S., Mishra, N., Motwani, R.: Clustering data streams. In: Proceeding(s) of 41st Annual Symposium on Foundations of Computer Science, pp. 359–366 (2000)
6. Ocallaghan, L., Meyerson, A., Motwani, R., Mishar, N., Gha, S.: Streaming data algorithms for high-quality clustering. In: Proceeding(s) of 18th International Conference, Data Engineering, pp. 685–704 (2002)
7. Aggarwal, C., Han, J., Wang, J., Yu, P.: A framework for clustering evolving data streams. In: Proceedings of the 29th International Conference on Very Large Data Bases, pp. 81–92 (2003)
8. Chen, Y., Tu, L.: Density-based clustering for real-time data stream. In: Proceeding of the ACM KDD 2007 Conference, pp. 133–142 (2002)
9. Matei, Z., Mosharaf, C., Tathagata, D., et al.: Resilient distributed datasets: a fault-tolerant abstraction for in-memory cluster computing. In: Proceedings of the 9th USENIX Conference on Networked Systems Design and Implementation, p. 2 (2012)

10. Martin, E., et al.: A density-based algorithm for discovering clusters in large spatial databases with noise. In: Proceedings of 2nd International Conference Knowledge Discovery and Data Mining (KDD 1996), pp. 226–231 (1996)
11. KDD dataset. http://kdd.ics.uci.edu/databases/kddcup99/kddcup99.html

Distributed Gaussian Mixture Model Summarization Using the MapReduce Framework

Arina Esmaeilpour[1,2]($\boxtimes$), Elnaz Bigdeli[2,3], Fatemeh Cheraghchi[2,3], Bijan Raahemi[2], and Behrouz H. Far[1]

[1] Department of Electrical and Computer Engineering, University of Calgary, 2500 University Dr. NW, Calgary, AB, Canada
esmaeia@ucalgary.ca
[2] Knowledge Discovery and Data Mining Lab, Telfer School of Management, University of Ottawa, 55 Laurier Ave. E, Ottawa, ON, Canada
[3] Computer Science Department, University of Ottawa, 600 King Edward, Ottawa, ON, Canada

Abstract. With an accelerating rate of data generation, sophisticated techniques are essential to meet scalability requirements. One of the promising avenues for handling large datasets is distributed storage and processing. Further, data summarization is a useful concept for managing large datasets, wherein a subset of the data can be used to provide an approximate yet useful representation. Consolidation of these tools can allow a distributed implementation of data summarization. In this paper, we achieve this by proposing and implementing a distributed Gaussian Mixture Model Summarization using the MapReduce framework (MR-SGMM). In MR-SGMM, we partition input data, cluster the data within each partition with a density-based clustering algorithm called DBSCAN, and for all clusters we discover SGMM core points and their features. We test the implementation with synthetic and real datasets to demonstrate its validity and efficiency. This paves the way for a scalable implementation of Summarization using Gaussian Mixture Model (SGMM).

Keywords: Distributed density-based clustering · Distributed cluster summarization · Gaussian mixture model · MapReduce

1 Introduction

Nowadays massive amounts of data are being generated by activities such as web surfing, using social networks and e-mail services, shopping and online bank services. In addition to increase in amounts of generated data, also the rate of growth is drastically high. The management and use of the accumulated large volume of data requires not only large storage devices but also substantial processing, analysis and retrieval methods. One approach to mitigate these problems is to summarize the data, reducing storage, processing and retrieval needs.

© Springer International Publishing Switzerland 2016
R. Khoury and C. Drummond (Eds.): Canadian AI 2016, LNAI 9673, pp. 323–335, 2016.
DOI: 10.1007/978-3-319-34111-8_39

Data summarization techniques are intended to produce summaries which are compact yet representative of the entire dataset [1]. Clustering can be used as a summarization technique. For large volumes of data, preservation of all cluster members is not feasible, so each produced cluster can be summarized as well. Cluster summarization aims to represent a cluster with fewer instances, while preserving the original shape and distribution of the cluster [2–5].

Summarization approaches are costly, requiring substantial processing time and memory. These computations need scalable methods to be able to process large amounts of data. In this case, distributing data and computation across a cluster of machines can be beneficial. The MapReduce framework provides a means to implement a model for automatically distributing and parallelizing processes in a cluster of commodity machines.

In this paper, this goal is met by implementing a density-based clustering and a cluster summarization technique using the MapReduce framework. Specifically, the work is focused on a MapReduce-based implementation of clustering with a Density-Based Spatial Clustering of Applications with Noise (DBSCAN) algorithm [6], and Summarization based on Gaussian Mixture Model (SGMM) [5]. Both of these techniques are advantageous in discovering and summarizing arbitrary-shape clusters [5,6].

The rest of this paper is organized as follows. Section 2 shows a short introduction of basic concepts and existing methods. In Sect. 3 the structure and phases of MR-SGMM are presented, and experimental results on synthetic and real datasets are shown in Sect. 4.

2   Related Works

Several algorithms have been developed to implement scalable density-based clustering methods using MapReduce [7–12]. In particular, distributed algorithms for DBSCAN using the MapReduce framework have been realized in [10–12]. All of these algorithms include four general steps of partitioning, local clustering, mapping and relabeling. These developments differ in their partitioning technique, and may not necessarily implement all steps in the MapReduce framework. In [12], authors propose an algorithm for a fully-distributed DBSCAN clustering technique using the MapReduce framework.

MapReduce: MapReduce is a programming framework which includes map and reduce functions. The map function receives data in the form of key/value pairs, and derives a set of intermediate key/value pairs. The reduce function groups all the intermediate values having the same intermediate key. A MapReduce program is scalable over hundreds or thousands of machines in a cluster, without need for involving the user in parallelization details [13,14].

DBSCAN: DBSCAN is a well-known density-based clustering algorithm that finds and connects dense regions to form clusters. Points in sparse regions between clusters are referred to as outliers or noise. DBSCAN is able to find arbitrary-shape clusters and, it handles noise and outlier very well [6].

SGMM: SGMM is an approach to summarize arbitrary-shape clusters which finds dense regions in a cluster, represents each dense region with a point called an SGMM core point, and delineates features related to the core points. The SGMM method is used to create a Gaussian Mixture Model (GMM), which is a combination of a set of normal distributions over all SGMM core points of the cluster, and can be used for a compact representation of the original data. SGMM preserves the original shape and the distribution of clusters, and has the ability to regenerate the cluster using core points and their features [5].

3   Distributed Gaussian Mixture Model Summarization Using the MapReduce Framework

In this section, the structure and phases of MR-SGMM are presented. The main purpose of the proposed method is to summarize a dataset with DBSCAN clustering algorithm, and then summarize each discovered cluster using the SGMM approach in a distributed manner.

The phases of the proposed method are shown in Fig. 1. First, the data set is partitioned and distributed to different machines. Then, process is performed on each machine locally. Finally, the results of all machines are merged. Detailed explanations of each phase are presented in Sect. 3.1 through Sect. 3.3, respectively.

Fig. 1. General phases of MR-SGMM

3.1   Partitioning

By default, the generic Hadoop framework partitions (splits) input data into 64 MB blocks. In the approach presented in this paper, we cannot rely on the default partitioning strategy of Hadoop, since the spatial position of each point in the space is important. If members of a cluster were assigned to different partitions, the clustering algorithm performed locally in each partition would not be able to see all members of the cluster. This means that the points would be clustered incorrectly.

To overcome this problem, the partitioning algorithm should consider the position of each point in the data space, and split the input data in such a way that each partition contains points which are spatially close. In this way, one cluster is either contained in one partition or spread across neighboring partitions.

The partitioning strategy in this paper is chosen to be simple and fast, creating partitions by dividing the interval of each dimension (feature) into equal portions, where the minimum and maximum values of each feature is used to determine its interval. This strategy assumes that the size of each partition could fit in the available memory of a single machine. The partitioning algorithm can be improved at any time, as it is an independent step in the entire process. Before partitioning, all features of the data are normalized between 0 and 1. Also the input data is indexed so that each point has a unique identification number.

The partitioning phase is implemented in the MapReduce framework. Two key elements of this phase are the map and reduce functions; the map function performs the following steps:

1. **Create a grid for the data space:** The interval of each dimension (feature) is divided into an arbitrary number of equal portions, with an overlap of width 2ϵ between the portions, where ϵ is the radius used by the DBSCAN algorithm. This overlap is needed because of the possibility that the data points of a cluster are spread across different partitions, so there should be a joint or boundary region between adjacent partitions. Selecting a 2ϵ-wide boundary region ensures sufficient information for the merge phase [11]. Figure 2 demonstrates an example, with each dimension divided into two portions. The shaded areas indicate the boundary regions.

2. **Find the grid number for each point:** The map function takes each point and determines its corresponding cell within the grid using its feature values. A border point will be one that is in more than one cell of the grid. Referring to Fig. 2, consider a point P that is in two different cells. The algorithm creates a set for each dimension to record the grid number(s). This set is referred to as the grid number set for the dimension. Referring again to Fig. 2, for point P the grid number set for $dimension_1$ (x) is $GN_{d_1} = \{1, 2\}$, and for $dimension_2$ (y) is $GN_{d_2} = \{2\}$. After creating the grid number sets for each dimension, a set will be created for the point P. This set is a Cartesian product of all dimensions' grid number sets, and is referred to as the grid number set for the point P. In Fig. 2, the grid number set for point P will be $GN_p = GN_{d_1} \times GN_{d_2} = \{(1, 2), (2, 2)\}$. Each pair shows the grid numbers in the two dimensions of the example. For a pair $(1, 2)$, the first number shows the grid number in dimension one, and the second number shows the grid number in dimension two.

3. **Determine border or non-border points:** In this step, the map function examines the cardinality of the grid number set for each point, to decide whether a point is a border point. If the cardinality of the grid number set for a point is greater than one, the point is deemed a border point, and if the cardinality is equal to one, the point is not a border point.

The reduce function receives a list of all points with the same grid number, collects them as a partition, and assigns a number to each partition. The output of the reduce function is a partition with its constitutive points.

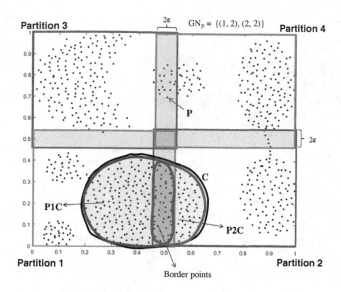

Fig. 2. A cluster scattered in two partitions

3.2 Local Processing

This phase is implemented in the MapReduce framework. Figure 3 shows the local processing phase.

Here, the partitioned data from the previous phase is given to the machines in the cluster for further processing. Local processing refers to all processes that are performed on a partition of the input data, which includes performing DBSCAN and SGMM locally. SGMM core points and their features are extracted for each discovered cluster. Finally, the clusters that should be merged in the subsequent merge phase are discovered.

The map function performs these steps:

1. **DBSCAN clustering:** DBSCAN [6] starts with an arbitrary point p in the dataset, and discovers all points that are density-reachable from p, using Eps and MinPts. For each cluster, all core points found by the DBSCAN algorithm are labeled, as they will be later required for the merge phase. For the current implementation, the DBSCAN clustering algorithm in the WEKA [15] is employed in the source code.
2. **Find SGMM core points:** In this step, the set of neighbors is found for all points in a cluster, based on a radius. The points are sorted based on the number of neighbors. The first point in the sorted list is deemed to be an SGMM core point; this core point and its neighbors are removed from the sorted list. This approach is continued on the remaining points, until all the points have been removed from the sorted list. In this way, all core points representing dense regions are discovered.

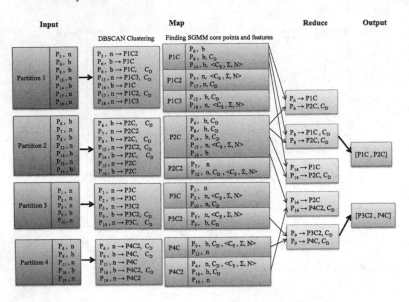

Fig. 3. Example local processing phase in the MapReduce framework

3. **Extract SGMM core point features:** Since only SGMM core points will be preserved to represent the cluster, a set of features will be required to summarize the characteristics of their neighbors. For each SGMM core point found in the previous step, Σ_i (the covariance calculated using the SGMM core point and its neighbors) and n (the number of neighbors of SGMM core point) are extracted.

In summary, the map function in this phase receives data points in a partition as input and sends only border points (points in boundary regions) to the reduce function. The reason for sending border points to the reduce function is that if the data points of a cluster are spread across different partitions, the border points can be used to help collect all the data points into one cluster, for example see Fig. 2.

The reduce function discovers the clusters that should be merged in the next phase (merge phase), based on a received list of border points. For example, in Fig. 2 border points belong to both cluster P1C and cluster P2C. The reduce function for each border point determines if the point is a DBSCAN core point in all its clusters. If so, these clusters should be merged to form a single cluster [10]. Also in Fig. 2, if there is at least one border point which is a DBSCAN core point in both P1C and P2C, the clusters should be merged. Finally, the reduce function creates a list of the clusters and sorts the list based on the partition number; the first item in the list will be selected as the label for the scattered cluster. Again referring to Fig. 2, the sorted list is [P1C, P2C] and P1C will be the selected label for cluster C. Clusters P1C and P2C cannot as yet be deemed final, since they still need to be merged. The resulting cluster after merging P1C and P2C will be a final cluster.

At the end of this phase, all points are clustered, for each cluster which is still not a final cluster, all SGMM core points and their features have been found. Also, all clusters that should be merged and the new label for these clusters have been determined.

3.3   Merge

This phase uses an algorithm implemented in the MapReduce framework. First, all clusters that should be merged to form the final clusters are merged, and then a GMM is created over a cluster using SGMM core points and features. The map function receives all points that are clustered in their partition, along with a list containing all clusters to be merged. It changes the label of the clusters to the new label, and creates the final clusters. All data points in each final cluster are then sent to the reduce function. The reduce function receives all data points which belong to a final cluster, and performs the following tasks:

1. **Obtain the size of cluster:** The reduce function counts the number of points in each final cluster, and calculates CS, the size of the final cluster.
2. **Calculate SGMM core point weights:** For each SGMM core point in the final cluster, the number of neighbors, n of the SGMM core point, has been saved during the local processing phase. By knowing n and CS from the previous step, the weight of each SGMM core point in the final cluster ($w_i = n/CS$) is calculated.
3. **Calculate the GMM for the clusters:** In this step, a GMM is calculated for each final cluster. A GMM is created over a cluster using all SGMM core points found for the cluster. Based on Eq. 2, each component of the GMM is calculated using an SGMM core point μ_i (the center of the component), the covariance of the SGMM core point Σ_i (representing the distribution information around the SGMM core point), and the weight of the SGMM core point w_i (which shows the relative contribution of the core point and its neighbors).

$$g(x) = \sum_{i=1}^{n} w_i N_{\mu_i \Sigma_i}(x),$$
(1)

where

$$N_{\mu_i \Sigma_i}(x) = \frac{1}{\sqrt{(2\pi)^d |\Sigma_i|}} e^{-\frac{1}{2}(x-\mu_i)\Sigma_i^{-1}(x-\mu_i)^T}$$
(2)

4   Experimental Results

In this section, results are reported for experiments using synthetic and real datasets to demonstrate and assess the performance of MR-SGMM. All datasets were analyzed and regenerated using MR-SGMM and single-node SGMM, enabling quantitative and qualitative comparison between these approaches.

The installed Hadoop cluster consisted of four nodes. Each node was a virtual machine with four Intel Xeon© 2.4 GHz CPUs, each with 8 GB of RAM, running on the Ubuntu Linux 12.04 operating system. The Hadoop version was 1.2.1, running on Java 1.7.0. The system used the default configuration settings.

Datasets: Synthetic and real datasets were used for the experiments described in this section. These included three synthetic datasets and two real geographic datasets. The three synthetic datasets each contained points in two dimensions, referred to in this chapter as SDS1, SDS2, and SDS3, having 8,000, 10,000 and 8,000 points, respectively. These datasets had been originally generated and used for clustering assessment in [16]. The real datasets were from the United States Census Bureaus MAF/TIGER database, containing features such as roads, railroads and rivers, as well as legal and statistical geographic areas [17]. The features that were selected to use in the experiments were latitude and longitude, but re-scaled to numbers between 0 and 1. These datasets are referred to here as RDS1 and RDS2, with 10,786 and 72,539 entries, respectively.

4.1 Data Regeneration

SGMM allows tools for generating random normal deviates for data regeneration. SGMM core points and their related variances can be used as input to Gaussian random number generators. For this step of the experiment, random normal deviates were produced based on the Box-Muller method [18]. Figure 4 shows results for MR-SGMM and single-node SGMM on the SDS1 datasets. Clustering results for original data, SGMM core points and the regenerated dataset are shown for SDS1 dataset. Visual comparison of the results for MR-SGMM and single-node SGMM shows encouraging resemblance.

4.2 Figures of Merit

A common approach for examining data summarization results is to compare calculated clustering analysis measures for the original (clustered) and regenerated data. In the present research, MR-SGMM and single-node SGMM were applied to the input datasets to cluster the datasets using DBSCAN, summarize the datasets based on the SGMM core points and their variances, and finally to regenerate the datasets. Clustering analysis measures were calculated for the original (clustered) datasets and for the regenerated data, using both methods of single-node and MR-SGMM. Two well-known clustering analysis measures, Dunn's and Davies-Bouldin (DB) indexes, were evaluated and compared for the single-node and MR-SGMM methods.

The Dunn's index [19] is a validity index which identifies compact and well-separated clusters for a specific number of clusters; it is given by:

$$D_c = \frac{\min_{i=1}^{c}\{\min_{j=i+1}^{c}\{dis(c_i, c_j)\}\}}{\max_{k=1}^{c} diam(c_k)}, \qquad (3)$$

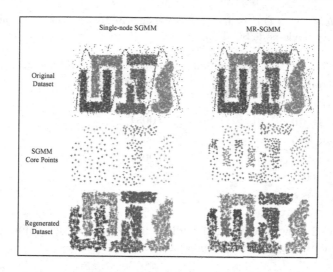

Fig. 4. Dataset SDS1: depictions of the original (clustered) dataset, SGMM core points and the regenerated dataset for the single-node SGMM and MR-SGMM.

where c is the number of clusters and $dis(c_i, c_j)$ is the dissimilarity function between two clusters c_i and c_j. Large values of this index indicate that clusters are compact and well-separated [19].

For experiments presented here, the data is summarized and regenerated using the core points of GMMs and the Dunn's indexes is calculated in both distributed and not distributed approaches.

Finally, the regeneration ability of MR-SGMM is examined using the difference between Dunn's indexes of regenerated data with MR-SGMM and original dataset compared to difference between Dunn's indexes of regenerated with single-node SGMM and original dataset. Difference in Dunn's indexes near zero indicates that how the MR-SGMM method regenerates the data which follows the shape and distribution of the original data similar to single-node SGMM. Table 1 shows the results for the five datasets discussed in this section (each value in the tables below is the average of ten independent runs).

A second measure considered for the present experiments was the DB index [20], which is the mean value of the ratio of within-cluster scatter to between-cluster separations. The DB index is defined as:

$$DB = \frac{1}{c} \sum_{i=1}^{c} \max_{i \neq j} \left(\frac{\delta_i + \delta_j}{\Delta_{ij}} \right), \tag{4}$$

where c is the number of clusters, δ_i is the average distance of all members of the cluster to their cluster center, and Δ_{ij} is the distance between centers of cluster i and cluster j. Small values of the index indicate compact and well-separated clusters [20].

Similar to the procedure for Dunn's index, the DB index difference was calculated for regenerated data versus original data for MR-SGMM and single-node

Table 1. Differences of Dunn's indexes for regenerated vs. original data for both the MR-SGMM and single-node SGMM approaches

Dataset	MR-SGMM regeneration vs. original data	Single-node SGMM regeneration vs. original data
SDS1	0.0073	0.0037
SDS2	0.0070	0.0054
SDS3	0.0029	0.0037
RDS1	0.0022	0.0032
RDS2	0.0021	0.0142

SGMM. Table 2 shows the results for all datasets (again, each value in the table represents the average of ten independent runs).

Table 2. Differences of DB indexes for regenerated vs. original data for both the MR-SGMM and single-node SGMM approaches

Dataset	MR-SGMM regeneration vs. original data	Single-node SGMM regeneration vs. original data
SDS1	0.0413	0.0307
SDS2	0.0681	0.1288
SDS3	0.0684	0.0471
RDS1	0.0499	0.0169
RDS2	0.0656	0.0532

Similar to the results for Dunn's index, the differences of DB indexes between the regenerated and original data with MR-SGMM were close to zero.

Results using Dunn's and DB indexes show that the performance of the MR-SGMM method was reliable, and managed to properly summarize and regenerate data. Its performance during testing was comparable to that of the single-node SGMM, and in some cases better results has been demonstrated.

4.3 Efficiency Comparison

To compare the efficiency of single-node SGMM and MR-SGMM, different sizes of datasets were executed using both approaches, and the execution time was recorded. Datasets with 8,000, 10,000, 18,000, 36,000 and 72,539 entries were used, with results shown in Fig. 5.

From Fig. 5, the MR-SGMM execution time was consistently shorter than that for single-node SGMM for the same processed data. As the number of data entries grew, the increase in execution time in single-node SGMM was

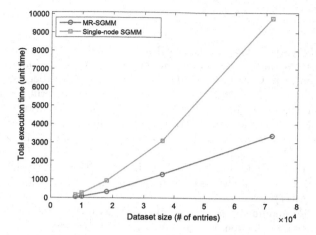

Fig. 5. Efficiency comparison between single-node SGMM and MR-SGMM

higher than that of MR-SGMM. Therefore, MR-SGMM became increasingly advantageous in handling increasingly large datasets thanks to its distributed implementation.

5   Conclusion and Future Works

We proposed and tested an implementation of a data summarization and regeneration approach using the MapReduce framework. This implementation (MR-SGMM), allows summarization of data based on a Gaussian Mixture Model (GMM) in a distributed manner. We used the information extracted from the SGMM process to regenerate the original data. Experiments comparing the regenerated data from the MR-SGMM to that of a single-node SGMM showed similar outcomes. We used two evaluation metrics to quantitatively compare MR-SGMM and single-node SGMM. These indicated that the data partitioning, local processing and data merging steps were performed properly, with similar results. Finally, we showed that the execution time for MR-SGMM grows more slowly with an increase in data size compared to the single-node method.

The present experimental results are focused on accuracy and efficiency of our method. Future evaluations would be considered on scalability analysis with larger datasets and performance analysis on different types of datasets. The current method relies on DBSCAN clustering, where the radius of the neighborhood for a point is fixed. For clusters of diverse densities, implementing MR-SGMM based on clustering techniques with different densities is desirable.

Acknowledgement. This research was partially supported by Mitacs.

References

1. Hesabi, Z.R., Tari, Z., Goscinski, A., Fahad, A., Khalil, I., Queiroz, C.: Data summarization techniques for big data-a survey. In: Khan, S.U., Zomaya, A.Y. (eds.) Handbook on Data Centers, pp. 1109–1152. Springer, Heidelberg (2015)
2. Cao, F., Ester, M., Qian, W., Zhou, A.: Density-based clustering over an evolving data stream with noise. In: SDM, vol. 6, pp. 328–339. SIAM (2006)
3. Yang, D., Rundensteiner, E.A., Ward, M.O.: Summarization and matching of density-based clusters in streaming environments. Proc. VLDB Endow. 5(2), 121–132 (2011)
4. Chaoji, V., Li, G., Yildirim, H., Zaki, M.J.: ABACUS: mining arbitrary shaped clusters from large datasets based on backbone identification. In: SDM, pp. 295–306. SIAM (2011)
5. Bigdeli, E., Mohammadi, M., Raahemi, B., Matwin, S.: Cluster summarization with dense region detection. In: Fred, A., Dietz, J.L.G., Aveiro, D., Liu, K., Filipe, J. (eds.) IC3K 2014. CCIS, vol. 553, pp. 68–83. Springer, Heidelberg (2014)
6. Ester, M., Kriegel, H.-P., Sander, J., Xiaowei, X.: A density-based algorithm for discovering clusters in large spatial databases with noise. In: KDD, vol. 96, pp. 226–231 (1996)
7. Yu, Y., Zhao, J., Wang, X., Wang, Q., Zhang, Y.: Cludoop: an efficient distributed density-based clustering for big data using Hadoop. Int. J. Distrib. Sens. Netw. 501, 579391 (2015)
8. Ma, L., Gu, L., Li, B., Qiao, S., Wang, J.: MRG-DBSCAN: an improved DBSCAN clustering method based on map reduce and grid. Int. J. Database Theor. Appl. 8(2), 119–128 (2015)
9. Kim, Y., Shim, K., Kim, M.-S., Lee, J.S.: DBCURE-MR: an efficient density-based clustering algorithm for large data using mapreduce. Inf. Syst. 42, 15–35 (2014)
10. He, Y., Tan, H., Luo, W., Mao, H., Ma, D., Feng, S., Fan, J.: MR-DBSCAN: an efficient parallel density-based clustering algorithm using mapreduce. In: IEEE 17th International Conference on Parallel and Distributed Systems (ICPADS), pp. 473–480. IEEE (2011)
11. Dai, B.-R.,, Lin, I., et al.: Efficient map/reduce-based DBSCAN algorithm with optimized data partition. In: IEEE 5th International Conference on Cloud Computing (CLOUD), pp. 59–66. IEEE (2012)
12. He, Y., Tan, H., Luo, W., Feng, S., Fan, J.: MR-DBSCAN: a scalable mapreduce-based DBSCAN algorithm for heavily skewed data. Front. Comput. Sci. 8(1), 83–99 (2014)
13. Dean, J., Ghemawat, S.: Mapreduce: simplified data processing on large clusters. Commun. ACM 51(1), 107–113 (2008)
14. Dean, J., Ghemawat, S.: Mapreduce: a flexible data processing tool. Commun. ACM 53(1), 72–77 (2010)
15. Hall, M., Frank, E., Holmes, G., Pfahringer, B., Reutemann, P., Witten, I.H.: The WEKA data mining software: an update. ACM SIGKDD Explor. Newsl. 11(1), 10–18 (2009)
16. Karypis, G., Han, E.-H., Kumar, V.: Chameleon: hierarchical clustering using dynamic modeling. Computer 32(8), 68–75 (1999)
17. United States Census Bureau. http://www2.census.gov/geo/tiger/tiger2010/
18. Box, G.E.P., Muller, M.E.: A note on the generation of random normal deviates. Ann. Math. Stat. 29, 610–611 (1958)

19. Dunn, J.C.: A fuzzy relative of well-separated clusters. J. Cybern. **3**(3), 32–57 (1973)
20. Davies, D.L., Bouldin, D.W.: A cluster separation measure. IEEE Trans. Pattern Anal. Mach. Intell. **2**, 224–227 (1979)

Erratum to: Advances in Artificial Intelligence

Richard Khoury[1](✉) and Christopher Drummond[2]

[1] Lakehead University, Thunder Bay ON, Canada
richard.khoury@lakeheadu.ca
[2] National Research Council Canada, Ottawa, Canada
christopher.drummond@nrc-cnrc.gc.ca

Erratum to:
R. Khoury and C. Drummond (Eds.)
Advances in Artificial Intelligence
DOI: 10.1007/978-3-319-34111-8

The affiliation in front matter Dr. John Oommen was not complete. The complete information is given below:

The author is also an Adjunct Professor with the University of Agder in Grimstad, Norway.

Erratum to:
Chapter 7: R. Khoury and C. Drummond (Eds.)
Advances in Artificial Intelligence
DOI: 10.1007/978-3-319-34111-8_7

In the original version, the references [4], [8], [14], and [15] were wrong. It should be read as follows:

1) Reference 4: It should be "De Melo, C.M., Carnevale, P., Read, S., Antos, D., Gratch, J.: Bayesian model of the social effects of emotion in decision-making in multiagent systems. In: Proceedings of AAMAS, vol. 1, pp. 55–62" instead of "De Melo, C.M., Carnevale, P., Read, S., Antos, D., Gratch, J.: Bayesian model of the social effects of emotion in decision-making in multiagent systems. Science 1, 55–62 (2012)"

2) Reference 8: It should be "Hoey, J., Schröder, T., Alhothali, A.: Affect control processes: Intelligent affective interaction using a partially observable Markov decision process. Artif. Intell. 230, 134–172 (2016)" instead of "Hoey, J., Schröder, T., Alhothali, A.: Affect control processes: intelligent affective interaction using a partially observable Markov decision process. Science 230, 134–172 (2016)"

The updated original online version for this Book can be found at 10.1007/978-3-319-34111-8
The updated original online version for this Chapter can be found at 10.1007/978-3-319-34111-8_7

R. Khoury and C. Drummond (Eds.): Canadian AI 2016, LNAI 9673, pp. E1–E2, 2016.
DOI: 10.1007/978-3-319-34111-8_40

3) Reference 14: It should be "Sequeira, P., Melo, F.S., Paiva, A.: Learning by appraising: an emotion-based approach to intrinsic reward design. Adapt. Behav. 22(5), 330–349 (2014)" instead of "Sequeira, P., Melo, F.S., Paiva, A.: Learning by appraising: an emotion-based approach to intrinsic reward design. Science 22(5), 330–349 (2014)"

4) Reference 15: It should be "Squazzoni, F., Jager, W., Edmonds, B.: Social simulation in the social sciences a brief overview. Soc. Sci. Comput. Rev. 32(3), 279–294 (2014)" instead of "Squazzoni, F., Jager, W., Edmonds, B.: Social simulation in the social sciences a brief overview. Science 32(3), 279–294 (2014)"

Author Index

Ali Akber Dewan, M.   115
Ali, Maqbool   89
Almeida, Hayda   168
Alsaeedan, Wojdan   162
Alves, Guilherme   96
Anderson, John   276

Bagheri, Ebrahim   192
Bagheri, Mohammad Ali   3
Bamber, Andrew   257
Banos, Oresti   89
Belitsos, Joseph   15
Bi, Yi   224
Bigdeli, Elnaz   323
Butz, Cory J.   207, 213

Chang, Chuan   21
Cheraghchi, Fatemeh   323
Choo, Hyunseung   89
Chowdhury, Md. Solimul   224
Chudzynski, Janusz   15
Csinger, Andrew   257

de Almeida, Claudianne M.M.   96
de Amo, Sandra   96
de Souza, Erico N.   33
Dirks, Matthew   257
Domenici, Paolo   40
Dong, Yue   304
dos Santos, André E.   207, 213
Du, Donglei   46
Du, Weichang   192

Escalera, Sergio   3
Esmaeilpour, Arina   323
Eude, Thierry   21

Fani, Hossein   192
Far, Behrouz H.   323
Felício, Crícia Z.   96
Feng, Zhiyong   224
Finnestad, Sonje   58
Freitas, Fred   243

Gao, Qigang   3
Gonzales, Christophe   213
Gotti, Fabrizio   150
Govil, M.C.   109

Hala, Stacey   40
Hamilton, Cameron   122
Hamilton, Howard J.   40
Heller, Robert   270
Hoey, Jesse   52
Hu, Baifan   33
Hunter, Aaron   263

Idris, Muhammad   89
Inkpen, Diana   131
Islam, Aminul   137

Jahjah, Vincent   180
Jaiswal, Ajay   109
Japkowicz, Nathalie   304
Jean-Louis, Ludovic   168
Jiang, Qian   285
Jiang, Xiang   33
Jin, Yong   46
Jung, Joshua D.A.   52

Khoury, Richard   180
Kinshuk   115
Kobti, Ziad   73
Kumar, Nitin   109

Lamontagne, Luc   180
Langlais, Philippe   150
Lee, Sungyoung   89
Li, Guichong   298
Lin, Fuhua   115, 270
Liu, Fangfang   224
Luo, Xuhui   102

Madsen, Anders L.   207
Matwin, Stan   33
Menai, Mohamed El Bachir   162
Meurs, Marie-Jean   168

Milios, Evangelos   137
Mohomed Jabbar, Mohomed Shazan   237
Morgan, Jonathan H.   52
Murray, Gabriel   64

Nagy, Geoff   276
Neufeld, Eric   58

Obando Carbajal, Luis Eduardo   21
Oliveira, Jhonatan S.   207, 213
Otten, Jens   243

Paixão, Klérisson V.R.   96
Pereira, Fabíola S.F.   96
Pilli, E.S.   109
Poole, David   257
Procter, Michael   270

Qiao, Dan   115

Raahemi, Bijan   323
Rahgozar, Arya   131
Rakib, Md. Rashadul Hasan   137
Ramanna, Sheela   83
Reichherzer, Thomas   15

Satterfield, Steven   15
Schröder, Tobias   52
Shahryari, Shervin   122
Siddiqi, Muhammad Hameed   89
Silver, Daniel L.   33
Singh, Ashmeet   83
Singh, Maheep   109
Soleimani, Ahmad   73

Watson, Lamar   15
Wen, Dunwei   115
White, Kenton   186
Wolf, Ingo   52

Xiang, Yang   285
Xu, Jinhua   102

You, Jia-Huai   224

Zaïane, Osmar R.   237
Zarrinkalam, Fattane   192
Zhang, Harry   46
Zhang, Jiongmin   316
Zhang, Yuan   316

Printed in the United States
By Bookmasters